D1116775

PHILOSOPHY OF CHEMISTRY

BOSTON STUDIES IN THE PHILOSOPHY OF SCIENCE

VOLUME 242

PHILOSOPHY OF CHEMISTRY

Synthesis of a New Discipline

Edited by

DAVIS BAIRD
University of South Carolina

ERIC SCERRI
University of California at Los Angeles

and

LEE McINTYRE
Boston University

A C.I.P. Catalogue record for this book is available from the Library of Congress.

ISBN-10 1-4020-3256-0 (HB) Springer
ISBN-10 1-4020-3261-7 (e-book) Springer
ISBN-13 978-1-4020-3256-1 (HB) Springer
ISBN-13 978-1-4020-3261-5 (e-book) Springer

Published by Springer,
P.O. Box 17, 3300 AA Dordrecht, The Netherlands.

Printed on acid-free paper

Printed in the Netherlands.

CONTENTS

VII. CHEMISTRY AND ONTOLOGY

SECTION I

CHEMISTRY AND THE PHILOSOPHY OF CHEMISTRY

CHAPTER 1

INTRODUCTION

The Invisibility of Chemistry

DAVIS BAIRD
South Carolina Honors College, University of South Carolina

ERIC SCERRI
Department of Chemistry and Biochemistry, University of California, Los Angeles

LEE MCINTYRE
Center for Philosophy and History of Science, Boston University

BUT *WHAT ARE* ALL THOSE CHEMISTS DOING?

Recently, one of us (Davis Baird) attended a meeting of historians of science and technology spanning all of the natural sciences and engineering and all (western) periods, ancient through contemporary. In the discussion of a paper on state-of-the-art history of modern (18th century forward) chemistry, a member of the audience made the claim that there was very little left to do in contemporary chemistry and that chemistry departments in his country were having trouble attracting graduate students. Baird found this perspective on contemporary chemistry both remarkable and implausible, and said as much. At the University of South Carolina (USC)—where he teaches—chemistry enrolls, and graduates, five times as many graduate students as physics. In this, USC is not unique.

The discipline of chemistry is, in fact, enormous and enormously productive. Joachim Schummer in this volume (Chapter 2) makes the point persuasively and concisely with data on the number of publications in various fields. With a grand total just shy of 900,000 papers indexed in chemical abstracts for the year 2000, chemistry is larger than all of the other natural sciences combined. After Baird suggested to the collected historians in the audience that contemporary chemistry was, in fact, a very active and productive discipline, a historian of mathematics sitting next to him leaned over and skeptically inquired, "But *what are* all those graduate students doing?"

"Making, measuring, and modeling . . . " is how Baird's colleague Catherine Murphy (Chemistry and Biochemistry, USC) would have replied. Recently, she and those in her lab have been making nanorods of silver with precisely controlled, diameters and lengths (Murphy and Jana 2002). It turns out that the aspect ratio (length/diameter) has a dramatic effect on the color of solutions of these rods. Short rods (20 nm by 30 nm) are orange, medium-length rods (20 nm by 100 nm) are red,

D. Baird et al. (eds.), Philosophy of Chemistry, 3–18.

and long rods (20 by 200) are blue.[1] Beyond the intrinsic scientific interest, various applications lie down the road. She also works on how strands of DNA flex, an important feature of how this long linear molecule wraps itself up into a tiny bundle for "storage" and unwraps when "being read."

Both research projects require each of her triumvirate: making, measuring, and modeling. Murphy has to *make*—synthesize—her nanorods, a multi-step process that will remind non-chemist readers of undergraduate chemistry labs carefully controlling temperature, pH, solution concentrations, reaction times, etc. The synthesized rods have to be *measured*—"characterized"—and here Murphy leans heavily on USC's Electron Microscopy Lab. Now she has rods of known aspect ratios, she knows how various properties—such as color in solution—depend on index ratios. But why? Here *models* for how the index ratio of the rods affects the absorption of radiation are necessary.

One cannot help but be struck by the ubiquity of chemistry. Everything in sight has the chemist's art written into it. Paints, varnishes, and other finishes all are the products of chemistry, some long established and (in this sense) "low tech," some new "high tech" paints with nanoparticles for striking visual and other functional qualities. There are batteries, light bulbs—reminding us of Oliver Sacks' recent "memoir of a chemical childhood," *Uncle Tungsten* (Sacks' uncle made better light bulbs through chemistry)—and liquid crystal computer displays. There is even chemistry hidden inside silicon chips, which are made by using ultra-precise lithography. Many of us use cheap "gel writer" pens with their new improved ink. And, of course, everywhere one looks, one sees plastics. In the 1960s film "The Graduate," Dustin Hoffmann was given this advice about the future—"one word, plastics." In this new century, that word might be "nanomaterials." Either way, it is chemistry.

Like the air we breathe, chemistry engulfs us. And like air, we take chemistry for granted. We do not see all of the making, measuring, and modeling that went into the world we inhabit. What are all those chemists doing, indeed! Yet somehow, despite its omnipresence, chemistry has remained largely invisible, denigrated by the physicist and, until recently, ignored by the philosopher. By way of an introduction to this volume, we briefly consider several of the more important reasons for chemistry's relative obscurity, and how the recent flourishing of work in the philosophy of chemistry—some documented in this volume—is helping to make this science more visible and accepted as an apt subject for critical reflection both within the laboratory and without.[2]

THEORETICAL ESCAPE FROM OUR MATERIAL BONDS

Until the last decade or so, historians, and especially philosophers, tended to focus on grand unifying theoretical visions instead of complicated local sights. Physics offers grand theories, and reflections on relativity, quantum theory, and physical theory generally have dominated 20th century philosophy of science. Chemistry offers a multiplicity of models drawing on both physical theory and experimental generalizations, each where appropriate. Yes, there is the periodic system, and we will see Ray Hefferlin's pursuit of a periodic system of molecules in this volume (Chapter 12). Joseph Earley (Chapter 11) discusses ways in which group theory can move chemical theory ahead, and Jack Woodyard (Chapter 13) develops a radically new way of interpreting quantum theory. However, much of the life of theoretical chemistry is better described by Murphy's "modeling," circumscribed attempts to make sense of synthesized local phenomena.

Theory, a philosophically acceptable substitute for fiction, like fiction, takes us away from the actual material world, away from the work-a-day world of the experimental chemist. There is a long tradition in philosophy that has tried to deny our clay-footed nature to see in us—and in our world—a more "important," more "fundamental," immaterial something, perhaps a soul and/or a set of first principles. Plato's peculiar ontological inversion, where the material world is, but an imperfect copy of such a realm of ideas, spells trouble for the philosophy of chemistry. Seeing chemistry—philosophically seeing chemistry—is perhaps like seeing our mortality. Denial is an enormously effective strategy.

Philosophers in denial have ample resources to draw on. Chemistry does sit right next to physics, with all its lovely unifying and foundational theory. Squinting our eyes up tight, it is possible to see chemistry as complicated applied physics. Even in denial, we say we are materialists, but the material world of our denial is the foundational world of physical theory, and so chemistry—in principle anyway—must be reducible to physics. But this has never been much more than an article of faith. The arguments for it have all relied on a highly sanitized picture of chemistry. It has been enough to persuade philosophers in denial that if one wants to understand science, it would be enough to understand physics. (To be fair, there is now a large group of philosophers who are not in denial, and many have pursued the philosophy of biology particularly vigorously. While there is also now a good group of philosophers of chemistry—several represented in this volume—this field has only very slowly, and recently, come to be accepted as part of the canon of philosophy of science.) However, we must be clear here: to understand science, it was often enough to understand physical theory. Through this work, we might gain some insight into chemical models and modeling, but the two other legs of Murphy's chemical tripod, making and measuring, are lost. Nonetheless, questions about the relations between chemistry and physics are central to understanding chemistry and chemistry's invisibility.

MAKING CHEMISTRY VISIBLE

A central aim of this volume is to continue to make chemistry more philosophically visible. Great strides already have been made in this and these are nicely documented in Schummer's contribution to this volume. Yet experience among the historians of science and technology suggests that more work is needed. Some of the fundamental sources for chemistry's invisibility can be found in the issues we have just quickly considered:

(i) the materiality of chemistry's objects;
(ii) the centrality and means of conceptualizing this materiality;
(iii) the nature and place of theory, and foundational issues in chemistry;
(iv) chemistry's relation to physics.

In reverse order, these issues establish the bases for the last four sections of the book:

Section 4, Chemistry and Physics;
Section 5, Chemical Theory and Foundational Questions;
Section 6, Chemistry and its Tools of Representation;
Section 7, Chemistry and Ontology.

Prior to these four sections are three more sections that mutually situate chemistry, the philosophy of chemistry, and the history and philosophy of science:

Section 1, Chemistry and the Philosophy of Chemistry;
Section 2, Chemistry and the History, and Philosophy of Science;
Section 3, Chemistry and Current Philosophy of Science.

Together, these seven sections take on chemistry's invisibility and make it—chemistry—more visible. We continue this introduction by briefly summarizing the work of the 18 authors of these seven sections.

CHEMISTRY AND THE PHILOSOPHY OF CHEMISTRY

Our first task is to situate the philosophy of chemistry. Joachim Schummer takes a more optimistic stance than our opening, but still he is struck by the failure until recently of philosophers to attend to chemistry. Given, chemistry's size, importance, and long and fascinating history, this fact cannot be ignored. It cries for explanation. Schummer provides excellent insight into the disciplinary forces that have driven philosophers' attention from chemistry. Indeed, remarkably, there is an inverse correlation between the amount of attention paid by philosophers to a field of study and the size of that field of study. Philosophers pay the most attention to their own history—which has the smallest literature—and the least attention to chemistry, which has the largest literature.

Schummer beautifully and concisely moves through the past, present, and future of the philosophy of chemistry. While there has been neglect in the past, the neglect has not been total. There has been a Marxist tradition of examining chemistry, vigorously pursued by Friedrich Engels, and carried on in Marxist countries. And, where philosophers have left the field vacant, chemists, chemical educators, and some historians of chemistry have moved to occupy it. There have been philosophical treatises on chemistry. Two important such contributions are considered in this volume, Aristotle—by Paul Needham (Chapter 3)—and Kant—by Jaap van Brakel (Chapter 4). Such was the not-completely neglected state of the philosophy of chemistry through the 1980s. There were important—but isolated—contributions, no sustained discussion of the subject, and no recognition by philosophers of science of the importance of the subject.

The situation began to change in the 1990s. This decade saw numerous conferences devoted exclusively to the philosophy of chemistry, the formation of two journals—*Hyle* and *Foundations of Chemistry*[3]—and the creation of the International Society for the Philosophy of Chemistry (ISPC). The ISPC has held annual international conferences on the philosophy of chemistry each summer, since 1997.[4] Indeed, several papers presented at the third, 1999, conference, held at the USC, form the basis for this volume.

Schummer outlines the spectrum of work that has made up this flourishing of the philosophy of chemistry. Reductionism has been a central issue (see Chapters 9 and 10, by Robin Hendry and G. K. Vemulapalli, respectively, in this volume). Also central are attempts to develop and adapt established concepts from the philosophy of science for chemistry—among others, naturalism, explanation, professional ethics, and themata in the history of science (see Chapters 5–8, by Otto Ted Benfey, Eric Scerri, Johannes Hunger, and Jeffrey Kovac, respectively). Philosophers have pursued conceptual analyses of key chemical concepts such as element, pure substance, compound, affinity (see Chapter 17, by Nalini Bhushan on natural kinds, and Chapter 18, by Michael Weisberg on pure substance).

Schummer closes his contribution speculating about, how philosophy of chemistry should develop. He suggests a variety of avenues of inquiry: What is the logic of chemical relations? What does the chemical system of classification (of chemistry's more than 20,000,000 known substances) tell us about ontology? What methods of discovery are peculiar to chemists, and how have they been so successful? How should we recast the relationship between science and technology in light of the long-standing history of the chemical industry and its close ties to academic chemistry? Here, we mention only a few of Schummer's suggestions. His entire list provides an exciting and compelling program for a fully developed mature philosophy of chemistry, one that rightly will force philosophy to rethink itself (see also Chapter 19, by Alfred Nordmann).

CHEMISTRY AND THE HISTORY AND PHILOSOPHY OF SCIENCE

The two contributions in the next section of the volume take up two important contributions to chemistry's relationship to the history and philosophy of science. First is Paul Needham's careful discussion of Aristotle's views about chemical reaction and chemical substance (Chapter 3). This is followed by Jaap van Brakel's eye-opening discussion of Kant's long neglected *Opus Postumum*, a work that only recently has come to wide attention. Its neglect allowed Kant's earlier dismissive views of chemistry to dominate philosophical considerations of chemistry. Here is a central source of chemistry's philosophical invisibility.

Atomism's rise, thanks to Dalton's resurrection of some ancient ideas of Democritus and Leucippus, has made atomists of all of us. However, Paul Needham (Chapter 3) makes clear that this conversion has been neither complete nor uncontested. Alternative approaches, most especially Aristotle's, remain embedded in contemporary chemical theory. Perhaps, the central theoretical problem that chemistry poses concerns combination: two distinct substances can combine to form a third yet distinct substance. Why does this happen? How does it happen? And, what does this say about how we conceptualize a substance's identity and properties? The atomists appear to have neat answers to these questions framed in terms of the combinations of individual indestructible atoms. However, Needham's more careful analysis shows, both how this approach is incomplete and how Aristotle's long ago criticisms of it have been important in the development of modern chemistry. Of particular interest here is how modal properties are used. Aristotle was more careful to recognize and deal with the manner in which the properties of elements in isolation are "lost," when in combination. However, since analysis can "bring them back," they must remain as a kind of potentiality. Here is one of several important features of Aristotle's theory that survives in modern chemistry (see Chapter 19, by Alfred Nordmann).

In his *Metaphysical Foundations of Natural Science*, Kant tells us that, "...chemistry can become nothing more than a systematic art or experimental doctrine, but never science proper; for the principles of chemistry are merely empirical and admit of no presentation a priori in intuition" (see Chapter 4, by Jaap van Brakel, for full quotation and references). Such was Kant's critical—and we use this word in both its common and its "Kantian" senses—assessment of chemistry. And this assessment, van Brakel shows us, has had an enormous impact. Even as late as 1949, the physicist and philosopher of science Herbert Dingle tells us that chemistry should not figure at all in the philosophy of science.[5] From Kant's critical point of view, chemistry suffered from two interconnected problems, it was insufficiently mathematical and its laws could only be found empirically, and hence these laws would always be subject to Humean skepticism. However, Kant's views on chemistry evolved in his post-critical period. Kant was specifically concerned with how his critical work had not come to terms with the diversity of substances chemists bring to our awareness and how his own metaphysical concept of matter did not deal well with this diversity (cf. Needham on Aristotle). This was a central project of Kant's *Opus Postumum.* Unfortunately, this work was first published—in a chaotic and unedited form—a century after Kant's death in 1804. The first English translation was published in 1993. Yet this is where Kant, in full awareness of the revolutionary developments and controversies that chemistry was undergoing at Lavoisier's hands and in the years following, begins to come to grips with chemistry as (at least an improper) science.

CHEMISTRY AND CURRENT PHILOSOPHY OF SCIENCE

The four contributions in the next section take up different long-standing concerns in the philosophy of science and look at them through the lens of chemistry. Although perhaps less frequently discussed in recent years, understanding how concepts play a central role in the historical development of science has much exercised philosophers of science. Otto T. (Ted) Benfey takes up this issue with work that develops Gerald Holton's notion of the role of themata in the history of science (Holton 1988). Eric

Scerri follows with a discussion of the role that reflections on chemistry can play in gaining deeper insight into the normative/descriptive distinction in the philosophy of science. Johannes Hunger shows how traditional concepts of explanation fail when transplanted to chemistry. And finally, Jeffrey Kovac develops an outline of an approach to professional ethics for chemists.

Holton is well known for arguing that general conceptual preferences—for example, for continuous theories or for theories that are symmetric in time—play an important role, over and above empirical data, in the historical development of science. Ted Benfey pursues this idea and develops a triad of opposing concept-pairs that provides a useful conceptual tool for looking at the history of science, and chemistry in particular. His three pairs are these: (1) reversible time–irreversible time; (2) continuity–discontinuity; (3) inner structure–outer structure. He illustrates how these six concepts work with numerous examples, many, although not all, taken from the history of chemistry. Thus, the ideal gas law posits an essentially uniform (or continuous) collection of point-mass (disregarding any inner structure) particles (reversibly) in motion. The theory works well, but not perfectly, and its imperfections can be traced to contraries of the three concepts employed. In fact, the particles are not point masses, but occupy space and have internal structure. They do not behave in a uniform manner at low temperatures, but will aggregate and exhibit discontinuities. And, of course, fundamental puzzles come from entropy and the irreversibility of time. With his three concept-pairs in hand, Benfey shows how fundamentally different traditions in science tend to focus on particular preferred collections of them. The mechanical sciences, which characterized the scientific revolution, tended to emphasize reversible time, a lack of inner structure, and discontinuity. The "organicist" (or "minority report" to the scientific revolution) emphasizes the opposites, directional time, inner structure, and continuity. And more generally, many scientific struggles have been struggles over how to incorporate these conflicting concepts.

Philosophy of science has long struggled to understand its position relative to the sciences. Is philosophy fundamentally about the—a priori—logic of science, dictating to scientists from epistemological foundations what can and cannot be an appropriate method? Or rather does philosophy fundamentally draw lessons about what are appropriate methods by observing how successful science goes about its business? We find more normative, a priori approaches in Kant, Frege and the logical positivists of the 20th century, and more descriptive, a posteriori approaches in Whewell, Mill and the recent "naturalist turn." Eric Scerri (Chapter 6) reminds us of this history and then presents a bit of his own history of coming to terms with chemical orbitals. Quantum mechanics tells us that orbitals do not exist, and early on Scerri drew strong normative conclusions from this, urging chemists from his stance in the foundations of chemistry, to recognize that these chemical workhorses were fictions. Chemists were not moved. And Scerri began to reconsider the strength of his normative position. He has arrived at a kind of compromise that draws on chemist Fritz Paneth's understanding that chemistry must at once recognize that quantum theory lies at chemistry's base, while also remaining an autonomous science that uses concepts such as orbitals and other even higher level macroscopic concepts. Scerri shows us how the philosophy of chemistry can teach the philosophy of science to be both descriptive and normative. Both attitudes—a term Scerri consciously takes from Arthur Fine—are important to our philosophical work and indeed to the work of practicing scientists.

Johannes Hunger (Chapter 7) takes on another of the standard topics in the philosophy of science, explanation. Hunger examines, in detail, various ways that chemists explain and predict the structural properties of molecules. We learn about ab initio methods, empirical force field models and neural network models, each of which have been used to explain and predict molecular structure. And we learn that none of these approaches can be subsumed under either hypothetico-deductive or causal models of explanation. Either chemistry does not offer proper explanations (the normative option) or our philosophical models for explanation are inadequate to cover explanation in chemistry (the descriptive option). Hunger takes the descriptive option and sketches a more pragmatic approach to the explanation that develops Bas van Fraassen's approach to explanation for chemistry. Once again, we find that the philosophy of science has much to learn from the philosophy of chemistry.

This section closes with a sharply normative chapter written by a chemist on professional ethics. Jeffrey Kovac argues that ethics lies at the very heart of chemistry, and science more generally. As an institution that grows our knowledge, science depends on the good moral character of its practitioners. Kovac has made good on these views in practice, incorporating ethics into his chemistry curriculum. Here (in Chapter 8), he outlines his approach to ethics for the sciences. He draws on the fact that the sciences are professions, and as professions, they depend on internal agreements among their practitioners to behave according to certain standards, and external agreements with the society in which they are embedded to provide a certain kind of product in return for support and monopoly status as purveyors of scientific knowledge. Kovac draws on Robert Merton's four ideals of science to articulate the nature of the profession of science:

(i) Universalism—science aims at universal truth based on universally accepted criteria;
(ii) Communism—scientific knowledge is a public or community good;
(iii) Disinterestedness—the advancement of science is more important than the advancement of its practitioners;
(iv) Organized skepticism—claims to scientific knowledge are provisional, based on accessible data and subject to revision
(Merton 1973).

Kovac concludes his chapter sketching a moral ideal for science. He focuses on two attributes: (1) scientists who make up the profession of science must have the habit of truth. Honesty with oneself, with one's colleagues, and with the society one serves is essential to the flourishing of science; (2) scientific goods should be exchanged as gifts, not commodities. Gift exchange is tied to all of Merton's ideals, but most closely to communism; gifts are part of how a community of inquirers establishes and preserves itself.

CHEMISTRY AND PHYSICS

The two chapters in the next section take up chemistry's relation to physics and, in particular, the bug-a-boo of reductionism. The chapters complement each other. The first, by Robin Hendry, looks at chemistry's relation to physics from the point of view of philosophy, bringing the careful analyst's conceptual scalpel to the minefield of reductionism and emergentism. The second, by G.K. Vemulapalli, looks at chemistry's relation to physics from the point of view of chemistry, simultaneously acknowledging the importance of the fundamental laws of physics for chemistry, while conceding none of chemistry's autonomy.

It is relatively easy to talk and gesture about how chemistry either does or does not reduce to physics. It is much harder to spell out exactly what is required to make good on the claim that chemistry does (or does not) reduce to physics. Philosophers have a concept of supervenience. In the case we are focused on here—chemistry putatively reducing to physics—supervenience requires that every chemical change be accompanied by a physical change. This is nearly universally held, for example, if two molecules are identical in all physical respects, they will not differ chemically. However, supervenience is not sufficient for the reduction of chemistry to physics. There could be "downward causation," where it is the chemical facts and laws that drive the physical facts and laws, not the other way around. Robin Hendry (Chapter 9) argues that those committed to the reducibility of chemistry to physics have not ruled out the possibility of downward causation, and moreover, he presents substantial evidence from the manner in which quantum mechanical descriptions for molecules are constructed and deployed by chemists in favor of downward causation. Quantum mechanical descriptions of molecules that have explanatory and descriptive power are constructed from chemical—not physical—considerations and evidence. Here in precise terms, we see chemistry supervenient on physics, but still autonomous, not reducible to physics.

So, if chemistry is not reducible to physics, what is the relationship between the two disciplines? This is the question that G.K. Vemulapalli takes up in Chapter 10. Drawing on a wealth of experience in the practice of chemistry, Vemulapalli looks at how chemists use physical theories, how they develop chemical theories over and above physical theories, and how these two theoretical domains relate to each other. It is clear from many examples that developments in physics profoundly influence chemistry. Relative molar masses—one of several examples that Vemulapalli presents—are now routinely measured by mass spectrometry working in combination with the ideal gas law. It is also clear from many examples that developments in chemistry influence physics. Faraday's work as an electrochemist led to Stoney's introduction of the concept of the electron, and Nernst's study of low temperature equilibria led to the third-law of thermodynamics. The development of quantum theory has played a major role in the chemical understanding of bonds. However, and here is one of Vemulapalli's main points, chemists do not *and cannot* simply plug chemical situations into the Schrödinger equation and get useful results. They must augment straight physical theory with chemical concepts, bond energies and bond lengths, for example. As Vemulapalli puts it, physical law provides fundamental conceptual insight and boundary conditions on what is possible, chemists have to add to this to find out how actual chemical species behave.

CHEMICAL THEORY AND FOUNDATIONAL QUESTIONS

Hendry and Vemulapalli nicely frame the space for the work taken up in the next section. Fundamental physical theories such as quantum mechanics raise difficult foundational questions that have demanded the efforts of many powerful minds in physics and the philosophy of physics. As chemistry is not reducible to physics, there is an autonomous space for chemical theory and for foundational issues in chemical theory. Three such issues are raised in this section. Joseph Earley examines the role of symmetry in chemistry and argues for closer attention to group theory on the part of his fellow chemists. Ray Hefferlin seeks to extend the idea of a periodic law from elements to compounds. Jack Woodyard takes on the fundamental obstacles that get in the way of a more straightforward application of quantum theory to molecules.

Joseph Earley (Chapter 11) argues for the importance of group theory both in chemistry and in the philosophy of chemistry. He shows how group theoretical concepts can shed needed light on several fundamental problems in chemistry. For one, there is the problem of chemical combination that we have already seen (Chapters 3 and 4, by Paul Needham and Jaap van Brakel, respectively). For another, there is the problem of reduction and emergence that we also have seen (Chapters 6, 9, and 10, Eric Scerri, Robin Hendry, and G.K. Vemulapalli, respectively). Earley looks at philosophical contributions to mereology (the study of parts and wholes). He shows that mereology, as it is currently developed, is incapable of dealing with chemical combinations, for, in general, the properties of elements alone are significantly changed when in combination. Earley suggests that group theory, and in particular the concept of closure under the group operation, provides the necessary conceptual apparatus to move mereology ahead, so that it can deal with chemical combination. Earley also suggests that group theory is exactly what is needed to provide a conceptually sharp way of articulating how a more complex system can emerge from a simpler system, while remaining supervenient on the simpler system—exactly the situation that Hendry discusses with respect to chemistry and physics.

Dimitri Ivanovich Mendeleev's periodic table of the elements has been tremendously important in chemistry and the philosophy of chemistry. It provides a powerful organizing principle that has led to the discoveries of new unsuspected elements and, until recently, it stands as one of the best examples of a genuinely chemical law. There have been—not completely successful—attempts to explain the periodic law with quantum theory (Scerri 2003b), but it remains, like Darwin's theory of natural selection in biology, a cornerstone of chemistry. It would be spectacular if a similar periodic system could be developed for molecules. It would help organizing the massive complexity that constitutes the huge number of known chemical species, and it would help predict new compounds that might be synthesized and developed for practical purposes. It would also be a wonderful further example of chemistry's autonomy. Ray Hefferlin (Chapter 12) has been a main player in two decades of attempts to develop such a system of molecules. Here, he provides an overview of the history of these attempts, their problems, and their prospects.

In our last chapter on the theoretical and foundational issues in chemistry (Chapter 13), Jack Woodyard constructively attacks the way classical quantum theory is fudged in quantum chemistry. Woodyard argues for fundamental change, abandoning Hilbert space representations in favor of a complex three-dimensional space in which "matter-waves" interact. We have already seen that the application of quantum theory to chemistry is not straightforward and requires additional input from experimental data and from techniques of approximation that work, but which cannot be motivated on theoretical grounds (Chapters 6, 7, 9, and 10, by Eric Scerri, Johannes Hunger, Robin Hendry, and G.K. Vemulapalli, respectively). Woodyard presents additional evidence that this traditional approach only works through the use of numerous band-aids to cover over assumptions that are known to be mistaken. The theory provides no means that is not *ad hoc* to visualize molecules, and perhaps most damning, as more terms are calculated in the theoretical series, the results move further from experimental values. Woodyard offers an alternative theory that retains the fundamentally correct core of quantum theory, but develops it as a theory of matter-waves in three-dimensional space, and he shows that his alternative provides results that agree with experiment better than the standard approach. Woodyard is aware that fundamental change, such as that which he offers, is almost never smoothly adopted, replacing current dogma, and he offers some engaging insights into revolutionary science from his point of view as a revolutionary.

CHEMISTRY AND ITS TOOLS OF REPRESENTATION

These issues about chemical theory and its relation to physical theory are central to the philosophy of chemistry, but they do not encompass the whole field. How chemistry goes about representing its objects is extremely important, both to the conduct of chemistry and to our philosophical understanding of chemistry. This section includes three chapters on chemistry's tools of representation. In the first, Ann Johnson describes the introduction of computers into chemical engineering. This is important because the rise of computer-aided design radically—incommensurately—changed the field. Johnson's contribution also is important because she reminds us that the philosophy of chemistry cannot ignore its technological end, chemical engineering. In the next chapter, Sara Vollmer considers seemingly simpler tools, the ways in which we symbolize chemical species "on paper." Her discussion probes the differences between pictorial representations and linguistic ones. Finally, the philosopher/chemist pair, Daniel Rothbart and John Schreifels, examines the pre-suppositions that go into the design of analytical instruments.

Ann Johnson's story about the introduction of computers into the practice of chemical engineering (Chapter 14) reminds us of several important features about science. Tools make a difference. Furthermore, it is through communities of practitioners that tools are introduced and these communities bring specific problems—indeed as Johnson has it, they are defined by their problems—to their work. By the end of the story, chemical engineering c. 1990, the work done by a chemical engineer has changed radically from that done c. 1950. Some changes might be expected (although, perhaps only in hindsight). Much chemical engineering involves working with partial differential equations. Prior to computers, which can iteratively produce numerical approximations, a chemical engineer had to work with analytically soluble equations. Computers relax this constraint, but, of course, they produce their own constraints. Initially, specialist programmers were necessary. This produced a disciplinary problem, which over time was solved by incorporating computer programing into the training of chemical engineers. When first introduced, computers were used to crunch numbers, while the surrounding engineering practice stayed more or less the same. However, as the possibilities of the tool became better understood, the practice itself changed. The computer was developed to simulate operations, and this opened up (and simultaneously closed off) a whole new field of operations for chemical engineers. In the end, the kinds of problems, acceptable solutions, necessary training, day-to-day practices, and disciplinary setting of chemical engineering all changed, producing a kind of incommensurability that stories of theory change miss.

Sara Vollmer's contribution (Chapter 15) concerns what seems a much simpler chemical tool, the manner in which we symbolize chemical species on paper. She contrasts John Dalton's approach with Jakob Berzelius's approach. Berzelius developed a forerunner to the now-common "H_2O" representation where compounds are represented in terms of the relative molar numbers of their elemental constituents. Dalton's approach was more pictorial. Sulfur trioxide (Dalton's understanding of sulfuric acid) was shown as a central circle (with some conventional markings to indicate its being sulfur) surrounded by three other "oxygen" circles, an approach that presages various "ball-and-stick" representations used today. Vollmer's interest here is to tease out how pictorial representations, such as Dalton's, differ from Berzelius's more linguistic representations. The answer lies somewhere in the geometry that the representation shares with its object, although, particularly when it comes to the illusion of three dimensions on a two-dimensional surface, this answer requires considerable care to explicate.

The final chapter in this section brings together Johnson's concerns with instrumentation and how communities interact with instrumentation, and Vollmer's concerns with pictorial representation. Daniel Rothbart and John Schreifels (Chapter 16) discuss a variety of instruments, from Hooke's microscopes to Binnig and Rohrer's scanning tunneling microscope. They are concerned to show that instruments are not passive "transparent" devices that merely "open a window" on a part of the world that is otherwise inaccessible. Instruments are active devices and in multiple senses. Instruments must be made, and this means that we have to rely on the extant collection of materials and techniques to work them. However, even before we put hand to lathe we have to design, and here Rothbart and Schreifels's add to Vollmer's discussion of pictorial representation. Rothbart and Schreifels think of instrument designs as thought experiments, where one can, through reading the diagrams, think through how the instrument will interact with its specimen. This reminds us that the passage through an instrument from specimen to observation is anything but passive. Instruments probe specimens generate a signal and modify this signal, all along the path to creating information about the specimen. Rothbart and Schreifels argue that it is in virtue of the mechanical operations that instruments and nature share, that we can trust what we learn from instruments. They draw both an epistemological moral—only instruments that properly share their modes of operation with their objects will produce genuine knowledge—and an ontological moral—the operations of the world are of the same sort as the operations of instruments. Speaking of "a clockwork universe" is more than mere analogy.

CHEMISTRY AND ONTOLOGY

Rothbart and Schreifel's ontological speculations bring us to the last section of the book, where we find chemistry's most profound lessons, ontology. All three chapters, here, deal with the lessons chemistry has for our understanding of substance. And, as Michael Weisberg reminds us, chemistry is the science of the structure and reactivity of substances. So, there is no surprise that there is much to be learned here. The first two chapters—Chapters 17 and 18, by Nalini Bhushan and Michael Weisberg—are about chemistry's lessons about natural kinds. In the philosophical literature, written by philosophers with at best a superficial knowledge of chemistry, chemistry is frequently cited as a bountiful source of paradigm natural kinds. Is not water H_2O, after all? Bhushan and, separately, Weisberg bring more chemical sophistication and skepticism to this simplistic notion of natural kinds. Water is not (simply) H_2O. Alfred Nordmann, in Chapter 19, draws on the work of Émile Meyerson and Gaston Bachelard—two philosophers with some genuine chemical knowledge—to articulate a more general, a richer, and a metachemical notion of substance. This metachemical notion of substance improves upon the metaphysical notion that has been plaguing us with philosophical pseudoproblems for centuries.

Nalini Bhushan argues that chemical kinds are not natural kinds. They are not natural kinds because many chemical kinds are human-crafted—synthesized—kinds and do not occur "naturally." However, they are also not natural kinds in the more important sense that chemistry does not offer a univocal way of carving up substances (synthesized or "natural") into kind categories. How a chemist classifies kinds, has to do with local chemical and functional needs, and is responsive to these needs. For some purposes, having a particular kind of reactivity will drive classification; for other purposes, structural issues will drive classification; and for yet other purposes, other classifications are appropriate.

Michael Weisberg's water example (Chapter 18) helps to make this point. "Ordinary," although purified (but how purified?), water found "naturally" on the Earth contains typical, percentage-wise predicable isotopic isomers. While most water is composed of "ordinary" "one proton–one electron" hydrogen, a small percentage is composed of a heavier isotopic isomer of hydrogen—"one proton–*one neutron*–one electron," "deuterium," or "D." We symbolize "heavy water," HDO, instead of H_2O. For many purposes, the water we want to talk about *must* contain the standard percentages of ordinary H_2O *and* "heavy" HDO. Properties such as freezing point and viscosity depend on this "water natural kind" having the "naturally occurring" percentages of isotopic isomers. Sometimes, it is crucial to draw distinctions between these isotopic isomers—and to separate them in the lab, preparing isotopically pure samples of the various "kinds" of water. Heavy water is used as a moderator in some nuclear reactors, but you would not want to drink it.

Bhushan and Weisberg both take pains to point out how chemical kinds are more complicated than philosophers untutored in chemistry might think (or wish). They take these facts from chemistry in slightly different directions. Bhushan argues that the realist conclusions that chemically naïve natural kind talk engenders are not supported by genuine chemistry. Instead, she opts for a more situated "particularist" approach to chemical kinds that takes from Nancy Cartwright's ontological views (Cartwright 1994, 1999). The position is realist, and the chemical kinds employed are "really out there," but only in locally constructed—perhaps synthesized—ways, not theoretically globally. Weisberg argues that work in the philosophy of language that causally ties reference to essences identified in an initial act of dubbing (Kripke 1980; Putnam 1975) rests on a false assumption. This assumption is what Weisberg calls the "coordination principle," and it asserts that ordinary kind talk (e.g., "water") can be mapped onto scientific kind talk (e.g., H_2O), where we actually discover the nature of these essences. However, as we have seen, water is not H_2O, and the coordination principle cannot be true. Weisberg urges us to consider a better coordination principle that allows references such as "water" to be responsive to the context of use. When a guest asks for a glass of water, it is the ordinary bulk liquid with its standard percentages of isotopic isomers that (no doubt, usually unconsciously, or even "metaphysically") "water" refers to. When a nuclear engineer asks for the valve for the water moderator to be opened, "water" refers to "heavy water."

So, chemistry teaches us that kinds are more complicated than we thought. Alfred Nordmann (Chapter 19) takes this conclusion further, the very notion of substance is more complicated than we thought. Our "metaphysical" notion of substance focuses exclusively on the millennially old "problem of change": what stays the same through change? However, to understand the material world we inhabit, we need a richer notion of substance, a notion that is sensitive to how substances come into being, how we identify them as substances, and how we project them forward into a world constantly in the making. Metachemistry gives us this notion of substance. While not denying the metaphysical notion of substance, metachemistry provides a broader notion that embraces the metaphysical "sub-stratum" notion, but adds important elements to it. Nordmann describes Bachelard's trio: "sub-stance"—that which lies behind observable phenomena; "sur-stance"—that which emerges in our engaging the material world; "ex-stance"—that excess of meaning that substance concepts have, and which allow us to project these substances beyond their context of creation. Nordmann connects this work by Bachelard with more recent work by Bruno Latour, where the focus is on the work scientists perform to bring substances into being. He also connects it to Peirce's notion that the real is what is arrived at, at the end of inquiry. We may start with a purely metaphysical notion of substance, but we end with a real metachemical notion.

As one can see, even through this brief summary of the papers that are contained in this volume, the philosophy of chemistry is now well launched as a discipline. After predictable growing pains over the last few decades, we feel that the field is now mature enough to offer a quorum of good work, such that the subtitle of this volume—"Synthesis of a New Discipline"—may live up to its promise. This, of course, cannot be judged by the merits of any *a priori* argument over the need for a new discipline, but only by the content of the work that already exists in the field. It is to this that we direct the reader in the pages that follow.

NOTES

1. Many institutes have recently sprung up to study nanoscience, all having a strong contingent of chemists. For example at UCLA, the institution of another of the editors of this volume (Scerri), the California Nano Systems Institute (CNSI) has recently been founded following the award of a grant amounting to approximately $50 million.
2. Even in the realm of popular science books, chemistry appears to be under-represented with physics and biology claiming far greater shelf space. Some exceptions include books on the periodic table and books by Peter Atkins, Philip Ball, and Roald Hoffmann all of whom have made valiant efforts to popularize chemistry.
3. The websites for the two journals are http://www.hyle.org/index.html and http://www.kluweronline.com/issn/1386-4238, respectively.
4. Several earlier meetings were also held in Germany and the UK, including one at the London School of Economics, in March 1994 and the first International Summer School in Philosophy of Chemistry, in July 1994. (Scerri 2003a). A detailed survey of the history of the field has been published (Van Brakel, 1999) and several books on philosophy of chemistry have also appeared (Bhushan and Rosenfeld 2000; Schummer 1996; Van Brakel 2000).
5. Nevertheless, Dingle who founded the *British Journal for the Philosophy of Science* was sufficiently interested in chemistry to act as the co-editor of the philosophical essays by the chemist Fritz Paneth (Dingle and Martin 1964).

REFERENCES

Bhushan, N. and Rosenfeld, S. 2000. *Of Minds and Molecules.* Oxford: Oxford University Press.

Cartwright, N. 1994. *Nature's Capacities and Their Measurement.* Oxford: Oxford University Press.

Cartwright, N. 1999. *The Dappled World: A Study of the Boundaries of Science.* Oxford: Oxford University Press.

Dingle, H. and Martin, G.R. 1964. *Chemistry and Beyond: Collected Essays of F.A. Paneth.* New York: Interscience.

Holton, G. 1988. *Thematic Origins of Scientific Thought: Kepler to Einstein.* Cambridge, Massachusetts: Harvard University Press (1st ed., 1973).

Kripke, S. 1980. *Naming and Necessity.* Cambridge, Massachusetts: Harvard University Press.

Merton, R.K. 1973. The Normative Structure of Science. In: Merton, R.K. (ed.), *The Sociology of Science.* Chicago: University of Chicago Press, 267–278.

Murphy, C. 2002. Nanocubes and Nanoboxes. *Science* 298(13 December): 2139–2141.

Murphy, C. and Jana, N. 2002. Controlling the Aspect Ratio of Inorganic Nanorods and Nanowires. *Advanced Materials* 14(1): 80–82.

Putnam, H. 1975. The Meaning of Meaning. In: *Mind, Language and Reality, Collected Papers*, Vol. 2. Cambridge: Cambridge University Press.

Scerri, E. 2003a. Editorial 13. *Foundations of Chemistry* 5: 1–6.
Scerri, E. 2003b. How Ab Initio is Ab Initio Quantum Chemistry. *Foundations of Chemistry* 6: 93–116.
Schummer, J. 1996. *Realismus und Chemie. Philosophische Untersuchungen der Wissenschaft von den Stoffen.* Würzburg: Königshausen & Neumann.
Van Brakel, J. 1999. On the Neglect of Philosophy of Chemistry. *Foundations of Chemistry* 1: 111–174.
Van Brakel, J. 2000. *Philosophy of Chemistry*. Leuven: Leuven University Press.

CHAPTER 2

THE PHILOSOPHY OF CHEMISTRY
From Infancy Toward Maturity

JOACHIM SCHUMMER
Department of Philosophy, University of Karlsruhe, js@hyle.org

INTRODUCTION

The time of complaining about the neglect of the philosophy of chemistry is over now. With more than 700 papers and about 40 monographs and collections since 1990, philosophy of chemistry is one of the most rapidly growing fields of philosophy.[1] Perhaps too rapidly, as it has become arduous for insiders to keep up-to-date, troublesome for newcomers to approach the field and virtually impossible for outsiders to survey the main ideas. Being involved since the late 1980s, I think it is appropriate to pause for a while and write a paper of the kind "Where do we come from?—Where are we now?—Where should we go to?"[2]

Thus, the chapter is divided into three parts. We come from philosophical neglect—that is, virtually from nowhere—which I try to explain in the first part by recalling the disciplinary history of philosophy. We are now in a state of rapid growth, of prolific publishing, to which I provide some structure, in the second part, by pointing out the major trends and topics.[3] "Where should we go to?" is a question to which I can give only a personal answer, based on a pragmatist judgment of topics of infancy and topics of maturity that I try to justify in the third part.

THE PHILOSOPHERS' NEGLECT OF CHEMISTRY IN CONTEXT

A rule of thumb about the philosophers' interest in the sciences

Let me start with a look at the amount of literature published in the various sciences. Such data provide a good estimate of the relative size of the disciplines, in contrast to the coverage in the media and other talk about science. Figure 1 presents the number of new publications (books, papers, patents, etc.) as indexed by the major abstract journals in 2000 and 1979.

The most striking point is that chemistry is not only the biggest discipline, but also bigger than the total of all the other natural sciences, including all their related flourishing technologies. The *INSPEC* database (formerly, and strangely, called *Science*

D. Baird et al. (eds.), Philosophy of Chemistry, 19–39.

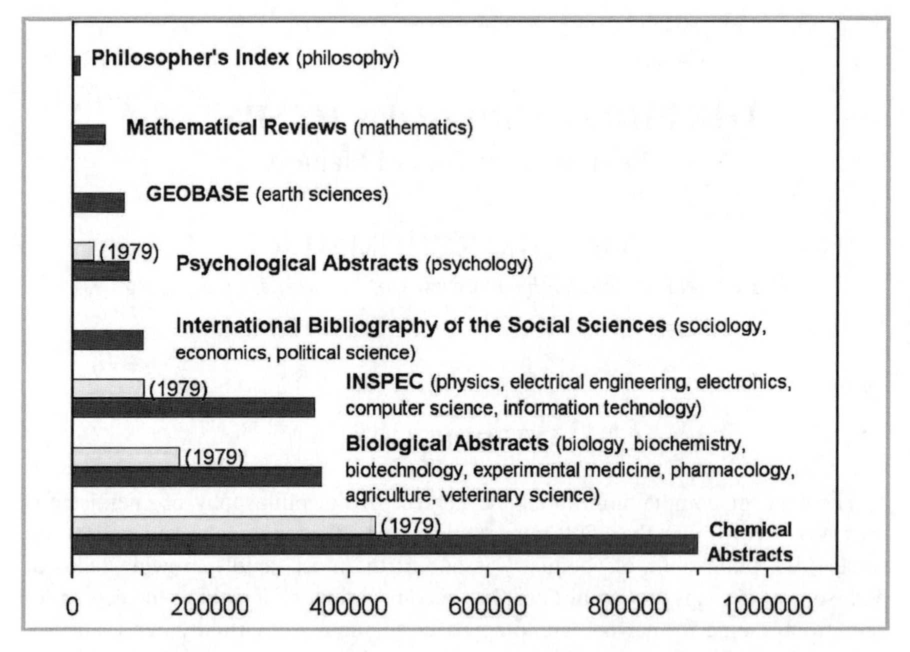

Figure 1: Number of new publications (papers, patent, books, etc.) indexed by major abstract journals in 2000 and 1979. 2000 data are from the journals' websites (in thousands: PI 10, MR 47, GB 74, PA 80, IBSS 100, INSPEC 350, BA 360, CA 899); 1979 data are from Tague et al. (1981).

Abstracts) has, besides physics, also "electrical engineering, electronics, communications, control engineering, computers and computing, and information technology" and a "significant coverage in areas such as materials science, oceanography, nuclear engineering, geophysics, biomedical engineering, and biophysics."[4] Yet, despite the rapid growth of computer sciences and information technology, all that now comes to less than 40% of the coverage of *Chemical Abstracts*. In addition, *Biological Abstracts* could greatly flourish in the past decade by covering, besides biology, also "biochemistry, biotechnology, pre-clinical and experimental medicine, pharmacology, agriculture, and veterinary science."[5] Despite the boom of the biomedical sciences and the overlap with chemistry, it is still only 40% of *Chemical Abstracts*. The earth sciences, less than a 10th of the size of chemistry, are even smaller than the social sciences and psychology.

The quantitative dominance of chemistry is no new phenomenon. To the contrary, many of the other abstract journals have grown more rapidly than *Chemical Abstracts* during the past three to four decades for various reasons. They could benefit from booming trends, as *Psychological Abstracts* from cognitive psychology; they absorbed new fields, as *Science Abstracts* did with computer science and information technology to become *INSPEC*; or they increased the overlap with chemistry, as *Biological Abstracts* did with biochemistry. By 1979, when no philosopher of science could even imagine the existence of philosophy of chemistry, *Chemical Abstracts* was more than four times as big as *Science Abstracts* (physics) and about three times as big as *Biological Abstracts*. Had those philosophers without prejudice gone into the laboratories, then they would have stumbled on chemistry almost everywhere.

Nowadays, philosophers overall write as many publications per year as chemists do in four days. Ironically, the figure suggests a rule of thumb about the philosophers' interest in the sciences: *the smaller the discipline, the more do philosophers write about it*, with the exception of the earth sciences. In the approximate order, philosophers write:

(1) about philosophy, as history of philosophy or, to be more correct, about what philosophical classics have published or left unpublished;
(2) about mathematics, as mathematical logic and philosophy of mathematical physics ("philosophy of science");
(3) about psychology, as philosophy of mind or naturalized epistemology;
(4) about the social sciences, as social and political philosophy and philosophy of social sciences;
(5) about experimental physics, as "philosophy of science";
(6) about biology as philosophy of biology; and
(7) to the smallest degree, about chemistry.

Thus, if philosophers produce general ideas about "science," there are good reasons to be mistrustful. On the other hand, if one really wants to understand the natural sciences, there are good reasons to start with chemistry.

A history of philosophy explanation

Many explanations have been advanced for the fact that philosophers have so stubbornly neglected chemistry as if it were virtually non-existent. Is it the lack of "big questions" in chemistry, its close relationship to technology, or the historically rooted pragmatism of chemists and their lack of interest in metaphysical issues? Or, is the alleged reduction of chemistry to physics (quantum mechanics) the main obstacle, so that, if chemistry were only an applied branch of physics, there would be no genuine philosophical issue of chemistry?

What all these approaches have in common is that they try to explain the neglect of philosophers by reference to chemistry, as if there were something wrong with chemistry. If there is only a bit of truth in our rule of thumb, however, it is the strange order of interest of philosophers that calls for explanation. In such an explanation, the neglect of chemistry would turn out to be only a special case, albeit an extreme one. I do not intend to provide a full explanation, but some hints from the *disciplinary* history of philosophy. Although we can, in retrospect, build a history of texts that we nowadays call philosophy, there is anything else than a continuous history of a discipline called philosophy, i.e., a history of a profession. The topical preferences of today's philosophers reflect the surprisingly young and awkward history of their discipline.

The relationship to mathematics goes back to a time, still at the turn to the 19th century, when "philosophy" was just the generic term for all the arts and sciences collected in the philosophical faculties, among which the mathematical arts made up the largest part since medieval times. Thus, professors in the philosophical faculties, i.e., professional "philosophers," had to teach much mathematics, including applied mathematics such as mechanics and geometrical optics, which should become part of modern physics in mid-19th century. Other natural sciences like chemistry and biology (natural history) were mainly taught in the higher faculties of medicine and, therefore, continued to be rather foreign to philosophers until today. When, during the 19th century, most disciplines including modern physics grew out of philosophy in the generic sense, psychology and the social sciences ("moral philosophy") still remained under the label of philosophy. Their final separation was not much before the early 20th century. Roughly speaking, the later a discipline became independent from philosophy (in the generic sense), the smaller is it nowadays, and the better are its historical ties to today's philosophers, in accordance with our rule of thumb.

The separation of the disciplines caused a serious crisis about the question if there are any topics left over for philosophy as an own discipline. While most disciplines grew independent by defining their own subject matter to be investigated by empirical methods, the remaining philosophers refused to do so. Many picked up Kant's 18th-century ideas, proposed prior to the disciplinary formation of the modern sciences, who had reserved metaphysical and epistemological foundations of the mathematical sciences as genuine philosophical topics, besides ethics and esthetics. That allowed them indeed to reconstruct a tradition that goes back to early modernity.

If we look at the tradition that modern philosophy of science considers as its own, it turns out that it is an extremely one-sided tradition, focused on mechanics that was formerly taught as "mixed" or "applied" mathematics in the philosophical faculties. A few points might illustrate that. First, the rise of early modern epistemology, both of the rationalist and the empiricist branch, with the exception of Francis Bacon, was closely connected to the rise of mechanical philosophy, which was strongly opposed to various kinds of chemical philosophies. Second, since modern physics has its theoretical roots in analytical or "rational" mechanics, which did not belong to the physical sciences but to mathematics still in the early 19th century, philosophical debates over "the scientific method" were, to a large part, about establishing mechanics as a *physical* science. Kant's former dictum that, unlike the experimental sciences, only mechanics is a proper science because it has an a priori foundation in mathematics, was an early and influential partisanship in these debates. That made it easy for Kantians to focus on mechanics and ignore the rest of the sciences. Finally, during the crucial phase of the professionalization of philosophy of science in the 20th century, it was first of all philosophically minded theoretical physicists who shaped the field with their numerous dissertations on the puzzles of quantum mechanics and relativity theory. They soon occupied most of the newly established chairs in philosophy of science—a situation that has not much changed since.

The long but historically incidental affinity to theoretical physics made "philosophers of science" neglect not only chemistry but also every other branch of the natural sciences, including experimental physics until recently. Relicts of the older meaning of "physics," as the generic term for the natural sciences still in the early 19th century and the ambiguity of the English term "physical" contributed to the confusion of philosophy of physics with philosophy of science. It was not before the early 1970s that biologists first reacted to the narrow focus and established their own groups together with biology-minded philosophers. It took another two decades that a similar movement occurred with respect to chemistry. In some sense, philosophy of science now lately repeats the 19th-century process of the ramification and professionalization of the natural sciences.

Philosophy of chemistry before the 1990s[6]

Although western mainstream philosophy of science has neglected chemistry, it is not true that there was no philosophy of chemistry before the 1990s. We will see later (pp. 26–27) that mainstream historiography of philosophy has simply ignored what philosophical classics had said about chemistry. Second, other scholars, particularly chemists and historians of chemistry, filled the gap left by professional philosophers. Third, philosophy of science in communist countries was broad enough to include chemistry, particularly in the period from the late 1950s to 1990.[7]

Dialectical philosophy of chemistry

Philosophy of chemistry in communist countries drew on Engels' dialectical materialism, where chemistry featured prominently as a case against what he called vulgar or French materialism, i.e., mechanical philosophy. Like Comte a few decades earlier but with reference to Hegel's distinction between "mechanism," "chemism," and "organism," Engels suggested a non-reductive hierarchy of the sciences. For the mechanical, chemical, and physiological level, he postulated different "forms of movement" each with own laws as well as general "dialectical laws" for the transformation from lower to higher levels.

While Engels' own treatment of chemistry remained fragmentary, 20th-century philosophers expanded on his ideas. They soon recognized that chemical phenomena could serve to illustrate universal laws of Engels' doctrine. For instance, acid–base reactions were used to exemplify his "law of contradictions" about counter-acting forces in nature. In addition, acid–base reactions, when performed by titration with indicators, could colorfully visualize his general "law of the change from the quantitative into the qualitative." Philosophers of science in communist countries had an established role in tertiary science education and were officially committed to interpret particular scientific facts, problems, and developments within the general framework of dialectical and historical materialism. Because Engels had reserved an own "form of movement" for chemistry, they were free to deal with chemistry as an autonomous field. Indeed, they produced a wealth of studies on modern chemical phenomena, laws, theories, theory dynamics, and sub-discipline formation. It is impossible to review the literature here, as there are studies on almost every philosophical issue, albeit of differing quality. At least it might be said that Engels' 19th-century framework was liberal enough to elaborate on such sophisticated topics as the relation between quantum chemistry and quantum mechanics, but epistemologically too naive to deal with quantum chemical concepts such as Pauling's resonance structures.[8]

The case of dialectical philosophy of chemistry proves that post-Kantian philosophy of science needed not have chemistry on its blind spot if it relied on later authorities. Engels published in the 1870s when the formation of modern scientific discipline was almost complete, with mechanics being only a sub-discipline of modern physics. Kant published a century earlier, before the modern discipline formation, in a pre-modern attitude to establish rational mechanics as the only real science and to discredit the then-growing experimental sciences. Although both have been made historical authorities of eternal validity in different philosophical ideologies, their views of science are only of historical significance nowadays, with Kant's view being surely much more anachronistic than Engels'. Yet, the Kantian legacy is still dominating in philosophy of science. For instance, when the eastern part of Germany was united to the western part in 1990, dialectical philosophy of chemistry immediately vanished in favor of the uninspired Kantianism that has pervaded the western part. Thus, the neglect of chemistry also results from the arbitrary choice of anachronistic authorities. In Part II, I will argue for an understanding of philosophy that gets rid of both anachronism and authorities.

Philosophy of chemistry without philosophers

As long as professional philosophers in the Western countries did not care about chemistry, scholars from various disciplines approached the field, each from their own perspective and with specific questions. Particularly, scholars of chemistry education have always recognized the need to reflect on methods and to work on the clarification of concepts, such that most of their journals are still a rich source for philosophers. Working chemists usually stumbled on philosophical issues when their own research challenged them to reflect on received notions or methodological ideas. Earlier prominent examples include Benjamin Brodie, Frantisek Wald, Wilhelm Ostwald, and Pierre Duhem. However, the series of philosophizing chemists did not stop in the early 20th century.[9] For instance, Paneth's (1962) work on isotopy made him think on the concept of chemical elements. Mittasch's (1948) reflection on the notion of causation in chemistry arose from his studies on chemical catalysis. Faced with the reluctance by contemporary scientists to accept his own theories and based on his detailed experience in laboratory practice, Polanyi (1958) challenged received rationalist methodologies of science by calling for social factors and the role of tacit knowledge. Caldin (1959, 1961), who as any other chemist primarily worked in the laboratory, argued that the then-prevailing Popperian methodology simply failed to grasp the role of experiments in the experimental sciences and the way scientists deal with theories.

Since the late 1970s, also theoretical chemists, who worked hard on the development of quantum chemical *models* for chemical purposes, began to question the naive reductionist view, albeit common among western philosophers of science, according to which chemical concepts and laws could simply be derived from quantum mechanical principles.[10] Woolley (1978), in a seminal paper, argued that the concept of chemical structure could not be deduced from quantum mechanics. Primas (1981, 1985) devoted a whole book on the issue of reductionism, arguing that quantum mechanical holism does not allow to derive statements about chemical objects without further assumptions. Del Re and Liegener considered chemical phenomena lying on a higher level of complexity that emerges from but does not reduce to the quantum mechanical level (Del Re 1987; Liegener and Del Re 1987a/b). Others began to question naive reductionism too (e.g., Theobald 1976; Lévy 1979; Bunge 1982; Weininger 1984).

Because the border between philosophy and history of science has never been sharply drawn, not surprisingly, many historians of chemistry approached the field by dealing with philosophical issues of the past, of which two for some time ranked so high among historical topics of chemistry that it is impossible to review the literature here. These are the metaphysical issue of atomism and the methodological issue of conceptual change and theoretical progress as exemplified by Thomas Kuhn's treatment of the "chemical revolution."[11] Of course, both topics attracted many philosophers as well. Particularly, the second topic caused, for a while, much fruitful collaboration and competition and a flood of case studies. Challenged by the historiographic rigor of their colleagues, the philosophers' case studies frequently did not much differ from historical work, except by their greater ambition to make them a case for or against a general methodological position, such as pro or con Popper, Kuhn, Lakatos, and so on. Yet, picking-up chemical stories as evidence for one or the other general methodology in science overall is hardly a conclusive argument, nor can it count as philosophy of chemistry proper. This has never been better criticized than by chemist-philosopher Ströker (1982) in one of the most detailed historical account of the "chemical revolution".

More than in other historical branches, historians of chemistry approached philosophical issues in a wealth of fine studies on the history of ideas, theories, and methods and the mutual impact between chemistry, on the one hand, and its neighboring disciplines, philosophy, humanities, religion, and the general society, on the other.[12] Insofar, as they did that with the aim of a better understanding of our present intellectual culture and the role of chemistry therein, they did a job that professional philosophers refused to do. Interestingly, the few western philosophers who dealt at book-length with chemistry, e.g., Bachelard (1932, 1953), Ströker (1963, 1982), and Dagonet (1969), were strongly historically minded. I will come back to this point, which is by no means pure chance (see the section on Analyzing the Structure of Chemistry).

The case of the pre-1990 history in western countries illustrates that there is a need for philosophical reflections on chemistry, whether professional philosophers take notice of that or not. The need comprises both the analysis of chemical details and integrating perspectives to locate chemistry in the overall culture and the history of ideas. Through their education, philosophers usually have particular skills to meet these needs, if accompanied by some understanding of and interest in chemistry. Both were lacking, however, with the exception of a few individuals who found themselves outside the established circles.

RECENT TRENDS AND TOPICS IN THE PHILOSOPHY OF CHEMISTRY

Establishing socially

The most obvious distinction of the emergence of the philosophy of chemistry in the 1990s to the previous period was its social establishment. Had former scholars worked in relative isolation, the new generation sought contact with each other and the exchange of ideas. Since the late 1980s, chemists, philosophers, and historians of chemistry began to gather in more or less formal working groups with regular meetings in many countries, such as the *Werkgroep Filosofie van de Chemie* in the Netherlands, the *Gruppo Nazionale di Storia e Fondamenti della Chimica* in Italy, and the *Arbeitskreis Philosophie und Chemie* in Germany. In addition, there was a call from the chemical industries for building bridges between chemistry and the humanities as an effort to improve the bad public image of chemistry.[13] In 1994, national meetings happened to grow to a series of international conferences in London (March), Karlsruhe (April), Marburg (November), and Rome (December). By 1997, international ties enabled the formal establishment of an *International Society for the Philosophy of Chemistry* with annual summer symposia. Two journals were launched, *HYLE: International Journal for Philosophy of Chemistry* (since 1995, edited by the author) and *Foundations of Chemistry* (since 1999, edited by Eric Scerri). The parallel rise of Internet technologies, which were soon employed for many purposes (e-journal, e-mail discussion forum, regularly updated bibliography, information boards for conferences, syllabi, etc.), essentially helped establish a community and attract a wider audience.

Rediscovering the philosophical classics

The historical neglect of chemistry mentioned earlier is, in part, also an artifact by historians of philosophy who simply ignored what philosophical classics had said about chemistry. This has been brought to light in a growing number of recent studies. A prominent example is Kant's *opus postumum* that was not published before the early 20th century (with an English translation as late as 1993), though it contained a complete revision of his former theoretical philosophy against the background of the new Lavoisian chemistry.[14] In addition, Hegel's extensive writing on chemistry, albeit placed in his most famous books, became subject to scholarly investigations only recently.[15] While Duhem's *La théorie physique. Son objet—sa structure* (1905–1906) has long been a classic in the philosophy of science and translated into many languages, his *Le mixte et la combinaison chimique* (1902) was translated into English not before 2002. Who would have thought that even Rousseau had written a book on chemistry (Bensaude-Vincent and Bernardi 1999)? It is up to historians of philosophy to explore further writings on chemistry in philosophical classics such as Leibniz, Schelling, Schopenhauer, Herschel, Comte, Peirce, Broad, Alexander, Mill, Cassirer, Bachelard to mention only a few who immediately spring to mind.

Struggling with reductionism

Still an important topic in today's philosophy of chemistry is reduction—not of biology to chemistry but of chemistry to physics. Criticism of reductionism plays different roles. First, it provides a more precise and technical understanding of the limits of quantum mechanical approaches to chemistry and thereby defines independent areas for the philosophy of chemistry. For instance, in a series of papers, Scerri (1991, 1994) has convincingly argued that quantum mechanical approaches are not able to calculate the exact electronic configurations of the atoms. Because that has been, since Bohr's early atomic theory, taken as the quantum mechanical explanation and reduction of the periodic system of the chemical elements, the latter is open to new philosophical analysis (see p. 28). Similar arguments can be found with regard to the concept of molecular structure, following-up the issues raised by Woolley (see p. 25). Second, the criticism of reductionism at the "lowest" level of chemistry to quantum mechanics challenges microreductionism as a general metaphysical, epistemological, or methodological position and thus contributes to general philosophy. In the most detailed philosophical study on various forms of reductionism (incl. supervenience and microstructural essentialism à la Putnam and Kripke), van Brakel (2000) has made chemistry a case to argue for a kind of pragmatism in which the "manifest image" of common sense and the empirical sciences has primacy over the "scientific image" of microphysics. For Psarros (1999), rejection of reductionism is even a necessary pre-supposition of his extensive work on the culturalist foundation of chemical concepts, laws, and theories, which he seeks in pre-scientific cultural practices, norms, and values. For many others, including myself, it supports a pragmatist and pluralist position about methods that distinguish clearly between fields of research where quantum mechanical approaches are poor or even useless when compared with other approaches, and those where they are strong and even indispensable. Third, once reductionism has lost its credit to secure the unity of the sciences, new relationships between autonomous sciences, like structural similarities and interdisciplinarity, have become subject to both philosophical and historical investigations.[16]

Adapting philosophical concepts

Because of their narrow focus on theoretical physics, concepts of mainstream philosophy of science frequently require considerable revision before they help shed some light on chemistry. It is the gap between what Kuhn (1976) has called the "mathematical" and the "Baconian sciences" that philosophers of chemistry must bridge, because modern chemistry comprises both. Since chemistry is by far the largest scientific discipline, with enormous impact on every other experimental science, philosophers of chemistry also make valuable contributions to our philosophical understanding of the sciences when they adapt classical concepts for an understanding of chemistry. Examples, which are scattered around in the two journals and in numerous general anthologies and collections,[17] include the concepts of experiment, law, model, prediction, explanation, natural kinds, substance, and process; the scientific approaches to concept building, model building and classification; the treatment of competing theories; methods of research in the sense of exploring the new; the role of instruments in research; the distinction and relationship between science and technology; and so on. In addition, the topic of scientific realism, sometimes misused to distinguish theoretical physics from the rest of the sciences that are thereby discredited as "immature sciences," appears in new light if applied to chemistry and even becomes a research methodological concept if applied to synthetic chemistry (Schummer 1996a). While philosophers of mathematical physics have confined methodology to the "context of justification," if not to proof theory, philosophers of experimental sciences like chemistry put more emphasis on the "context of discovery," i.e., on scientific research methodology.

Analyzing the structure of chemistry

Since each scientific discipline has its own fundamental concepts, methods, and theories, philosophy of chemistry reaches a state of maturity, in my view (see the last part of this chapter), when it focuses on peculiarities of chemistry. This requires not only a double competence in chemistry and philosophy, but also a deep understanding of the history of chemistry because our present scientific disciplines, with all their peculiarities, are historical entities, snapshots in a process of development. Thus, unlike general philosophers of science, with their eternal, albeit sometimes personal, ideas of "general science," many philosophers of chemistry do merge with historians of chemistry to analyze fundamental concepts, methods, and theories in modern chemistry. It is in these areas where many studies have been done in the past decade, such that I can give only a brief list of the most important topics.

As to fundamental concepts, philosophical and historical analyses include chemical concepts such as element, pure substance, chemical species, compound, affinity, chemical reaction, atom, molecular structure, and aromaticity.[18] Recent interest in chemical methods has focused both on practical methods, such as experimentation and instrumentation and chemical synthesis,[19] and on cognitive methods, such as the pictorial language of chemistry and the various forms of model building and representation.[20] Still neglected are methods of classification—probably a legacy of the traditional focus on the "classification-free" modern physics before the rise of the particle era—although recent studies on the periodic system combine classificatory and theoretical aspects.[21] With respect to chemical theories, the axiomatic mathematical structures of physics with their apparently universal validity made philosophers reluctant to accept what chemists, virtually without any difference in meaning, call theories, models, or laws. Thus, save the aforementioned studies on models in chemistry, most of the present works on chemical theories are strongly historically orientated or about quantum chemistry and physical chemistry.[22]

Transcending boundaries

Ironically, philosophy of chemistry emerged at a time when scientific activities increasingly transcended disciplinary boundaries toward problem-orientated research. From environmental science to nanotechnology (Schummer 2004c), chemists are heavily involved in these activities, such that philosophers of chemistry are challenged to take them seriously. Three recent books, which each of their own combine philosophical and historical analyses of transdisciplinary research, have taken up this challenge. Rheinberger (1997) has analyzed the experimental settings, epistemological conditions, and the transdisciplinary culture in which cancer research moved in the 1950s toward protein synthesis as the biochemical background of molecular biology. Applying ideas from ancient philosophy of nature and technology, Bensaude-Vincent (1998) has investigated modern material science, which has shifted from pure materials to composites that are individually designed for various technological purposes. With a critical view on classical approaches in the philosophy of science, Christie (2001) has examined the methodological basis on which theories of ozone depletion have actually been accepted in the atmospheric sciences since the 1970s.

Besides disciplinary boundaries of the sciences, there are also disciplinary boundaries within philosophy that philosophers of chemistry are about to transcend. If "philosophy of science" means philosophical reflection on science, there is no need to restrict that to epistemological, methodological, and metaphysical reasoning, as philosophers of physics have done. Philosophy is a much richer field, and sciences like chemistry have many more interesting, sometimes even more pressing, aspects that philosophers can deal with. Thus, recent and forthcoming work includes special issues on *Ethics of Chemistry* (by *HYLE*), on *Green Chemistry* (by *Foundations of Chemistry*), and on *Aesthetics and Visualization in Chemistry* (by *HYLE*). Once the full scope of philosophy is acknowledged, topics in the philosophy of chemistry spring up abundantly (see the section on Discovering Topics in the Philosophy of Chemistry: Some Examples). This might go at the expense of simple paradigms of the field, but the intellectual profit is incomparably richer.

FROM INFANCY TOWARD MATURITY: A PRAGMATIST POINT OF VIEW

Since every historical account orders the material according to certain preferences and values, also my review of the recent development of philosophy of chemistry in the last section is a personal one. It is based on certain ideas of what philosophy can and should do in my view to meet general societal needs and to avoid the shortcomings of stagnancy and ossification of which we have ample evidence in other fields of philosophy. In this part, which addresses the normative question "where should we go to?" I will now argue for these ideas by distinguishing between topics of infancy and topics of maturity from a pragmatist point of view.

Topics of infancy

To avoid misunderstandings, I emphasize that topics of infancy are very important topics and should not be neglected. They are important, however, only during a state of self-defining and structuring a field, and for preparing topics of maturity. I believe that philosophy of chemistry is, to a large part, still in that state.

Rediscovering the philosophical classics for a certain field is a typical topic of infancy or crisis (in the original Greek meaning of separation). It belongs to the general topic of disciplinary history writing. Instead of coming from nowhere, the new field is shown to be rooted in an old tradition, from which it receives authority and importance. Classical examples are Priestley's history of electricity from 1767 and Ostwald's history of electrochemistry from 1896. Similarly, complaining about or analyzing the previous neglect of a field is a topic of infancy, as everything I do in the present chapter. On the other hand, rediscovering the classics can be an inspiring enterprise, in that it shows us long forgotten perspectives to be followed-up in future philosophical research.

Second, struggling with reductionism is for the most part, but not always, a topic of infancy regarding chemistry, albeit a topic of maturity for the philosophy of physics. Once more, I emphasize, it is important for the philosophy of chemistry. It prepares the grounds for more relaxed and deeper studies of subjects whose logical independence has been proven before and thus moves toward topics of maturity. Furthermore, it places the philosophy of chemistry in the context of general philosophy and thus contributes to its broader acceptance. It helps us develop a better understanding of the much more complex relationships between the sciences, both historical and logical.

Third, adapting classical philosophical concepts to an analysis of chemistry is a preliminary topic too, whereas the analysis itself is not. For philosophers, concepts are tools like spectrometers are for chemists. If they use the wrong tools, the results are at best irrelevant. For instance, in the received philosophy of science view, an experiment is something to test, improve, or develop a theory. We can find this notion in chemistry too, and we even find instances for all the roles philosophers have ever assigned to experiments. Yet, if we look at what chemists mean by "experiment," it turns out that the great majority use the term "experiment" in a sense that philosophers are hardly aware of (Schummer 2004a).

A pragmatist definition of topics of maturity

The topics I have mentioned so far are topics of infancy only with respect to what I consider topics of maturity. I claim that a philosophical field reaches a state of maturity only if it defines its own issues with respect to the peculiarities of its object. In our case that means that topics of maturity are those that derive from peculiarities of chemistry. This is by no means a truism in philosophy. Indeed, many philosophers reject the idea and claim to the contrary that philosophical issues are prior to or independent of any particular objects and that true philosophical issues are only so-called perennial or general problems. For them, chemistry would be interesting only if and insofar as it provides examples or illustrations of their general problems. If the perennial view were right, there would never be a particular philosophy of chemistry, nor of biology, and so on, because the philosophical interest in such fields would be only instrumental to solving perennial issues. As I have indicated with my historical remarks on philosophy in the first part of this chapter, there are good reasons to doubt such perennial problems of philosophy. They rather result from an arbitrary historiography of philosophy with references to favorite authorities. Instead, I would defend a kind of pragmatism according to which philosophical issues should always derive from the specific objects (the *pragmata* in Greek).

Unlike the perennial view, the pragmatist view requires a detailed understanding of chemistry, not only in its present form but also in its historical development. That is why philosophy of chemistry needs to be closely linked to the history of chemistry. Moreover, because of the incredible size of contemporary chemistry (about one million publications per year), there is a need for new methods to grasp what chemists are actually doing. It does not help much to ask a chemist or two what their four million colleagues are doing. That is simply beyond the intellectual scope of individuals (Schummer 1999). Instead, one needs empirical methods for qualified statements. For instance, most people would not believe that much more than half of the chemists are synthesizing new substances on a regular basis; yet, that is what qualified statistical analysis says (Schummer 1997b). Last but not least, philosophical issues of chemistry should be related to the problems chemists are actually confronted with.

Once we have a better understanding of what chemists are really doing and concerned with, we can do the philosophical analyses that I consider topics of maturity. These include the conceptual, metaphysical, and methodological investigations mentioned earlier. However, it also includes topics beyond conventional philosophy of science.

An eye on the philosophy of biology

It is instructive to see what philosophers of biology consider as the major topics of their field nowadays, in a state of maturity as we can assume. For instance, take the table of contents of the recent anthology *The Philosophy of Biology*, ed. by David L. Hull and Michael Ruse, published by Oxford University Press in 1998 (see Table 1). It is a collection of 36 previous articles divided into 10 thematic sections. The first four sections are about metaphysical, methodological, and conceptual issues of evolutionary theory; Section V is devoted to ontological issues in taxonomy. The

Table 1: Table of Contents of *The Philosophy of Biology*, edited by David L. Hull and Michael Ruse, Oxford University Press, 1998; 36 articles grouped in 10 thematic sections.

Section headings	Related philosophical branches
I. Adaptation	
II. Development	Metaphysical, methodological, and conceptual issues of evolutionary theory
III. Units of Selection	
IV. Function	
V. Species	Ontology, classification
VI. Human Nature	Anthropology
VII. Altruism	Ethical theory
VIII. The Human Genome Project	Applied ethics
IX. Progress	Philosophy of history, epistemology
X. Creationism	Philosophy of religion

second half of the book relates various topics of biology to different branches of philosophy, beyond received philosophy of science, such as anthropology, ethical theory, applied ethics, philosophy of history, and philosophy of religion. There is no general scheme of how scientific topics are related to philosophical issues, such that we could simply transfer them to philosophical issues of chemistry. Instead, the philosophical issues derive from the peculiarities of the biological topics.

Discovering topics in the philosophy of chemistry: some examples

What philosophers of chemistry can learn, and in my view should learn, from the philosophy of biology is the discovery of philosophical issues that derive from the peculiarities of chemistry. There is no simple rule or recipe to do that, as philosophy is a creative enterprise. However, one can take the branches of philosophy as guides, as analytical instruments rather than as sets of perennial issues. To illustrate that, I finally provide a few examples from my own recent work, arranged according to different branches of philosophy and each with reference to, in my view, obvious peculiarities of chemistry.[23]

(1) *Logic*: Prior to formal or symbolic logic, philosophical logic explores conceptual structures that we use in representing and reflecting the world. If we look at the conceptual structure of chemistry, it turns out that it is built on a peculiar kind of relations (Schummer 1997c, 1998a). For instance, a chemical property describes a complex dynamic and context-dependent relation between various substances, and not something that is proper to an isolated thing. Some philosophical classics, like Hegel, Cassirer, and Bachelard, recognized the peculiar relational structure of chemistry earlier, which is also deeply rooted on our common sense, for instance, when we use metaphors from chemistry to describe social relations. However, with the exception of philosopher-chemist Peirce and his followers, logicians have badly neglected relations, such that we have only little understanding of how the complex conceptual structures of chemistry are built out of basic chemical relations (Schummer 1996, 1998a).

(2) *Ontology*: When conceptual structures are used to frame the world, we are entering the field of ontology. For its 20 million and more substances, chemistry has built the most advanced classification system of all science, for which there is no model in other fields. Taking chemistry as a classificatory science seriously, from an ontological point of view, requires the investigation of very abstract notions, such as chemical entity, species identity, similarity, class membership, distinctions, and hierarchies. Again, we are only at the beginning of an understanding that shows that the ontological structure of chemistry is presently in state of change (Schummer 2002a). For instance, while classical chemical classification has been based on a strict correspondence between pure substances and molecules, recent trends to include also quasi-molecular species challenges the traditional system and causes deep ontological issues about the criteria of species identity.

(3) *Methodology*: While philosophers of science have been telling us that scientists aim at a true theoretical description of the natural world, the great majority of chemists (which also means the great majority of scientists, see p. 20) have

actually been engaged in synthesizing new substances, i.e., changing the natural world. Chemical synthesis is, to be sure, the most obvious peculiarity of chemistry, albeit the most neglected one because it is foreign to any received idea of philosophy of science. Since methodology is concerned with scientific methods, a methodological understanding of synthetic chemistry requires an analysis of the goals, procedure, techniques, and dynamics of that endeavor (Schummer 1997a/b). In particular, we need a better understanding of the theoretical concepts of chemistry which, unlike quantum mechanical concepts, not only predict natural phenomena but also serve as guidance for the production of new substances a million times per year.

(4) *Philosophy of language and semiotics*: Chemists have their own sign language of structural formulas and reaction mechanism that calls for semiotic analysis (Schummer 1996c). Rather than being only a set of iconic, symbolic, or index signs (according to the classical semiotics of Peirce and Morris), it is a particular language system endowed with elements of theoreticity that allows chemists to communicate with each other in a concise and precise manner about chemical entities and relations; it is also the major theoretical device for predicting and producing new chemical substances (Schummer 1998a). As compared to its extraordinary success, we still have little philosophical understanding of how theory is encoded in the language and if the system is a new type of theory, different from what we know from other sciences.

(5) *Philosophy of technology*: Although synthetic chemistry stands out because of its productivity, it does not entail as such being a kind of technology. Indeed, synthetic chemistry is a good case to evaluate various standard distinctions between science and technology according to their underlying ideas of science (Schummer 1997d). It turns out that all these ideas are hardly in agreement with contemporary experimental science. The case of synthetic chemistry may help sharpen the concepts of science and technology for a better understanding of their relations. For instance, historians of technology now consider the chemical industry the only real science-based industry. If that is true, philosophers are challenged to explain the peculiar epistemological relation between chemical knowledge and technology without mixing them up.

(6) *Philosophy of nature*: While in all the other natural sciences, nature is, by definition, the object of their study, chemistry breaks the rule by the strange opposition of "natural versus chemical," held by both chemists and non-chemists. Here, philosophers are required to analyze on which peculiar notion of nature the opposition rests and whether such an opposition is well grounded or not. Historical analyses reveal that the opposition came up only after ancient Greek philosophy and has pervaded the Christian era from the earliest times up to the present (Schummer 2001a, 2003b, 2004b). Systematical analyses show that the opposition, while being descriptively meaningless, serves normative ends in implicit quasi-moral judgments (Schummer 2003b). Again, here is a task for philosophers to prepare the grounds for a normative discourse by making the implicit explicit.

(7) *Philosophy of literature*: Another important field for making the implicit explicit is the literature insofar as they mediate the public image of scientists. There are many complaints about the gap between the "two cultures" and about the bad image of scientists, expressed in such figures as the "mad scientist." It is also well known that scientists in the media frequently appear as a mixture of the medieval alchemist and the

modern chemist. Yet, little is known why writers shaped these figures. Investigations of the 19th-century literature reveal that writers indeed began pillorying chemistry of all sciences because of metaphysical, theological, and moral concerns (Schummer in preparation). Again, it is up to philosophers of chemistry to analyze those accusations of particularism, materialism, atheism, and hubris, which historically form the philosophical background of the public image of science nowadays.

(8) *Ethics*: Although applied ethics is now a flourishing field of philosophy, chemistry is almost neglected. That is more than surprising because many major moral issues, from environmental issues to pharmacological issues and chemical weapon research, are strongly related to chemistry. The fact that synthetic chemists do not only produce knowledge, but also change our material world has caused public concerns since at least two centuries. It requires sober ethical analyses that separate moral issues proper from quasi-moral concerns as mediated through the literature or normative notions of nature in order to prepare an ethical framework for a moral discourse (Schummer 2001b, 2001–2002, in preparation).

(9) *Esthetics*: More than any other scientists, chemists make heavy use of all kinds of means of visualization, from simple drawings to virtual reality. In addition, chemists have increasingly made claims to the beauty of their synthetic products, and there is clear empirical evidence that this is also an actual research motivation. Both call for systematic investigations of the role of esthetics in chemical research (Schummer 1995, 2003a; Spector and Schummer 2003). In particular, esthetic analysis may help understand crucial issues of research creativity and innovation. If beauty is an accepted research value, we need to understand on what esthetic theory that notion of beauty is based and how esthetic values relate to other research values, both epistemological and moral.

The list of topics could be easily extended further. Yet, I do not want to put my own research into the focus. If the chapter has a message, then it is to encourage philosophers of chemistry to think for their own, be skeptical about perennial problems, and discover new philosophical issues of chemistry. There are plenty of them waiting for discovery.

Let me finally come back to Figure 1. If one compares the size of philosophy with that of chemistry, it is evident that the philosophy of chemistry will never be even visible in such a figure. However, the sheer mass of chemistry and its omnipresence does make the philosophy of chemistry one of the most important and difficult fields of philosophy.

NOTES

1. Interested readers may find a regularly updated online bibliography maintained by the author at http://www.hyle.org/service/biblio.htm.
2. A first draft of this chapter was presented as the opening lecture of the *Sixth Summer Symposium on the Philosophy of Chemistry, Washington, DC, 4–8 August 2002.*
3. Earlier review articles include van Brakel and Vermeeren (1981); van Brakel (1996, 1999, 2000, chap. 1); Ramsay (1998); Brock (2002); Schummer (2003c).

4. Quoted from the *INSPEC* website.
5. Quoted from the website of *Biological Abstracts*.
6. This section and Part II borrows from Schummer (2003c). For brevity reason, references are largely confined to monographs and collections; more references may be found in the online bibliography quoted in Note 1.
7. Unfortunately, the philosophy of chemistry literature in communist countries is not reviewed yet, except for the German Democratic Republic, see Laitko (1996). For a bibliography, see Schummer (1996b).
8. For the historical background of the debates on resonance structures, see Rocke (1981). For philosophical analyses, see Laitko and Sprung (1970), pp. 80–109; Vermeeren (1986); Schummer (1996a, Sec. 6.5.2).
9. Interested readers may find more on the chemists mentioned in this section in the *HYLE* series "Short Biographies of Philosophizing Chemists."
10. An earlier paper by a theoretical chemist that includes many later critical ideas is Hartmann (1965).
11. Hoyningen-Huene (1998) has argued that the chemical revolution was even Kuhn's paradigm case for his notion of scientific revolutions.
12. To name but a few historians with obvious philosophical interests: J.H. Brooke, W. Brock, M.P. Crosland, A.G. Debus, E. Farber, R. Hooykaas, D. Knight, T.H. Levere, A.N. Meldrum, H. Metzger, M.J. Nye, A. Rocke, and many more.
13. Two valuable publications from these initiatives are Mittelstraß and Stock (1992) and Mauskopf (1993).
14. Carrier (1990), Vasconi (1999), van Brakel (2000, chap. 1.2).
15. Engelhardt (1976), Burbidge (1996), Ruschig (1997).
16. For example, Danaher (1988), Janich and Psarros (1998), and Reinhardt (2001).
17. In addition to the ones mentioned elsewhere, these include Janich (1994); Psarros, Ruthenberg, and Schummer (1996); Mosini (1996); McIntyre and Scerri (1997); Psarros and Gavroglu (1999); Sobczynska and Zeidler (1999); Bhushan and Rosenfeld (2000); Earley (2003); and Sobczynska, Zeidler, and Zielonacka-Lis (2004).
18. Book-long philosophical or historical studies on chemical concepts include Klein (1994), Schummer (1996a), Psarros (1999), Görs (1999), Brush (1999), van Brakel (2000), and Neus (2002).
19. Baird (1993), Rothbart and Slayden (1994), Schummer (1996a, 1997a/b, 2002a), Holmes and Levere (2000), Morris (2002), and Baird (2004).
20. Laszlo (1993), Janich and Psarros (1996), Francoeur (1998), Schummer (1999–2000), van Brakel (2000), and Klein (2001).
21. Scerri (1998, 2001, forthcoming) and Cahn (2002).
22. For example, Nye (1993), Schummer (1998b), and Gavroglu (2000).
23. A comprehensive treatment of such topics will be Schummer (2005c).

REFERENCES

Bachelard, G. 1932. *Le pluralisme cohérent de la chimie moderne*. Paris: J. Vrin.

Bachelard, G. 1953. *Le matérialisme rationnel*. Paris: Press Universitaire de France.

Baird, D. 1993. Analytical Chemistry and the Big Scientific Instrumentation. *Annals of Science* 50: 267–290.

Baird, D. 2004. *Thing Knowledge: A Philosophy of Scientific Instruments*. Berkeley & Los Angeles, CA: University of California Press.

Bensaude-Vincent, B. 1998. *Eloge du mixte. Matériaux nouveaux et philosophie ancienne*. Paris: Hachette.

Bensaude-Vincent, B. and Bernardi, B. (eds.). 1999. *Jean-Jacques Rousseau et la chimie*. Special issue of *Corpus* 36.

Bhushan, N. and Rosenfeld, S. (eds.). 2000. *Of Minds and Molecules. New Philosophical Perspectives on Chemistry*. New York: Oxford University Press.

Brock, W. 2002. The Philosophy of Chemistry. *Ambix* 49: 67–71.

Brush, S.G. 1999. Dynamics of Theory Change in Chemistry. *Studies in History and Philosophy of Science* 30A: 21–79, 263–302.
Bunge, M. 1982. Is Chemistry a Branch of Physics. *Zeitschrift für allgemeine Wissenschaftstheorie* 13: 209–223.
Burbidge, J.W. 1996. *Real Process: How Logic and Chemistry Combine in Hegel's Philosophy of Nature*. Toronto: University of Toronto Press.
Cahn, R.M. 2002. *Historische und philosophische Aspekte des Periodensystems der chemischen Elemente*. Karlsruhe: Hyle Publications.
Caldin, E.F. 1959. Theories and the Development of Chemistry. *The British Journal for the Philosophy of Science* 10: 209–222.
Caldin, E.F. 1961. *The Structure of Chemistry in Relation to the Philosophy of Science*. London–New York: Sheed & Wards [reprinted in *Hyle* (2002), 8(2): 103–121].
Carrier, M. 1990. Kants Theorie der Materie und ihre Wirkung auf die zeitgenössische Chemie. *Kantstudien* 81(2): 170–210.
Christie, M. 2001. *The Ozone Layer: A Philosophy of Science Perspective*. Cambridge: Cambridge University Press.
Dagognet, F. 1969. *Tableaux et langages de la chimie*. Paris: du Seuil.
Danaher, W.J. 1988. *Insight in Chemistry*. Lanham, MD: University Press of America.
Del Re, G. 1987. The Historical Perspective and the Specificity of Chemistry. *Epistemologia* 10: 231–240.
Duhem, P. 2002. *Mixture and Chemical Combination and Related Essays*, ed. and trans. P. Needham. Dordrecht: Kluwer.
Earley, J.E. (ed.). 2003. *Chemical Explanation: Characteristics, Development, Autonomy*. New York: New York Academy of Science (Annals of the New York Academy of Science, Vol. 988).
Engelhardt, D.V. 1976. *Hegel und die Chemie: Studie zur Philosophie und Wissenschaft der Natur um 1800*. Wiesbaden: Pressler.
Francoeur, E. 1998. *The Forgotten Tool: A Socio-historical Analysis of Mechanical Molecular Models*. PhD Thesis, McGill University, Montreal.
Gavroglu, K. (ed.). 2000. *Theoretical Chemistry in the Making: Appropriating Concepts and Legitimising Techniques*. Special issue of *Studies in History and Philosophy of Modern Physics* 31(4).
Görs, B. 1999. *Chemischer Atomismus. Anwendung, Veränderung, Alternativen im deutschsprachigen Raum in der zweiten Hälfte des 19. Jahrhunderts*. Berlin: ERS-Verlag.
Hartmann, H. 1965. Die Bedeutung quantenmechanischer Modelle für die Chemie. In: *Sitzungsberichte der Johann Wolfgang Goethe-Universität*. Wiesbaden: F. Steiner, 151–168.
Holmes, F.L. and Levere, T.H. (eds.). 2000. *Instruments and Experimentation in the History of Chemistry*. Cambridge, MA: MIT Press.
Hoyningen-Huene, P. 1998. Kuhn and the Chemical Revolution. In: SILFS (ed.), *Atti del Convegno Triennale della SILFS. Roma, 3–5 Jan. 1996*. Pisa: Editioni ETS, 483–498.
Janich, P. (ed.). 1994. *Philosophische Perspektiven der Chemie*. Mannheim: Bibliographisches Institut.
Janich, P. and Psarros, N. (eds.). 1996. *Die Sprache der Chemie*. Würzburg: Königshausen & Neumann.
Janich, P. and Psarros, N. (eds.). 1998. *The Autonomy of Chemistry*. Würzburg: Königshausen & Neumann.
Klein, U. 1994. *Verbindung und Affinität: Die Grundlegung der Neuzeitlichen Chemie an der Wende vom 17. zum 18. Jahrhundert*. Basel: Birkhäuser.
Klein, U. (ed.). 2001. *Tools and Modes of Representation in the Laboratory Sciences*. Dordrecht: Kluwer.
Kuhn, T.S. 1976. Mathematical vs Experimental Traditions in the Development of the Physical Sciences. *The Journal of Interdisciplinary History* 7: 1–31.
Laszlo, P. 1993. *La Parole des choses ou le langage de la chimie*. Paris: Hermann.
Laitko, H. 1996. Chemie und Philosophie: Anmerkungen zur Entwicklung des Gebietes in der Geschichte der DDR. In: Psarros, N., Ruthenberg, K., and Schummer, J. (eds.), *Philosophie der Chemie: Bestandsaufnahme und Ausblick*. Würzburg: Königshausen & Neumann, 37–58.
Laitko, H. and Sprung, W.D. 1970. *Chemie und Weltanschauung*. Leipzig, 80–109.
Lévy, M. 1979. Les relations entre chimie et physique et le problème de la réduction. *Epistemologia* 2: 337–369.

Liegener, Ch. and Del Re, G. 1987a. Chemistry vs. Physics, the Reduction Myth, and the Unity of Science. *Zeitschrift für allgemeine Wissenschaftstheorie* 18: 165–174.

Liegener, Ch. and Del Re, G. 1987b. The Relation of Chemistry to Other Fields of Science: Atomism, Reductionism, and Inversion of Reduction. *Epistemologia* 10: 269–283.

Mauskopf, S.H. (ed.). 1993. *Chemical Sciences in the Modern World*. Philadelphia: University of Pennsylvania Press.

McIntyre, L. and Scerri, E.R. (ed.). 1997. *The Philosophy of Chemistry*. Special isssue of *Synthese* 111(3).

Mittasch, A. 1948. *Von der Chemie zur Philosophie*. Ulm: Ebner.

Mittelstraß, J. and Stock, G. (eds.). 1992. *Chemie und Geisteswissenschaften: Versuch einer Annäherung*. Berlin: Akademie-Verlag.

Morris, P.J.T. (ed.). 2002. *From Classical to Modern Chemistry. The Instrumental Revolution*. Cambridge: The Royal Society of Chemistry.

Mosini, V. (ed.). 1996. *Philosophers in the Laboratory*. Roma: Accademia Nazionale di Scienze—Lettere ed Arti di Modena.

Neus, J. 2002. *Aromatizität: Geschichte und Mathematische Analyse eines fundamentalen chemischen Begriffs*. Karlsruhe: Hyle Publications.

Nye, M.J. 1993. *From Chemical Philosophy to Theoretical Chemistry. Dynamics of Matter and Dynamics of Disciplines 1800–1950*. Berkeley: University of California Press.

Paneth, F.A. 1962. The Epistemological Status of the Chemical Concept of Element. *The British Journal for the Philosophy of Science* 13: 1–14; 144–160 [first published in German in 1931].

Polanyi, M. 1958. *Personal Knowledge: Towards a Post-Critical Philosophy*. Chicago: University of Chicago Press.

Primas, H. 1981. *Chemistry, Quantum Mechanics and Reductionism. Perspectives in Theoretical Chemistry*. Berlin–Heidelberg–New York: Springer.

Primas, H. 1985. Kann Chemie auf Physik reduziert werden? *Chemie in Unserer Zeit* 19: 109–119; 160–166.

Psarros, N. 1999. *Die Chemie und ihre Methoden. Ein philosophische Betrachtung*. Weinheim: Wiley-VCH.

Psarros, N. and Gavroglu, K. (eds.). 1999. *Ars Mutandi: Issues in Philosophy and History of Chemistry*. Leipzig: Leiziger Universitätsverlag.

Psarros, N., Ruthenberg, K., and Schummer, J. (eds.). 1996. *Philosophie der Chemie: Bestandsaufnahme und Ausblick*. Würzburg: Königshausen & Neumann.

Ramsey, J.L. 1998. Recent Work in the History and Philosophy of Chemistry. *Perspectives on Science* 6: 409–427.

Reinhardt, C. (ed.). 2001. *Chemical Sciences in the Twentieth Century: Bridging Boundaries*. Weinheim: Wiley-VCH.

Rheinberger, H.J. 1997. *Toward a History of Epistemic Things. Synthesizing Proteins in the Test Tube*. Stanford: Stanford University Press.

Rocke, A.J. 1981. Kekulé, Bulterov, and the Historiography of the Theory of Chemical Structure. *The British Journal for the History of Science* 14: 27–57.

Rothbart, D. and Slayden, S.W. 1994. The Epistemology of a Spectrometer. *Philosophy of Science* 61: 25–38.

Ruschig, U. 1997. *Hegels Logik und die Chemie. Fortlaufender Kommentar zum 'realen Mass'*. Bonn: Bouvier.

Scerri, E.R. 1991. The Electronic Configuration Model, Quantum Mechanics and Reduction. *The British Journal for the Philosophy of Science* 42: 309–325.

Scerri, E.R. 1994. Has Chemistry Been at Least Approximately Reduced to Quantum Mechanics? *PSA* 1: 160–170.

Scerri, E.R. 1998. The Evolution of the Periodic System. *Scientific American* 279(3): 56–61.

Scerri, E.R. (ed.). 2001. *The Periodic System*. Special issue of *Foundations of Chemistry* 3(2).

Scerri, E.R. (forthcoming). *The Story of the Periodic System*. New York: McGraw-Hill.

Schummer, J. 1995. Ist die Chemie eine Schöne Kunst? Ein Beitrag zum Verhältnis von Kunst und Wissenschaft. *Zeitschrift für Ästhetik und Allgemeine Kunstwissenschaft* 40: 145–178.

Schummer, J. 1996a. *Realismus und Chemie. Philosophische Untersuchungen der Wissenschaft von den Stoffen*. Würzburg: Königshausen & Neumann.

Schummer, J. 1996b. Bibliographie chemiephilosophischer Literatur der DDR. *Hyle* 2: 2–11.

Schummer, J. 1996c. Zur Semiotik der chemischen Zeichensprache: Die Repräsentation dynamischer Verhältnisse mit statischen Mitteln. In: Janich, P. and Psarros, N. (eds.). *Die Sprache der Chemie*. Würzburg: Königshausen & Neumann, 113–126.

Schummer, J. 1997a. Scientometric Studies on Chemistry I: The Exponential Growth of Chemical Substances, 1800–1995. *Scientometrics* 39: 107–123.

Schummer, J. 1997b. Scientometric Studies on Chemistry II: Aims and Methods of Producing new Chemical Substances. *Scientometrics* 39: 125–140.

Schummer, J. 1997c. Towards a Philosophy of Chemistry. *Journal for General Philosophy of Science* 28: 307–336.

Schummer, J. 1997d. Challenging Standard Distinctions between Science and Technology: The Case of Preparative Chemistry. *Hyle* 3: 81–94.

Schummer, J. 1998a. The Chemical Core of Chemistry I: A Conceptual Approach. *Hyle* 4: 129–162.

Schummer, J. 1998b. Physical Chemistry: Neither Fish nor Fowl? In: Janich P. and Psaros, N. (eds.). *The Autonomy of Chemistry*. Würzburg: Königshausen & Neumann, 135–148.

Schummer, J. 1999. Coping with the Growth of Chemical Knowledge: Challenges for Chemistry Documentation, Education, and Working Chemists. *Educación Química* 10(2): 92–101.

Schummer, J. (ed.). 1999–2000. *Models in Chemistry*. Special issue of *Hyle* 5:2 ("Models in Theoretical Chemistry"); 6:1 ("Molecular Models"); 6:2 ("Modeling Complex Systems").

Schummer, J. (ed.). 2001–2002. *Ethics of Chemistry*. Special issue of *Hyle* 7(2) and 8(1).

Schummer, J. 2001a. Aristotle on Technology and Nature. *Philosophia Naturalis* 38: 105–120.

Schummer, J. 2001b. Ethics of Chemical Synthesis. *Hyle* 7: 103–124.

Schummer, J. 2002a. The Impact of Instrumentation on Chemical Species Identity: From Chemical Substances to Molecular Species. In: Morris, P.J.T. (ed.). *From Classical to Modern Chemistry. The Instrumental Revolution*. Cambridge: The Royal Society of Chemistry, 188–211.

Schummer, J. 2003a. Aesthetics of Chemical Products: Materials, Molecules, and Molecular Models. *Hyle* 9: 73–104.

Schummer, J. 2003b. The Notion of Nature in Chemistry. *Studies in History and Philosophy of Science* 34: 705–736.

Schummer, J. 2003c. The Philosophy of Chemistry. *Endeavour* 27: 37–41.

Schummer, J. 2004a. Why Do Chemists Perform Experiments? In: Sobczynska, D., Zeidler, P., and Zielonacka-Lis E. (eds.), *Chemistry in the Philosophical Melting Pot*. Frankfurt et al.: Peter Lang, 395–410.

Schummer, J. 2004b. Naturverhältnisse in der modernen Wirkstoff-Forschung. In: Kornwachs, K. (ed.), *Technik—System—Verantwortung*. Münster: LIT Verlag, 629–638.

Schummer, J. 2004c. Multidisciplinarity, Interdisciplinarity, and Patterns of Research Collaboration in Nanoscience and Nanotechnology. *Scientometrics* 59: 425–465.

Schummer, J. (forthcoming). Forschung für die Armen versus Forschung für die Reichen: Verteilungsgerechtigkeit als moralisches Kriterium zur Bewertung der angewandten Chemie. In: Sedmak, C. (ed.), *Option für die Armen in den Wissenschaften*. Freiburg: Herder.

Schummer, J. Historical Roots of the 'Mad Scientist': Chemists in 19th-century Literature. *Ambix* (in preparation).

Schummer, J. *Chemical Relations. Topics in the Philosophy of Chemistry* (book manuscript in preparation).

Sobczynska, D. and Zeidler, P. (eds.). 1999. *Chemia: Laboratorium, Mysli i Dzialan*. Poznan: Wydawnictwo Naukowe Instytutu Filozofii UAM.

Sobczynska, D., Zeidler, P., and Zielonacka-Lis, E. (eds.). 2004. *Chemistry in the Philosophical Melting Pot*. Frankfurt et al.: Peter Lang.

Spector, T. and Schummer, J. (eds.). 2003. *Aesthetics and Visualization in Chemistry*. Special issue of *Hyle* 9.

Ströker, E. 1967. *Denkwege der Chemie. Elemente Ihrer Wissenschaftstheorie*. Freiburg: Alber.

Ströker, E. 1982. *Theoriewandel in der Wissenschaftsgeschichte: Chemie im 18. Jahrhundert*. Frankfurt: Klostermann.

Tague, J., Beheshti, J. and Rees-Potter, L. 1981. The Law of Exponential Growth: Evidence, Implications, and Forecasts. *Library Trends* 30: 125–150.

Theobald, D.W. 1976. Some Considerations on the Philosophy of Chemistry. *Chemical Society Reviews* 5(2): 203–214.

van Brakel, J. 1996. Über die Vernachlässigung der Philosophie der Chemie. In: Psarros, N., Ruthenberg, K., and Schummer, J. (eds.), *Philosophie der Chemie: Bestandsaufnahme und Ausblick*. Würzburg: Königshausen & Neumann, 13–26.

van Brakel, J. 1999. On the Neglect of the Philosophy of Chemistry. *Foundations of Chemistry* 1: 111–174.

van Brakel, J. 2000. *Philosophy of Chemistry. Between the Manifest and the Scientific Image*. Leuven: Leuven University Press.

van Brakel, J. and Vermeeren, H. 1981. On the Philosophy of Chemistry. *Philosophy Research Archives* 7: 501–552.

Vasconi, P. 1999. *Sistema delle scienze naturali e unità della conoscenza nell'ultimo Kant*. Firence: Olschki.

Vermeeren, H.P.W. 1986. Controversies and Existence Claims in Chemistry: The Theory of Resonance. *Synthese* 69: 273–290.

Weininger, S.J. 1984. The Molecular Structure Conundrum. Can Classical Chemistry be Reduced to Quantum Chemistry. *Journal of Chemical Education* 61: 939–944.

Woolley, R.G. 1978. Must a Molecule Have a Shape? *Journal of the American Chemical Society* 100: 1073–1078.

SECTION II

CHEMISTRY AND THE HISTORY AND PHILOSOPHY OF SCIENCE

CHAPTER 3

ARISTOTLE'S THEORY OF CHEMICAL REACTION AND CHEMICAL SUBSTANCES

PAUL NEEDHAM
Department of Philosophy, University of Stockholm, SE-106 91 Stockholm, Sweden

INTRODUCTION

The aim of this chapter is to formulate the Aristotelian idea that a substance is potentially, and not actually, present in a mixture or combination, and to consider some problems to which it gives rise.[1] This conception was adopted and adapted by Duhem (1902) as the basis of his understanding of the diversity of chemical substances systematized by chemical formulas and for providing what he thought was the natural interpretation of matter as treated by thermodynamics. Aristotle's theory is pursued here from this Duhemian perspective, with an eye to its relation to contemporary macroscopic conceptions of chemical substance. Despite Duhem's many suggestions and allusions, however, questions remain which need to be resolved if the interpretation is to be at all definite.[2] Much the same might be said of Aristotle's own writings, and leading commentators have sought to fill out perceived lacunas with reconstructions which, as for example Freudenthal readily concedes, are "not explicit in Aristotle's writings" (1995, 200).

The aspects emphasized by Duhem suggest an approach somewhat different from Freudenthal's, not least in being, within a perspective of two and a half millennia, distinctly modern. This consciously anachronistic viewpoint may not be of interest to Aristotelian scholars, but may succeed in arousing interest on the part of those concerned with the roots of modern chemistry in an alternative to more familiar atomistic conceptions. The general approach of the theory focuses on what can still be regarded as the right problem, and the general lines of criticism of the atomists' approach to this problem could, in their essentials, be validly applied throughout the 19th century. One such line of criticism fixes on the idea of indivisibility, and was deflected only when atoms came to be treated as composed of several other entities. But this was by no means the only line of attack. The criticisms have been obscured by undue emphasis on the positivist objection entirely based on the premise that atoms could not be seen, which plays no role in the following discussion, just as it did not for the interesting 19th-century opponents of atomism. The Aristotelian conception therefore has greater claim to be considered continuous with modern views than seems to be generally recognized, and a critical assessment of its merits illuminates the development of the modern macroscopic conception of substance. An exposition

D. Baird et al. (eds.), Philosophy of Chemistry, 43–67.

and analysis of Aristotle's theory of mixture from this perspective is developed in this chapter.

ARISTOTLE'S WAY OF SETTING UP THE PROBLEM

In the course of spelling out the problem which the theory of substance should address, Aristotle considers various theories of his predecessors. The atomic theory of Leucippus and Democritus is singled out as of particular merit:

> Plato only investigated the conditions under which things come-to-be and pass-away; and he discussed not *all* coming-to-be, but only that of the elements. He asked no questions as to how flesh or bones, or any of the other similar things, come-to-be... In general, no one except Democritus has applied himself to any to these matters in a more than superficial way. Democritus, however, does seem not only to have thought about all the problems, but also to be distinguished from the outset by his method.

Whereas others fail to "give any account of combination... offering no explanation, e.g., of action and passion—how in natural actions one thing acts and the other undergoes action," Aristotle recognized that the atomists had a clear idea of how to deal with these issues:

> Democritus and Leucippus, however, postulate the "figures," and make alteration and coming-to-be result from them. They explain coming-to-be and passing-away by their dissociation and association, but alteration by their grouping and position. And since they thought that the truth lay in the appearance, and the appearances are conflicting and infinitely many, they made the "figures" infinite in number (*DG* I.2, 315^a29-315^b11).[3]

Nevertheless, his critique is scalding. It might be considered to involve two strands, though intimately intertwined—that no coherent account is given of the ultimate atoms or "indivisibles," and that the theory fails to deal with the central problem, namely of explaining how new compounds, with new properties, can appear as a result of mixing. I will concentrate on the latter issue, of explaining the appearance and disappearance of substances, and remark on the former only in passing. In the light of Lewis's (1998) claims that the presocratic atomists did not hold atoms to be indivisible, but merely undivided, it should be pointed out that Aristotle's case against atoms is by no means exhausted by arguments against the impossibility of separating their spatial parts. Even when the translation has him referring to atoms as indivisibles, indivisibility may not be at issue.[4]

On the atomic conception of combination, "constituents will only be combined relatively to perception; and the same thing will be combined to one percipient, if his sight is not sharp—while to the eye of Lynceus nothing will be combined" (*DG* I.10, 328^a13f.), so that substances only appear mixed to us because we cannot distinguish the individual juxtaposed particles. If the ingredients were thus preserved, combination would not be an objective physical state of matter, which Aristotle thought there was no reason to accept. In fact, the atomic view has no real conception of chemical substances, but merely postulates shapes of atoms "infinite in number," as we saw, to account for "conflicting and infinitely many" appearances. Aristotle considers, on the contrary, that "there is a limit to the number of differences" (*Cael* III.4, 302^{b}31) exhibited by different substances, and an explanation of the variety of compounds must be based on a finite, and preferably small, number of properties. Democritus and Leucippus "have never explained in detail the shapes of the various elements . . . [and] the principles they assume are not limited in number, though such limitation would necessitate no other alteration in their theory" (*Cael* III.4, 303^{a}12-19). These primary properties, as they may be called, would explain how the mixing process can generate new substances, endowed with new properties, with the passing away of the properties of the original ingredients. No solution to what, as one translator puts it, "the paradox as to how flesh and bones and any other compounds will result from the elements," is to be found by conceiving of the process as mere "composition, just as a wall comes-to-be from bricks and stones . . . [and that] this 'mixture' will consist of the elements preserved intact but placed side by side with one another in minute particles" (*DG* II.7, 334^{a}20-1, 334^{a}26f.; Forster's trans.). Rather, the mixing process would, by Aristotle's lights, involve the affecting of one piece of matter by another, and it is the primary properties which endow matter with the capacity to affect and be affected by other matter.

Aristotle's general views on the capacities and susceptibilities of matter are reflected in his difficulties in understanding the existence of atoms. For whereas the postulation of atoms was to explain the changes involved in mixing by different combinations of particles which persist unchanged, Aristotle could not see how they could be coherently described without ascribing to them properties which would prevent them persisting unchanged:

> it is impossible that they [the indivisibles] should not be affected by one another: the slightly hot indivisible, e.g., will suffer action from one which far exceeds it in heat. Again, if any indivisible is hard, there must also be one which is soft; but the soft derives its very name from the fact that it suffers a certain action—for soft is that which yields to pressure (*DG* I.8, 326^{a}11-15).

Aristotle is posing a dilemma: either atoms are preserved intact, in which case no combination occurs, or combination occurs, in which case the atoms are not preserved intact. The notion of "combination" at issue here includes that of cohesion, but involves more than the mere sticking together of bits of matter; it includes the idea of the creation of a new *kind* of stuff, featuring properties not possessed by the original material. How could the mere juxtaposition of inert bodies generate new properties?

Since matter affects and is affected by matter in virtue of being of one kind of substance or another, questions arise of what indivisibles are made of and how that dictates their behavior. One of Aristotle's difficulties with atoms stems from this idea that they must be made of some substance and his rejection of the idea that a single substance is naturally separated, like a heterogeneous mixture of oil and water which are two substances. Pieces of the same kind of matter would form a uniform whole, as do drops of water when they come into contact: "if all of them [the indivisibles] are uniform in nature, what is it that separated one from another? Or why, when they come into contact, do they not coalesce into one, as drops of water run together when drop touches drop . . . ?" (*DG* I.8, 326^{a}31-3). Yet if they were made of different kinds of stuff, comprising a heterogeneous mixture, they would evidently be divided (and not merely divisible) into separate parts. They must be uniform. But then, why should uniformly constituted bodies divide into particles of one size and not another? As Aristotle puts it, "why should indivisibility *as such* be the property of small, rather than of large, bodies?" (*DG* I.8, 326^{a}23f.).

Aristotle simply takes the view that some quantities of matter are uniform or homogeneous, whereas others are heterogeneous mixtures, exhibiting a natural separation into homogeneous parts. The uniformity implied by homogeneity seems to be taken as an indication of a single substance, and compounds resulting from mixing are required to be homogeneous, just like the elements themselves: "if combination has taken place, the compound *must* be uniform—any part of such a compound is the same as the whole, just as any part of water is water" (*DG* I.10, 328^{a}10f.), where again water is taken as a paradigm. Passages such as these clearly express the view that single substances are homogeneous, or as Aristotle says, homoeomerous (literally, whose parts are like the whole). This is borne out by what he says in favor of rejecting the natural separability of a single substance. What would now be understood as a heterogeneous, two-phase mixture of a single substance—of ice and (liquid) water, say—all of whose parts are not like one another, is foreign to the Aristotelian view. Different phases were understood, it seems, to comprise different substances, so that a single substance constitutes a single phase. Moreover, the general tenor of his argument certainly suggests that he subscribed to the converse thesis—that homogeneous quantities comprise a single substance. He says, for example, that anyone who adopts "the view of Anaxagoras that all the homoeomerous bodies are elements . . . misapprehends the meaning of element. Observation shows that even mixed bodies are often divisible into homoeomerous parts; examples are flesh, bone, wood, and stone" (*Cael* III.4, 302^{b}13-7). The fact that Aristotle takes it that there obviously is a problem of explaining the variety of compounds presupposes that different substances can be easily recognized as such, and the homogeneity criterion of sameness of substance would provide such a means.

The distinction between homogeneous and heterogeneous matter derives from observable features of what would now be called macroscopic phenomena. But it would be wrong to understand Aristotle as dismissing atomism on the grounds that the ultimate inhomogeneity it ascribes to mixtures is not directly observable. Aristotle's respect for observation did not preclude theoretical treatment. But he saw nothing that would count in favor of the atomist view which would justify contravening the apparent homogeneity of bodies, and much that counts against it. He devised his own approach as the only way of accommodating and explaining what could be observed. Later, the Stoics came to see an opening for a third general view, but this required denying a thesis common to both the atomist and Aristotelian conceptions of the impenetrability of matter—a denial which many philosophers have balked at since. In any case it is questionable, although this cannot be argued here, to what extent they succeed in putting Aristotle's position aside.[5]

THE ARISTOTELIAN THEORY OF MIXING

The requirements of the Aristotelian conception of mixing, which play a major role, we saw, in his criticism of the atomic theory in *DG* I.8, are first laid out in *DG* I.6. The new properties of mixts arise from the interplay of contraries when the bodies that bear them come into contact. Contrary properties capable of endowing matter with the ability to act and the susceptibility to be acted upon are thus restricted to tangible properties—those which act by virtue of spatial contact of the bodies which bear them. In *DG* II.2 it is argued that these contraries can be reduced to two pairs—moist and dry, hot and cold—from which other tangible contraries, such as solidity and fluidity, can be derived, but which are not themselves further reducible to one another.

Chemists have continued in considerably more recent times to base their explanations of the transformations of substances on the interplay of contraries. Acids are contrasted with bases, electropositive with electronegative substances, and so on. But by modern lights the range of properties Aristotle thought it adequate to consider is restricted, to say the least; and even within the confines of this range, the arguments for the reduction to the four primary qualities of fundamental phase properties and such like are, where comprehensible, unconvincing. Even from within Aristotle's purview, the status, within this scheme of reduction, of lightness and heaviness exhibited by the natural motions of the elements as they move to different parts of the sublunar region, remains something of a mystery. Nevertheless, the general strategy should be clear enough. A body is warmed when it comes into contact with a hot body, provided the former is cold, or at least less hot; and a moist body wets another when in contact. The primary qualities endow matter with the capacity of acting and being acted upon—the "action" and "passion" in virtue of which something happens on mixing, involving the disappearance of old substances and the appearance of new.

The broad strokes of Aristotle's theory of mixing must strike the modern chemist as remarkably familiar. General considerations of stability determine whether reaction is possible at all; and where possible, separate considerations are relevant to the circumstances determining the rate of reaction. As for the general "thermodynamics,"

> Amongst those things . . . which are both active and passive, some are easily divisible. Now if a great quantity (or a large bulk) of one of these materials be brought together with a little (or with a small piece) of another, the effect produced is not combination, but increase of the dominant; for the other material is transformed into the dominant. (That is why a drop of wine does not combine with ten thousand gallons of water; for its form is dissolved, and it is changed so as to merge in the total volume of water.) On the other hand, when there is a certain equilibrium between their powers, then each of them changes out of its own nature towards the dominant; yet neither becomes the other, but both become an intermediate with properties common to both (*DG* I.10, 328^{a}24-30).

Immediately following this passage, Aristotle points out that such circumstances are not sufficient to determine the "kinetics" of mixing:

> Thus it is clear that only those agents are combinable which involve a contrariety—for these are such as to suffer action reciprocally. And further, they combine more freely

> if small pieces of each of them are juxtaposed. For in that condition they change one another more easily and more quickly; whereas this effect takes a long time when agent and patient are present in bulk (*DG* I.10, 328^{a}31-5).

Two kinds of interaction between substances are distinguished in the former passage. The first strikes modern readers as distinctly odd, and they will search Aristotle's texts in vain for an answer to the question immediately arising of exactly when a difference in bulk is just sufficient to tip the balance from combination to the overwhelming, as Aristotle sometimes says, of the less abundant ingredient by the other. Even the notion of the relative amounts of material, and the property whose magnitude provides the measure of this relative amount, is, as we will see, deeply unclear. But here it seems that "bulk" means volume, in which case we can ask how the existence of processes of this first kind squares with the initial claim that contact is a prerequisite for all interactions. Two bodies come into contact, it would seem, at their common surface. A common surface is the same for both bodies, however, whatever the ratio of their volumes. Accordingly, if exceeding a certain difference in (or ratio of?) volume is necessary and (in the absence of hindrance) sufficient for a certain kind of interaction, and for precluding effects of another kind of interaction, then the contact condition cannot be taken quite literally. This seems to be Aristotle's view, since he does loosen up the contact condition.

The important point introduced by the contact condition is that "contact *in the proper sense* applies only to things which have position" (*DG* I.6, 322^{b}32f.), and the imposition of the property of occupying a region of space precludes, for example, numbers exerting any influence. (Aristotle thought that the atomist conception of Leucippus and Democritus "in a sense makes things out to be numbers or composed of numbers. The exposition is not clear, but this is its real meaning" [*Cael* III.4, 303^{a}9f.].) Although Aristotle goes on immediately to say that touching is "defined" as having extremities together, it soon transpires that he does not hold himself to this definition at all. Rather, touching "in general applies to things which, having position, are such that one is able to impart motion and the other to be moved" (323^{a}23f.). Had he strictly adhered to the original definition, it would follow that touching is a symmetric relation. But he allows that, although "as a rule, no doubt, if A touches B, B touches A ... Yet it is possible ... for the mover to merely touch the moved, and that which touches need not touch a thing which touches it" (323^{a}25ff.). So touching has become a general relation of affecting, and possibly occasioning being affected, standing between objects which occupy positions in space and, perhaps, are not too far apart. The passage just quoted raises the possibility of touching without being touched, which may be interpreted as a statement of Aristotle's general thesis that all changes can be traced to an unmoved mover. [Gill (1989, 199ff.) gives the example of the art of medicine, "located" in the soul of the doctor, touching the sick patient without itself being touched.] Interactions of the first kind, in which one body overwhelms another, would seem to be cases of touching without being touched. For in overwhelming another, a body is not itself affected by its action like one which combines with another body by a process of the second kind described in the passage from *On Generation and Corruption* I.10 above.

Combination entails a mutual affection in which the primary qualities are affected in both interacting bodies. As the idea is more succinctly summarized later,

> flesh and bones and the like come-to-be when the hot is becoming cold and the cold becoming hot and they reach the mean, for at that point there is neither hot nor cold . . . In like manner also it is in virtue of being in a "mean" condition that the dry and the moist and the like produce flesh and bone and the other compounds (*DG* II.7, 334^{b}25ff.; Forster's trans.).

When the relation of the bulks of the two quantities of matter is below the threshold for overwhelming, the aggregated powers endowed by the primary qualities strive toward a mean, intermediate between the original values. Opposing contraries correspond to different degrees on a single scale. Hot and cold, for example, are two extremes of the same magnitude, let us call it warmth, and moist and dry two extremes of humidity. In modern terms, the primary notions are basically relational, things not as warm as one another being distinguished in virtue of the one being warmer than, or conversely, colder than the other, and the other primary notion of humidity can be similarly understood in terms of a relation "is at least as humid as." Not only does what is hot, or supremely warm, affect what is minimally warm, or cold; it is in general possible that "the slightly hot . . . , e.g., will suffer action from [that] which far exceeds it in heat" (*DG* I.8, 326^{a}12f.), or indeed whatever has some degree of a primary quality by whatever has a different degree. Thus, any two bodies of the same order of size, the one warmer than the other, can combine in a mixing process of the second kind to generate yet a further degree of warmth, intermediate between the original two and therefore different from both. This seems to be Aristotle's conception of the new properties resulting from combination.

An important corollary now follows from these considerations. The compounds generated by such mixing processes are homogeneous in the sense explained above: "if combination has taken place, the compound *must* be uniform—any part of such a compound is the same as the whole, just as any part of water is water." Homogeneity also applies to the original ingredients. (If not, on the supposition introduced above, the ingredient will be a heterogeneous mixture, comprising homogeneous parts, which are single substances.) Consequently, *each part* of the resulting compound will display a new property—reducible, on Aristotle's view, to new degrees on the scales of warmth and humidity. No part will exhibit any of the original properties, and therefore none of the original substances is preserved in the compound. They have all passed away.

Although this important corollary of Aristotle's theory is an inevitable consequence of the homogeneity requirement, it is not dependent on that requirement. Combination, according to Aristotle, is a process arising from the interaction of bodies in virtue of their powers and susceptibilities to affect and be affected by one another. It is precisely the properties conferring these powers and susceptibilities which are characteristic of the initial ingredients, and the changes wrought by mixing ensure that the properties initially characterizing quantities of matter as one kind of substance or another no longer obtain, and are replaced by others characterizing a different substance. His charge against the atomists, we saw, was that even the small particles they envisage must be made of some substance, in virtue of which they will be susceptible to modification and thus not survive intact in the creation of new substances. Such changes might conceivably be incorporated into a theory which does not require the products of mixing to be strictly homogeneous.

This corollary is a central pillar of Aristotle's theory, and a fundamental bone of contention on competing theories. He prepares for it in the lead up to the twofold theory of mixing by introducing his famous distinction between actual and potential:

> Since, however, some things *are potentially* while others *are actually*, the constituents can be in a sense and yet not-be. The compound may *be actually* other than the constituents from which it has resulted; nevertheless each of them may still *be potentially* what it was before they were combined, and both of them may survive undestroyed (*DG* I.10, 327^{b}23ff.).

So the original ingredients "may survive" in so far as they *can* be recovered from the compound. This is sharply distinguished from separability of ingredients as envisaged by the atomists: "re-separation . . . differs from reconstitution in implying that in some way the ingredients are there all along" (Sorabji 1988, 67). Something has to bear the property of being possibly of one or other of the original substance kinds, and this is the material of the compound which, as he says, "can be" of the original constituent kinds yet (actually) is not. This is not the ludicrous claim that the matter of the compound does not have these original kind properties (corollary above) and yet does. It is the claim that this matter does not have these properties but could have. The precise formulation of this modal feature calls for more attention, however, and is taken up again in section on What Bears the Aristotelian Potentialities? below.

ARISTOTLE'S CONCEPTION OF ELEMENTS

The second book of *On Generation and Corruption* gets under way with the reminder that "we have still to consider the elements" (*DG* II.1, 328^{b}32). This notion has not played any role in Aristotle's development of the general conditions governing the coming-to-be and passing away of substances in the first book. In particular, combination was not introduced as a matter of simple bodies coming together to form complex bodies, but of bodies endowed with the potential to act and the susceptibility to be acted upon coming together and realizing their potential. The elements are eventually defined in II.3, after certain preliminaries have been dealt with.

The preliminaries establish the general pattern to which the definitions of the elements are to conform. To begin with,

> Our own doctrine is that although there is a matter of the perceptible bodies (a matter out of which the so-called elements come-to-be), it has no separate existence, but is always bound up with a contrariety (*DG* II.1, 329^{a}24f.).

Aristotle goes on say that matter "underlies, though it is inseparable from, the contrary qualities; for the hot is not matter for the cold nor the cold for the hot, but the *substratum* is matter for them both" (329^{a}30f.). Scholars are divided about whether the passage from which these quotations are taken commits Aristotle to a doctrine of prime matter or not. I confine myself to the observation that Aristotle is saying that what has to be defined are properties—the properties of being such-and-such a kind of element for the various elemental substances—in terms of other properties, and this in turn imposes certain constraints. Properties have to be predicated of something—the subject that bears them, or as he says, underlies them, or is a substratum for them. There is no question of isolating the underlying matter in the sense of entirely "separating" it from the properties it bears. At any particular time, then, a quantity of matter can be designated as that which bears such-and-such a property at that time—that which is hot, for example, or the hot for short. This does not entail that the properties borne by matter are necessary features of the matter; properties may be acquired and relinquished. But at no time is matter entirely devoid of properties. The elemental properties, in particular, are transient. Aristotle subscribed to a doctrine of the transmutability of the elements, and the matter of the recently destroyed element is that "out of which" the new one comes-to-be. In principle, all the properties (of a certain group) possessed by a quantity of matter might conceivably change simultaneously without there ever being a time when no property applies. But Aristotle seems to impose restrictions. Elements at least cannot all undergo transmutation by modifying all their element-characterizing properties at once, and some must remain unchanged while the others are modified. But under no circumstances could the hotness of a body, say, be preserved at the same time as it is modified into its contrary. During a process in which the warmth of a body is decreased, there must be something bearing each of the different degrees of warmth. And although such changes might be loosely described by saying that the properties of bodies may be changed, it is "bodies [that] change into one another . . . , whereas the contrarieties do not change" (329^{b}1f.). Change is change in matter involving the gaining of properties it lacks; properties are not what change.

"Matter" will be used here to refer to the bearer of substance properties such as the elemental ones, with no implication of the features sometimes ascribed to the so-called prime matter, such as imperceptibility and pure potentiality. Matter is observable unless of microscopic dimensions, and endowed with potentialities in virtue of the properties it bears.

As for the properties in terms of which the elemental properties are to be defined, Aristotle assumes they are to be taken from contraries, and goes on to ask "What sorts of contrarieties, and how many of them, are to be accounted principles of body? . . . [O]ther thinkers assume . . . them without explaining why they are *these* or why they are just *so many*" (329^{b}3f.).

Aristotle seeks the explanation for the particular defining properties of the elements in some distinctive feature of what generally explains coming-to-be and passing away by enabling them to enter into one of the two kinds of reaction described in I.10. As with any substance capable of entering into reaction, he claims in II.2 that "the elements must be reciprocally active and susceptible, since they combine and are transformed into one another" (329^{b}23f.). Moreover, "hot and cold, and dry and moist, are terms, of which the first pair implies *power to act* and the second pair *susceptibility*" (329^{b}24f.). Aristotle goes on to argue that other relevant features are all derived from warmth and humidity, "but that these admit of no further reduction. For the hot is not *essentially* moist or dry, nor the moist *essentially* hot or cold; nor are the cold and the dry derivative forms, either of one another or of the hot and the moist. Hence these must be four" (330^b25ff.). Thus, other relevant features are reduced to warmth and humidity, and the distinctive feature of elemental properties is that they are defined in terms of what he later calls "contrary extremes" (*DG* II.8, 335^a8) of warmth and humidity—what is hot is what is as warm as can be, and so forth. This explains, then, why the principles are four in number.

Substances in general are to be characterized by properties conferring an active power to affect others and a passive susceptibility to be affected, and these are derivable, according to Aristotle, from degrees of the primary qualities, warmth and humidity. Four special cases are the extreme values of these primary determinables: maximal and minimal degrees of warmth—hot and cold—and of humidity—moist and dry—presumably conferring particularly potent powers and susceptibilities. The extreme degrees are, at any rate, what distinguish the elements as Aristotle defines them in II.3. Of these,

> any four terms can be combined in six couples. Contraries, however, refuse to be coupled; for it is impossible for the same thing to be hot and cold, or moist an dry. Hence it is evident that the couplings of the elements will be four: hot with dry and moist with hot, and again cold with dry and cold with moist (*DG* II.3, 330^a30ff.).

So there will be exactly four elements, the pairs corresponding, respectively, to fire, air, water, and earth.

Aristotle goes on immediately to speak of the elements as simple, and in so doing appeals to a generally accepted idea on the part of "all who make the simple bodies elements" (330^b7). Just as he questioned whether his predecessors had any explanation to offer of the characterizations they offered of elements, so he might have raised some questions about their conception of simplicity. But surprisingly enough he does not do so in *On Generation and Corruption*, where he goes on to speak of the elements as simple despite the fact that the manner of defining them just presented makes no appeal to any such notion. This must strike the contemporary reader as rather strange. When Lavoisier introduced the modern notion of an element as the "last point which analysis is capable of reaching," he stressed that there can be no further restrictions on the number and character of the elements; "we must admit, as elements, all the substances into which we are capable, by any means, to reduce bodies by decomposition" (1789, p. xxiv). He criticized the Aristotelian doctrine of the four elements accordingly. If we adopt the one procedure—of defining the elements as what cannot be further decomposed, or of defining them of the basis of other considerations such as the possession of extreme contraries as Aristotle does in *On Generation and Corruption*—then we cannot, it seems, reasonably adopt the other. Not that to do otherwise necessarily involves a contradiction; but it should be an open question whether what satisfies the condition imposed by one definition also satisfies the other, and not something that can be decided a priori. Nevertheless, Aristotle does precisely that, actually proposing a definition of the general idea of an element in a manner akin to Lavoisier's definition in another work, *On the Heavens*. The Aristotelian notion of simplicity is deeply problematic, however. The issue is taken up in section on Simple Substances below.

This is not the only point of conflict with the Lavoisian conception. Having defined the elements, Aristotle goes on immediately to ascribe them a feature which radically distinguishes them from what Lavoisier understood as an element. The theme of the next chapter is introduced with the claim that "it is evident that all of them are by nature such as to change into one another" (*DG* II.4, 331^{a}13), and Aristotle takes it upon himself to "explain what is the manner of their reciprocal transformation, and whether every one of them can come-to-be out of every one" (331^{a}11f.). Bodies with contrary properties will naturally affect one another, and Aristotle quickly turns to considerations bearing on the rate of change:

> it is easier for a single thing to change than for many. Air, e.g., will result from Fire if a single quality changes; for Fire, as we saw, is hot and dry while Air is hot and moist, so that there will be Air if the dry be overcome by the moist. Again, Water will result from Air if the hot be overcome by the cold . . . (*DG* II.4, 331^{a}25-30).

In general, any element can change into another with a defining feature in common. Any element can, in fact, be transmuted into any other. But:

> the transformation of Fire into Water and of Air into Earth, and again of Water and Earth into Fire and Air, though possible, is more difficult because it involves the change of more qualities. For if Fire is to result from Water, both the cold and the moist must pass-away . . . This second method of coming-to-be, then, takes a longer time (*DG* II.4, 331^{b}4-11).

Gill suggests that transmutations of the latter kind are slower because continuity requires that the two contraries at issue must change consecutively and not simultaneously: "the dryness of fire can first be overpowered to yield an interim air, and then the heat can be overpowered finally to yield water; alternatively, the heat of the fire can first be overpowered to yield an interim earth . . . " (1989, 74). Rather than requiring that all interim stuff be elemental, the idea of continuity might equally be thought to require successive change through all intermediate degrees. A quantity of water, say, would then be changed to air by all of the quantity successively passing through all intermediate degrees of warmth, so no part of this quantity is of either element kind within the transition period. Which of the two kinds of transformation process distinguished in I.10 is involved in transmutation is settled by these passages from II.4, however. Transmutation is a process of the first kind, in which one piece of matter of a given kind is sufficiently larger in size than the other piece of matter of another kind that the primary qualities of the former overwhelm those of the other, and no intermediate degree is realized. (In I.10 it seems that the conjunction of properties constituting a substance kind, elemental or not, simultaneously overwhelm; but in II.4 it is the overwhelming of extremes of warmth and humidity one at a time that is at issue.) It is not clear how such a discontinuous transformation could be anything but instantaneous if there are no intermediate stages to be traversed. But this also seems to say something about particle size, which we saw in the Aristotle's Conception of Elements section is a factor governing the rate of transformation. Aristotle says, for example, that "if water has first been dissociated into smallish drops, air comes-to-be out of it more quickly; while, if drops of water have first been associated, air comes-to-be more slowly" (*DG* I.2, 317^a27-30). Perhaps certain parts of the overwhelming quantity which are sufficiently larger than particular parts of the overwhelmed quantity and in sufficiently close contact with these parts overwhelm first. Then other parts can come into suitable position and overwhelm, so that the entire transformation does take place in a temporally ordered sequence of stages. On this account, it is the division of quantities that explains the rate of transformation. But if the process proceeds by stages, why cannot the quantity converted to an interim element in the first stage be transformed into the final element at the same time as the second stage of transformation of another part to the interim element is under way, and so on? In that case, transformations of fire to water and air into earth need hardly be much slower that other elemental transmutations.

It is not clear to me that a choice can be made between alternatives such as these on the basis of Aristotle's texts, or that it is possible to see on what Aristotle based his various claims about differing rates of transformation in *DG* II.4. Nevertheless, strange as all this may seem to us, Aristotle took himself to be addressing an "evident" phenomenon seen in everyday occurrences. The conversion of water to air is how Aristotle conceives of evaporation, which according to his theory proceeds more quickly than the conversion of water to fire when exposed to an excess of the latter; solidification involves the coming-to-be of earth; combustion the conversion of earth to fire, which is again quicker than conversion of water to fire.

On the post-Lavoisian conception, elements are usually understood to be preserved during chemical transformation. But the general corollary of the Aristotle's Conception of Elements section implies that, on Aristotle's theory, when elements combine to form compounds by a mixing process of the second kind, they are not preserved in the resulting compound. And now we have seen that they are susceptible to another kind of transformation, transmuting into other elements as a result of a mixing process of the first kind. Elements are as ephemeral as any other kind of substance on the Aristotelian view. Whatever might be said about this doctrine, it should not be confused with the denial of the persistence of matter. Although the ideas of preservation of elemental kinds and the persistence of matter are intimately associated on the post-Lavoisian view, Aristotle has made ample provision for distinguishing these ideas, and so cannot be criticized for incoherently denying persistence on the grounds of denying preservation of the elements. This is central to the present interpretation and worth spelling out properly.

When water becomes air, for example, that which was the water, and so was cold and moist, *is identical with* that which is later air, and so hot and moist. The primary qualities hot, cold, moist, and dry are therefore properties which may apply at one time, t, and not at another, t', to pieces, or following usage in the discussion of the logic of mass terms, quantities,[6] of matter $\pi, \rho, \sigma, \ldots$ The primitive relations of warmth and humidity in terms of which the extreme contraries are defined are, accordingly, also to be understood as time-dependent relations, π being as warm as ρ for example, at a time t.[7] Thus, defining water is a matter of defining a property by:

π is water at t if and only if π is cold at t and π is moist at t, and air by

π is air at t if and only if π is hot at t and π is moist at t,

and so on. A transmutation from water into air will, accordingly, involve the truth of a conjunction of the kind "π is water at t and π is air at t'," where t' is later than t but the same object, π, bears the different properties.

A TEXTUAL PROBLEM

Persistence of matter, thus interpreted in terms of the identity of what undergoes change, is arguably what is at issue in the passages van Brakel (1997, 258 and fn. 14) cites in support of his claim that the general view since Aristotle has been that "[w]ater in all its modifications (liquid, solid, vapour) is the same substance." With one exception (from *Problems*), all the several passages van Brakel mentions are taken from *Meteorology*. The one he actually quotes most clearly seems to conflict with the conception of the transmutation of the elements described above: "the finest and sweetest water each day ... dissolve[s] into vapor and rise[s] into the upper region, where it is condensed by the cold and falls again to the earth." But placing the passage in context shows that Aristotle thinks of the change as involving generation and destruction:

> the sun ... by its movement causes change, generation and destruction—it draws up the finest and sweetest water each day and makes it dissolve into vapour and rise into the upper region, where it is condensed by the cold and falls again to the earth (*Meteor.* II.2, 354^b6ff.; Lee's translation).

The "it" in "makes it dissolve into vapour" is naturally taken to refer to that *thing* which is "finest and sweetest water," and this does not necessarily entail preservation of the *property* of being water. It is this thing which loses these properties and becomes vapor. That he speaks of this thing dissolving fits with what has been said about the mechanism of transmutation involving a mixing process of the first kind.[8] More generally, Aristotle argues that "[e]ternal they [the elements] cannot be, for both fire and water and indeed each of the simple bodies are observed in process of dissolution" (*Cael* III.6, 304^b26f.). The issue is not quite so cut and dried, however, and draws attention to some of the tensions which arise in comparing Aristotle's various texts.

Van Brakel could respond by pointing out that two chapters later Aristotle says what he means by vapor, namely that "vapour ... is moist and cold" (360^a22f.), i.e., has the defining features of water. Once again, however, the picture is tainted by the larger context. Air in *Meteorology* is not the simple element described in *On Generation and Corruption*, but a mixture of vapor and smoke. The latter substance is called upon because "it is absurd to suppose that the air which surrounds us becomes wind simply by being in motion" (*Meteor.* II.4, 360^a26f.; Lee's trans.). The details of Aristotle's theory of the winds need not be pursued here. Even in *On Generation and Corruption*, however, there is an unelaborated comment in II.3, where the definitions of the four elements are given, hinting at further restrictions on the number of elements—"Fire and Earth . . . are extremes and purest; Water and Air, on the contrary, are intermediates and more combined" (II.3, 330^{b}34f.)—although this puts water and air on a par as far as their elemental status is concerned. As it stands, however, this might just be a badly formulated way of saying that what actually occupies the intermediate region between the center occupied by earth and the upper region occupied by fire is not in fact water and air, but mixtures which are less watery (like water) or airy than the matter below is earthy and the matter above fiery.

But *Meteorology* definitely goes a step further. There it is maintained that, "Air . . . is made up of these two components, vapour which is moist and cold . . . and smoke which is hot and dry; so that air, being composed, as it were, of complementary factors, is moist and hot" (*Meteor.* II.4, 360^{a}21ff.; Lee's trans.). Why in particular the two of the contraries cold and dry should be annulled, and not the other pair, remains unexplained, however. It would also seem that the theory of winds requires that the air is not homogeneous, and the two components form a juxtaposition rather than a compound. The annulment of contraries just mentioned would not be a local phenomenon, then, holding of every part of the air, but a large-scale phenomenon, noticeable only to someone unable to discern the distinctly qualified parts. This doctrine seems radically different from the theory developed in *On Generation and Corruption*. And in *On the Heavens* III.7, in the course of arguing against the atomist conception, Aristotle actually disputes the thesis that "water is a body present in air and excreted from air" by arguing that "air becomes heavier when it turns into water" (305^{b}9f.). Any suggestion that the air of *On Generation and Corruption* and *On the Heavens* is a theoretical notion, as against the observable substance treated in *Meteorology*, must reckon with the frequent appeal to observation to dismiss alternative views in the former two works, such as that at *Cael* IV.5, 312^{b}34ff. and passages discussed in connection with this topic above.

Meteorology is a study of the phenomena which occur in the sublunar region—of "everything which happens naturally, but with a regularity less than that of the primary element [translator's fn.: The fifth element of which the heavenly bodies and their spheres are made] of material things" (*Meteor*. I.1, 338^{b}20f.; Lee's trans.). It is clearly conceived at the outset as building on the general theory of matter and change laid down in *Physics* and *On Generation and Corruption*. But it seems that in striving to accommodate these phenomena, Aristotle has stretched his original theory beyond the limits of consistency. So it is doubtful that whatever counts in favor of regarding water as preserving its waterhood when changing to air in *Meteorology* can be treated as a straightforward elaboration of the doctrines of *On Generation and Corruption* and *On the Heavens*. There are divergences in the latter two works too, to which we now turn; but these are more easily regarded as complementing one another.

SIMPLE SUBSTANCES

Following the development of Aristotle's ideas in *On Generation and Corruption*, the notion of an element has been characterized in terms of properties determining how substances react with one another to generate new substances. According to the

Aristotelian conception, such properties are realized in varying degrees, and can all be reduced to two primary magnitudes. The elements distinguish themselves by taking on extreme or limiting values of these underlying determinables. Otherwise, they share with all other substances capable of entering into mixing processes or reactions the feature that they are not present in a compound, so that no part of a compound bears the element-defining characteristics. Thus, the elements are characterized by features exhibited only in isolation. Other theories of the elements might treat them as more widespread, but for Aristotle something is an element only when in isolation. Another notion Aristotle associates with the idea of an element, that of simplicity, has been mentioned without being integrated into the account. It is now time to consider how this might be done.

Simplicity is normally taken to be a feature distinguishing elements from other substances in virtue of an idea of composition—simple substances are those which are not composed of any other substances. Is such a notion available to Aristotle given what his theory already commits him to as summarized in the previous paragraph? It might seem that the idea is only available on theories like those of the atomists and the Stoics, which have it that the original components are preserved even in the most intimate of mixtures. Since the original ingredients are not present in a compound on Aristotle's conception, it might seem that he is not in a position to talk of components of a compound, and so not in a position to distinguish between simple and complex substances by contrasting substances composed or not of other substances.

Aristotle clearly did not agree. He goes so far as to declare that "every compound will include all the simples bodies" (*DG* II.8, 335^{a}9). This immediately raises the further question of what, if all compounds are alike in being compounds of all four elements, distinguishes one particular compound from another. They are distinguished by their different degrees of warmth and humidity on the conception of mixing developed in *On Generation and Corruption*, and these features should be paralleled by some difference in constitution if this new idea is to be upheld. In I.10 Aristotle contrasts a compound with an inhomogeneous juxtaposition, whose parts do not "exhibit the same ratio between its constituents as the whole" (328^{a}9), suggesting that proportions of component elements can be distinguished and are characteristic of particular kinds of compound. The statement from II.8 occurs toward the end of *On Generation and Corruption*, where Aristotle summarizes his theory without—the fleeting reference to ratios in I.10 notwithstanding—carefully introducing the new idea into his scheme as he had done with other aspects of his theory. Is it even compatible with the rest of the theory? Clearly, the notion of composition at issue cannot be that of actually comprising various substances as parts, which is precluded by the corollary discussed in section on The Aristotelian Theory of Mixing. It must involve the notion of potentiality. A component must be a substance that can be obtained from a compound. But there must be more to it than that since the elements do not distinguish themselves from compounds in respect of disposition to become other substances. Any substance can enter into a mixing process (of the first or second kind) and be entirely converted into another; elements are no exception.

In Aristotle's discussion of the elements in *On the Heavens*, the leading idea is precisely the distinction between simple and complex, rather than what governs mixing processes as in *On Generation and Corruption*. "An element, we take it, is a body into which other bodies may be analyzed, present in them potentially or in actuality (which of these is still disputable), and not itself divisible into bodies different in form. That, or something like it, is what all men in every case mean by element" (*Cael* III.3, 302^{a}15ff.). This is not the way the notion of an element was introduced in *On Generation and Corruption*. Nevertheless, given the reference there to simple bodies, and the reference here to all men, but more importantly, to potential presence, it is presumably thought to be consistent with it. The problem at the end of the last paragraph is addressed in *On the Heavens* with the claim that "flesh and wood and all other similar bodies contain potentially fire and earth, since one sees these elements exuded from them; and, on the other hand, neither in potentiality nor in actuality does fire contain flesh or wood, or it would exude them" (*Cael* III.3, 302^{a}20ff.). Presumably the idea is that a quantity of pure flesh can be converted into a quantity which is part fire, part earth, these parts occupying spatially separated regions. Starting with a quantity of fire, on the other hand, and *without the addition of any other matter*, it is not possible to obtain flesh or wood. A quantity of fire can be converted into flesh and wood, just as Gill points out that "earth is potentially clay and potentially wood" (1989, 150), but only by mixing with some other quantity of matter. For the fact remains that on the Aristotelian conception, what is flesh and wood can become fire and earth, and what is fire and earth can become flesh and wood. It seems, however, that the theory of mixing is to be interpreted to say that conversion of one kind of substance to another is only possible by a *mixing* process (whether of the first or the second kind). At all events, to the extent that it is unclear whether Aristotle has a notion of a physical process of heating which is not a mixing process of the first or second kind, we should be wary of reading our notion of decomposition by heating into his theory. In that case, failure of the parts of a quantity of some kind to be converted to any other without the addition of any other quantity of matter hardly distinguishes the elements, and the burden of explaining the Aristotelian notion of components rests heavily on the idea of a ratio of components. But without Lavoisier's use of the notion of mass to keep track of the amounts of different elements constituting a quantity of a compound, the practical implications of this idea are not easy to see. Of course, none of the views which Aristotle opposed were any better in this respect.

Practical matters aside, Aristotle's theoretical conception of components might be pictured along the following lines. A quantity which is a compound of a given kind is a homogeneous body with specific degrees of warmth and humidity lying between the extreme values found for these magnitudes. A compound kind has a characteristic degree of warmth, w, and degree of humidity, h, and can be denoted by a pair (w, h), any quantity of which is represented by a point O in Figure 1 below. A body of this kind might have resulted from a mixing process of the second kind involving initial ingredients of kinds (w', h') and (w'', h''), represented by points O′ and O″, respectively. The degree w is intermediate between w' and w'', and h between h' and h'', presumably[9] in relation to how much of each of the initial kinds of substance were mixed. How, exactly, this is determined—indeed, what the measurable quality is in respect of which the amount of each quantity might be described—is not specified in Aristotle's texts, and we can only guess at what, if anything, he had in mind.

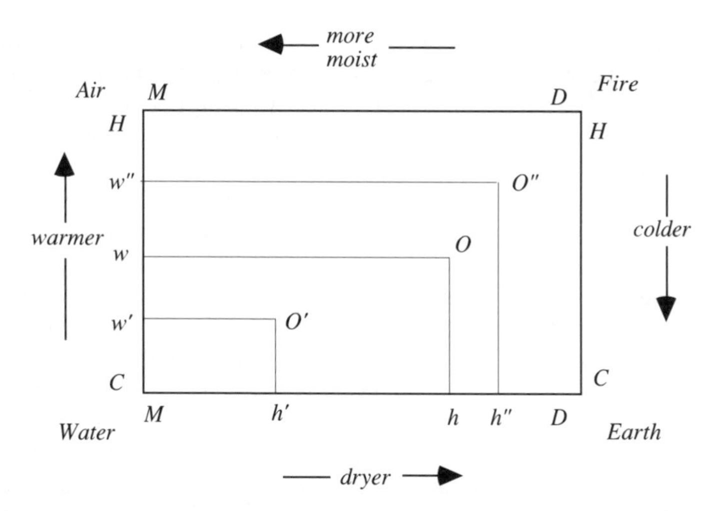

Figure 1: Aristotelian compounds

But let us consider how the problem might be approached within the Aristotelian framework.

Perhaps the original ingredients can be recovered from the quantity of kind (w, h). But there is no reason to suppose that only one kind of conversion into different substances is possible. This is the case with the modern view of compounds, which might be decomposed into different sets of compounds in different ways, and it is only decomposition entirely into elements that is unique. So too, Aristotle's idea seems to be that any compound can be split entirely into substances bearing extreme degrees of the primary magnitudes, i.e., the elements, represented by the corners of the diagram. But is there any reason, even in this case, to think that the analysis would be unique? Suppose mixing water with earth in certain proportions yields a quantity with degree of humidity h. Further addition of air might bring it to degree of warmth w. But this same combination of degrees of warmth and humidity might also be realized by taking less air and some fire, together with less than the original amount of earth to compensate for the extra dryness introduced by the fire. In short, two independent variables w and h are insufficient to determine three independent ratios of four elements.

It is difficult to see how the account might be filled out retaining all of Aristotle's theses about elements and compounds. Moreover, any such elaboration would run counter to the widely accepted view that Aristotle had no notion of a measure of the amount of matter (Jammer 1997, 19). But what, then, is to be made of the explicit (if little developed) reference to the "ratio between its constituents" in *On Generation and Corruption* I.10 (and several references in *Meteorology* IV to substances comprising predominantly, or more or less of, one or other of the elements)? Without some such notion, it is difficult to understand how there could be a variety of compounds if all simple substances are to be found in every compound. Note that this otherwise arbitrary restriction might find some kind of rationalization in terms of the situation portrayed in Figure 1. At least three elements must be involved if both warmth and humidity are to take on intermediate degrees, and perhaps it is natural to think that a unique resolution of the intermediate degrees characterizing a compound into extreme degrees would involve all four elements. But then there must be some additional magnitude contributing to the analysis.

There are pointers in Aristotle's texts to a solution to the problem of elemental proportions which are worth pursuing because of the issues they would resolve, despite the tensions raised for the interpretation of Aristotle. These pointers have, however, been understood differently. Consider a problem raised by Freudenthal (1995) who assumes from the outset that Aristotle has a definite notion of elemental components being *preserved* in a compound. Freudenthal says this immediately poses a problem of stability. Preservation of a compound is threatened, given this assumption, because of the conflict between the contrary defining features of the elements and because of the inherent tendencies of the elements to move off toward different parts of the universe.[10] As a leading exponent of thermodynamics, Duhem was well aware of the problem of stability, which this science tackles in terms of describing conditions for equilibrium under prevailing constraints. This problem gives some insight into why Duhem may have been attracted by the Aristotelian conception according to which, as he understood it, "elements... cease to exist at the moment when the mixt is created, and the homogeneous mixt whose smallest part contains the elements potentially... can regenerate them by the appropriate corruption" (1902, 182). On this understanding, there can be no conflict between contrary defining features of the elements in a compound, as Freudenthal would have it, the extreme contraries having given way to an intermediate degree. The second source of instability in Freudenthal's scenario introduces features which are new to the present discussion, namely the intrinsic heaviness or lightness of the elements. These features are expressed by the tendency, as the Aristotelian view has it, of earth to fall toward the center of the universe, of fire to rise toward the upper reaches of the region bounded by the sphere defined by the lunar orbit, and of water and air to move to intermediate regions. Earth is said to have an absolute heaviness because it falls in every other medium, whereas water has a relative heaviness because it falls through fire and air but not earth, as does air because it falls through fire. Similarly, fire is absolutely light, and air and water relatively light (*Cael* IV.1, 308^{a}29ff.).[11] The notion of the simplicity of the elements is introduced in *On the Heavens* with reference to these inherent qualities of heaviness and lightness:

> Bodies are either simple or compounded of such; and by simple bodies I mean those which possess a principle of movement in their own nature, such as fire and earth with their kinds, and whatever is akin to them. Necessarily, then, movements also will be either simple or in some sort compound—simple in the case of the simple bodies, compound in that of the composite—and the motion is according to the prevailing element (*Caelo* I.2, 268^{b}27-269^{a}3).

It therefore seems reasonable to look here for a third independent relation governing the ratio of constituents in a compound. This is at least in agreement with Freudenthal in so far as, in treating lightness and heaviness as a second source of instability, he takes these properties to be irreducible to warmth and humidity. And as already indicated in section on The Aristotelian Theory of Mixing, whether or not the general strategy of *DG* II.2 is to be understood as encompassing such a reduction, there is not the slightest hint there of how such a reduction might be conceived.

In Needham (1996, 267) I made the simple suggestion that the equilibrium position of a quantity of a compound along a radial line terminating at the inner boundary of the superlunary region is the result of the earth content pulling downwards and the fire content pulling upwards. Although this suffices, in conjunction with equations obtained from the relation of the intermediate degrees of warmth and humidity to the ratios of sums of amounts of elements carrying the corresponding extreme degrees, to determine the ratios of the elements, it oversimplifies what Aristotle says. Why should not the relative heaviness and lightness of the other two elements be taken into account? This would involve complications of considering the medium in which the body comes to rest in order to determine whether, for example, air's relative lightness or relative heaviness comes into play. But before getting involved with such details, the entire proposal might be considered off track because Aristotle speaks of motion rather than rest in this last passage, and the relevant observable feature should be speed of descent or ascent rather than position. [There is disagreement, however, on whether this is speed of free fall (O'Brien 1995) or of one arm of a counterweighted balance (Lewis 1995).] Moreover, it seems that Aristotle means that "motion is according to the prevailing element" in the sense of according to its *preponderance*, i.e., the amount by which it exceeds the other, based on differences between, rather than proportions of, the elemental constituents (Lewis 1995, 94; O'Brien 1995, 69; cf. *Caelo* IV.4, 311^a30-3). This second line of approach involves considerable complications of its own, however. But note that it also implies a certain shift in assumptions about what is at issue. The preponderance idea calls into question Jammer's claim, that "[w]eight, in ancient thought, was . . . intensive rather than extensive" (1997, 18). Aristotle is confusing on this point. Some passages clearly suggest that the relations of heavier and lighter are intensive notions, as when he says "[b]y lighter or relatively light we mean that one, of two bodies endowed with weight and equal in bulk, which is exceeded by the other in the speed of its natural downward movement" (*Caelo* IV.1, 308^a31f.). Again, reference to kinds of matter combined with no reference to bulk, as when he says "[o]f two heavy things, such as wood and bronze, we say that the one is relatively light, the other relatively heavy" (*Caelo* IV.1, 308^a8-9), also suggests intensive notions. But he certainly had extensive conceptions: "[t]he larger the quantity of air the more readily it moves upward" (*Caelo* IV.2, 308^b27).

Not only the proportions of the elements, but also the amount of each in a given quantity of a compound are involved when taking into account the idea that preponderance determines motion. This is more than the three independent ratios required to determine composition, calling for more information than the original problem required. O'Brien and Lewis provide no solution to that problem, but simply take the idea of elemental proportions for granted when explaining how the preponderance of some elements over others might be understood to give rise to different speeds of descent or ascent (Lewis 1995, 96–97; O'Brien 1995, 69). To give some idea of the kind of relation envisaged without going into their differences of opinion, suppose we have two quantities, π and ρ, of different kinds of compound (such as bone and flesh). If E_π were the earth content of π, W_ρ the water content of ρ, and so on, then consider how elemental differences might determine which of π and ρ falls faster in a given medium, say air. Let us suppose that air does not fall in air, and so the air content makes no contribution. In that case, the speed of fall in air of π is greater than that of ρ if

$$(E_\pi + W_\pi - F_\pi) > (E_\rho + W_\rho - F_\rho)$$

i.e., if the net excess of heavier over light constituents in π is greater than that in ρ. Clearly, the "if" cannot be strengthened to "if and only if" since different combinations of constituent amounts can give rise to the same net excess.

We might speculate that a calibration of speeds of fall is possible on the basis that, for a given volume, pure earth falls fastest (although greater volumes of pure earth fall faster), and a quantity π (of the appropriate volume) for which $(E_\pi + W_\pi) = F_\pi$ remains at rest in air. [If $(E_\pi + W_\pi) < F_\pi$, π would rise in air.] This brings us back to my original suggestion, with the modification that the attainment of an equilibrium position depends both on the medium and the volume of the quantity of the compound substance. For the particular volume of the compound substance which floats unhindered at a certain height in air, we have a third equation, independent of those governing the degrees of warmth and humidity,[12] sufficient for determining the ratios of the constituent elements. Following Lewis's suggestion (1995, 90) that the speeds concern differences in amounts of material on two arms of a balance, the calibration of speeds of fall might be considered to provide a measure of the total amount of matter in different volumes of the same substance.

Resolving the problem of composition (ratio of constituents) along these lines involves a good deal of speculation, and consequently may not provide a solution to the problem of interpreting Aristotle's view. There is therefore little point in discussing variations on the theme. But it is important to show that some definite sense can be made of the notion of the elemental proportions in a compound conceived along Aristotelian lines, at least in principle. Armed with such a notion, an issue left open at the end of section on The Aristotelian Theory of Mixing can now be taken up again.

WHAT BEARS THE ARISTOTELIAN POTENTIALITIES?

Aristotle's notion of the potential presence of the elements seems to involve a claim to the effect that a compound is possibly earth, water, air, and fire. There is an unclarity here. But to simplify the discussion an unAristotelian example—though appropriate for anyone like Duhem, interested in applying Aristotle's ideas in a more modern context—is chosen in terms of which the essential points will be made. Water will be considered as a compound with hydrogen and oxygen as components. The initial claim is then that:

Water is possibly hydrogen and oxygen.

This would seem to mean that of any given quantity of water, say π, a certain modal property is true of it, as expressed by:

$$\pi \text{ is possibly hydrogen and oxygen.}$$

But what, exactly, bears the property of being possibly hydrogen and oxygen, and how should this modal conjunction be understood? It is surely not possible that the whole of π is possibly hydrogen, nor that the whole of π is possibly oxygen. An appropriate insertion of "partly" would seem to be called for. We might try:

$$\pi \text{ is partly possibly hydrogen and } \pi \text{ is partly possibly oxygen.}$$

i.e., there is a part of π, say ρ, which is possibly hydrogen, and there is a part σ of π which is possibly oxygen, these two parts being mutually exclusive and jointly exhaustive. This could be put by saying that for any given quantity, π, of water

$$\exists\rho(\rho \subset \pi \wedge \Diamond (\text{Hydrogen}(\rho) \wedge \text{Oxygen}(\pi - \rho))) \tag{1}$$

where $\subset$ is the mereological relation of proper part (excluding the possibility of identity) and $\pi - \rho$ is the mereological difference or remainder of π less ρ; i.e., what remains of π when ρ is removed. [Formally, this difference is defined as the sum of all those quantities which are a part of π and separate from (do not overlap) ρ.] $\exists\rho$ is the existential quantifier, read, "there is a ρ such that...," $\Diamond$ is the modal operator, read "it is possible that...," and $\wedge$ means "and."

(1) expresses a sense in which the elements are potentially present in a compound. But, it might be objected, any proper part ρ' of a part ρ said to be possibly hydrogen is in fact water, and so some part, ρ'', of it (ρ') is, by the same count, possibly oxygen. Now if the whole of ρ is possibly hydrogen, by virtue of homogeneity presumably every part of ρ is too, so the part ρ'' of our arbitrary part of ρ is not only possibly oxygen, but possibly hydrogen too. This does not threaten contradiction, however, since

$$\Diamond\varphi \wedge \Diamond\psi \cdot \supset \Diamond(\varphi \wedge \psi)$$

where $\supset$ denotes implication, is a notoriously invalid principle for an acceptable modal logic, and so

$$\Diamond \text{ Hydrogen}(\rho'') \wedge \Diamond \text{ Oxygen}(\rho'')$$

does not imply

$$\Diamond ((\text{Hydrogen}(\rho'' \wedge \text{Oxygen}(\rho''))$$

[Compare an earlier quote from Gill stating that (a quantity of) earth is potentially clay and potentially wood; this does not imply that the quantity is potentially both clay and wood.] But as it stands, (1) hardly does justice to the homogeneity of the compound, which requires that there is no definite or unique division of a quantity of water into two parts, one of which is possibly hydrogen while the other is possibly oxygen. As

Aristotle put it, the idea that a compound is "composed of the elements, these being preserved in it unaltered but with their small particles juxtaposed each to each," raises the problem that "Fire and Water do not come-to-be out of any and every part of flesh." He was looking for an account which, just as "a sphere might come-to-be out of *this* part of a lump of wax and a pyramid out of *some other* part, it was nevertheless possible for either figure to have come-to-be out of either part indifferently," would describe "the manner of coming-to-be when both come-to-be out of any and every part of flesh" (*DG* II.7, 334^{a}29ff.). Duhem expresses some such idea when, in the passage already quoted at greater length, he speaks of a "homogeneous mixture whose smallest part contains the elements potentially . . . [and] can regenerate them by the appropriate corruption" (1902, 182).

(1) does not actually say that there is a unique partition of π into what is possibly hydrogen and what is possibly oxygen; an additional clause would have to be added to the effect that no distinct such partition exists. But (1) is compatible with such a uniqueness claim and should therefore be elaborated with an additional clause inconsistent with the uniqueness claim and more in line with the Aristotelian idea that, in some sense, any part of water might be partly hydrogen. There is a restriction on this idea, however: there is no unique partition, but any particular partition stands in the same proportion as any other. Let "Same Ratio(π, ρ, σ, τ)" abbreviate the relation of the ratio of π to ρ being the same as that of σ to τ. Then, for any given quantity, π, of water, (1) can be elaborated to read:

$$\begin{aligned} &\exists \rho \subset \pi (\Diamond\, (\text{Hydrogen}(\rho) \wedge \text{Oxygen}(\pi - \rho)) \wedge \\ &\quad \forall \sigma \subset \pi (\text{Same Ratio}(\rho, \pi - \rho, \sigma, \pi - \sigma) \supset \\ &\qquad \Diamond\, (\text{Hydrogen}(\sigma) \wedge \text{Oxygen}(\pi - \sigma)))) \end{aligned} \tag{2}$$

where $\forall\sigma$ is the universal quantifier, read "for all σ, ..."

If a large and a small lead ball are melted and mixed, and the resulting quantity divided into two balls, a larger and a smaller in the same volume ratio as the originals, there is no saying whether the matter of the resulting larger ball is identical with that of the original or not. Similarly on the Aristotelian conception of a compound, quantities of hydrogen and oxygen may be combined to form water which is subsequently reduced to its elements, but there is no saying whether the matter of the original quantity of hydrogen is identical with (as distinct from being of the same kind as) that of the resulting quantity.

CONCLUSION

Aristotle's theory of substances and their interconversion is certainly not without its problems. It by no means follows that any of the opposing views from antiquity were in any better shape, and it is doubtful whether they should be considered any closer to modern views. I have tried to present Aristotle's contributions to the development of ideas in chemistry in a more positive light than is usually done.[13] Some of his insights can be summarized as follows:

1. The fundamental issue in chemistry is the generation of new substances from old by chemical reactions. Aristotle recognized that this required an account of how initial ingredients in a mixing process can affect and be affected by one another so as to bring about the formation of new products.

2. What is it that distinguishes one substance from another? It is not clear that the atomists had any idea of substance, but merely postulated vast numbers of small particles displaying in principle infinite variations of a more or less radical kind with no general principle of collection under chemical kinds. Aristotle saw an appropriate criterion in the distinction between homogeneous and heterogeneous matter: homogeneous material comprises a single substance—either an element or a compound—whereas heterogeneous matter comprises a mixture of several substances.

This principle was evidently abandoned with Lavoisier's analysis of water, which proceeded by reducing steam, and cannot meaningfully be thought of as in any particular phase when ascribed a composition of hydrogen and oxygen. But the alternative view was first systematized with the discovery of the phase rule one century later.[14] Even if Aristotle got some important features wrong, the idea of a fundamental relation between phase and substance is retained and elaborated in modern science rather than being entirely overthrown.

3. What we call a phase change involves a discontinuity. Aristotle recognized a discontinuity in the first of the two kinds of mixing that he distinguished, which are not such that the one goes over into the other continuously. Rather, there is an abrupt change when one body is so much larger than another as to overwhelm it. Although Aristotle's specific theory of phase is rejected, it should be acknowledged that he grappled with a phenomenon that we still recognize as standing in need of an explanation.

4. It may seem that the characteristically Aristotelian idea of potential presence has been abandoned. It was said here that the post-Lavoisian view is usually taken to view elements as preserved in compounds. But the modal element has not been entirely dispensed with. Clearly, the actual presence of elements in compounds cannot be upheld in virtue of the properties exhibited by elements in isolation, so that quantities of matter characterized by certain properties exhibited in isolation must be understood to potentially possess other features exhibited by compounds; and conversely, the matter of compounds must be understood to potentially exhibit certain properties in isolation. Characterization of the elements by the periodic table complements any lists of properties exhibited in isolation with potential properties which the elements would exhibit were they combined in compounds. Otherwise, definition in terms of actual properties would exclude what is ordinarily counted as matter of the same element kind in a different state of chemical combination.

Fixing on atomic structure does not circumvent this general feature. The electronic structure characteristic of isolated atoms, or of atoms in molecules of the isolated element, is changed in compounds. Preservation of nuclear structure, even if necessary, is clearly not sufficient to explain the presence of elements whose chemical features are supposed to specifically depend on electronic properties. The fact that considerations of this kind led Paneth (1962) to question whether modern chemistry has dispensed with the Aristotelian conception of an element is a good indication that the conception does not stand or fall with the continuous conception of matter.[15]

NOTES

1. Unfortunately, several English terms common to philosophers' discussions of Aristotle and everyday chemical idiom are not used in the same way. The word "substance" is understood throughout in the sense of chemical substance. Because I am obliged to follow English translations of Aristotle, however, "combination" and "compound" are used in the philosopher's sense—indicating a homogeneous mixture but without recognition of any distinction between solutions and compounds—for the sake of continuity with quoted passages.
2. See Needham (2002a) for an interpretation of Duhem's view of mixtures.
3. Quotations from Aristotle are taken from Barnes' edition unless otherwise indicated. *DG* abbreviates *On Generation and Corruption*, *Cael* abbreviates *On the Heavens* and *Meteor* abbreviates *Meteorology*.
4. Lewis (1998, 2) says that whether the Greek adjective is to bear the modal (dispositional) construal or not is a matter of context, and therefore a matter of interpretation, rather than a feature that can be directly read off the word itself.
5. For a discussion see Needham 2002a.
6. See, e.g., Cartwright (1970) and Roeper (1985).
7. It might be suggested that they should in fact be indexed with two times in order to accommodate statements of the kind "π is now warmer than ρ was." But such details are not pursued here.
8. The translator points out that Aristotle usually makes no distinction between dissolution and melting (*Meteor* 318, fn. a; Lee's trans.).
9. By analogy with the "method of mixtures" in calorimetry.
10. Even Gill, who interprets Aristotle to say that the elements are not preserved in a compound (1989, 79, 149, 151, 212), and are merely potentially recoverable rather than mechanically separable like oil floating on water, develops much the same idea. Organisms preserve their identity by means of an active principle which counteracts this inherent instability of the compound substances of which they are made, much as a static force enables a pillar to hold up a bridge, without incurring any change: "if the soul were not present to prevent dispersion, the parts of composite bodies would scatter according the nature of their elements" (1989, 212). But the thesis of the inherent instability of compounds is deeply perplexing given the general theory of chemical reaction developed in *On Generation and Corruption*, according to which determinate degrees of the primary qualities lead bodies of the same order of size to attain values intermediate between the original degrees. If this is not the description of a process driven towards equilibrium, what is it? Why should contrary degrees of warmth and humidity be said to make the original ingredients act and be susceptible to action if it now turns out that they do not lead to a state of equilibrium but ulterior forces are required to constrain the product of such a reaction to remain in the same state? A commentator faced with the problem of interpreting the entire corpus might have to acknowledge that Aristotle says some such thing; but in that case, the question of the coherence of the whole might also be addressed. Coherence is threatened by the *extraneous* cause of instability due to contact (Aristotle does not allow a vacuum) with other homogeneous bodies of different kinds, which should eventually lead to the elimination of all differences of substance with the attainment of an intermediate degree of warmth and of humidity throughout the sublunar region—the soggy, tepid death of the universe. Organisms count for only a small part of the material.
11. Note that the converse of the heavier than relation, the relation of being less heavy, is not to be identified with being lighter. Similarly for the converse of the lighter than relation.
12. And perhaps a fourth, of the form $E_\rho = (A_\rho + F_\rho)$, for a different volume which comes to rest in water.
13. See, for example, Horne (1966).
14. Cf. Needham (2000) and (2002b) for discussion.
15. Research on which the chapter is based was supported by The Bank of Sweden Tercentenary Foundation.

REFERENCES

Aristotle. 1984. Barnes, J. (ed.), *The Complete Works of Aristotle*, Vol. 1. Princeton: Princeton University Press.

Aristotle. 1955. *On Coming-to-be and Passing-Away* (trans. by Forster, E.S., Loeb Classical Library), Vol. III. London: Heinemann.

Aristotle. 1939. *On the Heavens* (trans. by Guthrie, W.K.C., Loeb Classical Library), Vol. VI. London: Heinemann.

Aristotle. 1952. *Meteorologica* (*Meteor*) (trans. by Lee, H.D.P., Loeb Classical Library). London: Heinemann.

Cartwright, H.M. 1970. Quantities. *Philosophical Review* 79: 25–42.

Duhem, P. 1902. *Le mixte et la combinaison chimique: Essai sur l'évolution d'une idée*. Paris: C. Naud. Translated in *Mixture and Chemical Combination, and Related Essays*, Needham, P. (trans. and ed.). Dordrechet: Kluwer Academic Press, 2002.

Freudenthal, G. 1995. *Aristotle's Theory of Material Substance: Heat and Pneuma, Form and Soul*. Oxford: Clarendon Press.

Gill, M.L. 1989. *Aristotle on Substance: The Paradox of Unity*. Princeton: Princeton University Press.

Horne, R.A. 1966. Aristotelian Chemistry. *Chymia* 11: 21–27.

Jammer, M. 1997. *Concepts of Mass in Classical and Modern Physics*. New York: Dover Press.

Lavoisier, A.-L. 1789. *Traité élémentaire de Chimie*. Paris. *Elements of Chemistry in a New Systematic Order, Containing All the Modern Discoveries*, Kerr, R. (trans.). New York: Dover, 1965.

Lewis, E. 1995. Commentary on O'Brien. *Proceedings of the Boston Area Colloquium in Ancient Philosophy* 11: 87–98.

Lewis, E. 1998. The Dogmas of Indivisibility: On the Origins of Ancient Atomism. *Proceedings of the Boston Area Colloquium in Ancient Philosophy* 14: 1–21.

Needham, P. 1996. Aristotelian Chemistry: A Prelude to Duhemian Metaphysics. *Studies in History and Philosophy of Science* 26: 251–269.

Needham, P. 2000. What is Water? *Analysis* 60: 13–21.

Needham, P. 2002a. Duhem's Theory of Mixture in the Light of the Stoic Challenge to the Aristotelian Conception. *Studies in History and Philosophy of Science* 33: 685–708.

Needham, P. 2002b. The Discovery that Water is H_2O. *International Studies in the Philosophy of Science* 16: 205–226.

O'Brien, D. 1995. Aristotle's Theory of Movement. *Proceedings of the Boston Area Colloquium in Ancient Philosophy* 11: 47–86.

Paneth, F.A. 1962. The Epistemological Status of the Chemical Concept of Element (original 1931). *British Journal for the Philosophy of Science* 13: 1–14; 144–160.

Roeper, P. 1985. Generalisation of First-Order Logic to Nonatomic Domains. *Journal of Symbolic Logic* 50: 815–838.

Sorabji, R. 1988. *Matter, Space and Motion: Theories in Antiquity and Their Sequel*. London: Duckworth.

van Brakel, J. 1997. Chemistry as the Science of the Transformation of Substances. *Synthese* 111: 253–282.

CHAPTER 4

KANT'S LEGACY FOR THE PHILOSOPHY OF CHEMISTRY

J. VAN BRAKEL
Institute of Philosophy, Catholic University of Leuven

INTRODUCTION

Kant's relation to chemistry is ambiguous. Did he make detailed contributions to the chemistry of his time (Carrier 1990)? Or was he a dilettante (Heinig 1975)? Did he revise his transcendental and/or metaphysical system under the influence of the work of Lavoisier (Dussort 1956)? Or is his statement that chemistry is not a proper science his final judgment? The latter view has had a large impact on both scientists and philosophers. Scarcely known is that in later life Kant may have changed his mind. Research on Kant's *Transition* or *Übergang*—his thinking after he had written the three *Critiques* and the *Metaphysical Foundations of Natural Science*—has only recently come off the ground. In the first part of this chapter, I shall show how a few lines from Kant have had a disproportionate influence on how philosophers and scientists have tended to see the relative status and relation of physics to chemistry. In the second part, I shall try to make sense of Kant's later views on chemistry and how that ties in with his philosophy.[1]

THE STANDARD RECEPTION OF KANT'S VIEW ON CHEMISTRY

Though Kant made brief comments about science in his *Critique of Pure Reason* and in the *Prolegomena to Any Future Metaphysics*, his most developed views can be found in the preface of his *Metaphysische Anfangsgründe der Naturwissenschaft* of 1786 (further *MAdN*).[2] The title of the latter already indicates that science is only possible because of certain metaphysical foundations.

Kant's use of the term "science" is somewhat ambiguous; moreover, we have to be aware of the difference in use of terms such as "doctrine," "theory," "science," *Wissenschaft*, "physics," "natural science," "natural history" in Kant's time and modern (equally ambiguous) terminology. For Kant, science comes under the more inclusive concept of doctrine (Plaass 1994, 232). The "doctrine of nature" can be divided into "historical doctrine of nature" and "natural science" (4: 468). Under "doctrine" (or description or history) of nature would presumably come biology, empirical psychology, geography, and so on.

D. Baird et al. (eds.), Philosophy of Chemistry, 69–91.

According to Kant, natural science is "proper" (pure) to the extent that it "treats its object wholly according to a priori principles" (4: 468): "only that whose certainty is apodeictic can be called science proper." In contrast, improper science instead draws on laws of experience (mere regularities, subject to Hume's scepticism); improper science is also called "systematic art or experimental doctrine" (4: 473). However, no matter how "undeveloped" a doctrine of nature, they all draw on the faculty of judgment and in that sense have an ultimate orientation toward the goal of natural science proper (4: 469).

Natural science proper is physics. It has a pure and applied part. The pure part is strictly apodeictic (proper in the strong sense). The applied part needs the "assistance of principles of experience" (4: 469) and is proper in the weak sense. An example of strong/pure and weak/applied proper science would be pure and applied geometry. Physics is based on *a priori* principles, both from mathematics and philosophy (a metaphysics of nature):[3] "natural science proper presupposes metaphysics of nature" (4: 469).[4] The use of mathematics introduces the pure part of science, and at the highest level of abstraction there is the metaphysical *a priori*.

Chemistry is based on *a posteriori* principles only. Sciences like chemistry are rational (because they use logical reasoning), though not *proper* sciences, because they miss the basis of the synthetic *a priori*. Chemistry can be "a systematic art or experimental study" (4: 468), but never a proper science, because chemical phenomena do not lend themselves to the mathematical treatment connecting them to the *a priori*: "a doctrine of nature will contain only so much science proper as there is applied mathematics in it."[5] Probably the most famous quote on chemistry from Kant is the following:[6]

> So long, then, as there is for the chemical actions of matter on one another no concept which admits of being constructed, i.e., no law of the approach or withdrawal of the parts of matters can be stated according to which (as, say, in proportion to their densities and suchlike) their motions together with consequences of these can be intuited and presented a priori in space (a demand that will hardly ever be fulfilled), chemistry can become nothing more than a systematic art or experimental doctrine, but never science proper; for the principles of chemistry are merely empirical and admit of no presentation a priori in intuition. Consequently, the principle of chemical phenomena cannot make the possibility of such phenomena in the least conceivable inasmuch as they are incapable of the application of mathematics (4: 470–471).

Chemistry is not even applied theoretical science, in the sense that the geometry a carpenter uses is applied science. Chemistry rests *only* on empirical principles. A chemical science, which would be constituted on *a priori* metaphysical and mathematical principles is, in the present state of the discipline, not possible.[7] Chemistry would only achieve the status of proper science if its concepts of given objects were arrived at by means of the presentation of the object in *a priori* intuition.[8] Hence, an empirical science such as chemistry uses a rational method of inquiry, but is not a *proper* or pure science; it is *not* an *eigentliche Wissenschaft* (proper science). Hence:

> the most complete explication of certain phenomena by chemical principles always leaves dis-satisfaction in its wake, inasmuch as through these contingent laws learned by mere experience no a priori grounds can be adduced (4: 469).

Moreover, in the crucial passage, Kant not only says that chemistry does not count as a proper science because it uses no mathematics, but that this requirement would be difficult *ever* to fulfil.[9]

Hence, for the "received" Kant a necessary requirement of "proper" science is its tie to metaphysics and mathematics. The idea that "proper" science uses mathematics has continued to the present day, though the metaphysics has been dropped. To put it crudely, the logical positivists threw out metaphysics, replacing it by logic and aligning the latter with mathematics as the "metaphysical foundation" of all "proper" science.[10]

IMPACT OF KANT'S VIEW THAT CHEMISTRY IS NOT AN *EIGENTLICHE WISSENSCHAFT*

Kant's earlier views on science have been extremely influential, far beyond his followers and the narrow circle of Kant specialists. In the English-speaking world, Kant's view that chemistry is not a genuine science is "exemplified by its place in William Paley's *Natural Theology*, the mandatory textbook read by every Cambridge gentleman throughout the nineteenth century" (Nye 1993, 5). The chemist Meyer (1889, 101) in an address delivered to *The Association of German Naturalists and Physicians*,[11] later published in translation in the *Journal of the American Chemical Society*, mentions in passing that Kant's view on chemistry was referred to in the *Deutsche Rundschau* of November 1889. Paneth [1931 (1962, 7–8)] argues that Kant's definition results in an extremely narrow and inappropriate conception of science and he declares "chemistry, too, [is] a true science, even in those branches where it contains little or no mathematics."[12] But such occasional opposition from chemists to Kant's views has never become influential.[13]

More recent examples of the ubiquitous influence of Kant's view are found in the philosophy of nature; for example, Hartmann (1948) says: "all of chemistry that is lawlike is pure physics." In 1949, the physicist and philosopher of science Dingle put it in stronger terms:[14]

> The truth is that chemistry indeed has no place in the strict scientific scheme. . . . Chemistry rightly figures prominently in the history of science; in the philosophy of science it should not figure at all.

Even more recently, in 1994, the quantum chemist Bader, echoing Kant, but explicitly adding a reductionistic picture, wrote:[15] "A scientific discipline becomes exact, in the sense that predictions become possible, as soon as the classification represents the physics that underlies an observation." And at a philosophy of chemistry conference in 1996,[16] Frenking (1998, 106–107), a theoretical chemist, discussing the autonomy of chemistry, says:

> Chemistry would become revolutionised in 1927, now that the very basis of all chemical phenomena, i.e. the chemical bonding, was understood for the first time. Chemistry as true science is still in a developing stage because quantum chemical research of the many chemical phenomena is still in an infant stage.

Although he does not refer to Kant, it seems more than plausible that his "true science" is a descendent of Kant's *eigentliche Wissenschaft*. Similarly, Arabatzis (1998, 155) suggests that the conflict between physicists and chemists over the electron was "fully resolved" with "the advent of the exclusion principle, spin, and eventually quantum mechanics." With the advance of quantum chemistry, according to which chemistry would be fully reducible to physics, chemistry would finally have reached the status of a proper science. Finally, Mainzer argues (at the same conference) that chemistry has become a *science* in the sense of Kant, because it uses ever more mathematics:

> Chemistry is involved in a growing *network of mathematical methodologies and computer-assisted technologies* with increasing complexity. Thus chemistry, is a *science* in the sense of Kant, but with changing frontiers (Mainzer 1998, 49, emphasis in original).

BRIEF DIGRESSION ON DIRAC

Since the advent of quantum mechanics, in addition to Kant's "chemistry is not an *eigentliche Wissenschaft*," an obligatory reference to Dirac's authority has been used to justify the reduction of chemistry to physics, i.e., to the proper mathematics of quantum mechanics.[17] Dirac (1929) said:

> The underlying laws necessary for the mathematical theory of a large part of physics and the whole of chemistry are thus completely known, and the difficulty is only that exact applications of these laws lead to equations which are too complicated to be soluble.

One could consider this the 20th century culmination of Kant's view.[18] Like Kant's view, Dirac's remark is referred to in lectures for a general audience of scientists in a way that suggests everybody knows it. Usually the reference is uncritical, e.g., by Noble laureate Mulliken in *Physics Today* (1968);[19] only rarely is the reference to Dirac more critical.[20] Dirac's view fit "the times," because logical positivism not only replaced metaphysics by logic, but stressed the unity of science: all sciences were to be reduced to physics.

BRIEF DIGRESSION ON PSYCHOLOGY

In the preface of the *MAdN* Kant also says that psychology is even further removed from proper science than chemistry.[21] Even an experimental psychological science is not possible. Hence, psychology may not even be an "improper science." The relevant passages are difficult to interpret, and there are some ambiguities if one compares it with older texts of Kant. In the 19th century, this led to extensive debates among interpreters. Some scholars even went so far as to say that the often-quoted passage where psychology is demoted as a possible science, is an oversight, inconsistent with the rest of Kant's writings (Drews 1894, 259). Instead, it was suggested that in fact psychology might score better as a "proper science" than chemistry: Kant worked with the distinction of "bodily nature" (*res extensa*) and "thinking nature" (*res cogitans*). And in his lectures on metaphysics of 1765/1766 Kant described psychology as the "metaphysical empirical science of people"[22] and used the expression *mathesis intensorium*, suggesting that even if it were not so at the moment, psychology nonetheless could become a proper science. However, this refers to Kant's "pre-critical" years and hence does not bear on the issue of proper science as discussed in the *MAdN*.

Moreover, Kant's later negative remarks concerning psychology becoming a science are directed at a particular type of psychology, viz. that which is introspection-based.[23] That is not science at all; not even improper science. In contrast, Kant's writings and lectures on anthropology contain much that might now be called psychology, and anthropology is for Kant a respectable empirical science.[24] Moreover, the *mathesis intensorum* remained of crucial importance for Kant in his critical period.[25] One can find in Kant a foreshadowing of psychophysics, which quantifies and mathematizes perception; this is the *mathesis intensorium*.[26] So what is now called psychophysics might even become a proper science. However, Kant did not consider that part of psychology.[27]

ANOTHER KANT

In retrospect, looking at the whole of Kant's philosophical career, Lequan (2000, 5) has suggested that chemistry plays a double role for Kant; on the one hand, it is part of physics or natural science; on the other, it serves as an analogy, metaphor, or paradigm for the method of critical philosophy.[28] Though in the pre-critical period, chemistry plays a rather peripheral role, in his later work it gets its revenge. Kant's increasing interest in chemistry can be discerned from 1780 onwards, coming more and more to the fore towards the end of his life. In 1786, chemistry was still a simple empirical art—a "counter model" to proper science. However, in the *Reflections on Physics and Chemistry*[29] of the late 1790s and in the *Opus postumum* (further *OP*)[30] of the same period, his interest in the work of contemporary chemists moved center stage and became an integral part of his philosophical project.[31]

At first, the double role of chemistry created for Kant a kind of paradox. On the one hand, as part of natural science, chemistry was not a proper science (because it was "merely" empirical); on the other hand, chemistry stands as a paradigm for the method of critical philosophy (e.g., the methods of analysis, separation, purification, and synthesis are shared by chemistry and philosophy).[32] Perhaps neither philosophy nor chemistry can be mathematized, but they have something else: their constructive method. Chemistry as a *practical* science or art is more like moral science than like physics. For example, Kant says that "a procedure which resembles chemistry," is a procedure needed in the analysis of moral common sense—what Körner (1991) has referred to as "the quasi-chemical method."[33]

On the other hand, chemistry presents the philosophy of nature with new problems, "unbeknownst to mathematical physics," viz. to give an account of the variety of substances. Not being able to give a philosophical account of the variety of substances, Kant started to see as a gap in his philosophy.[34] An important reason for the "gap"

Kant identified in his philosophy in later life, is that if the synthetic *a priori* has to guarantee the possibility of all experience, much is missing if one is stuck with only physics (i.e., Newtonian mechanics). As he already says in his physics lectures of 1785, the *Danziger Physik*:[35]

> Chemistry has raised itself to greater perfection in recent times; it also rightfully deserves the claim to the entire doctrine of nature: for only the fewest appearances of nature can be explained mathematically—only the smallest part of the occurrences of nature can be mathematically demonstrated. Thus, e.g., it can, to be sure, be explained according to mathematical propositions when snow falls to the earth; but why vapors transform into drops or are able to dissolve—here mathematics yields no elucidation, but this must be explained from universal empirical laws of chemistry . . .

This "gap" leads to Kant's *Übergang* or *Transition* project (see below).

REREADING THE "RECEIVED" KANT

If one takes seriously Kant's lifelong occupation with chemistry, the older passages can be read in a less dismissive light. Dismissing the passage in the preface of the second edition of the *Critique of Pure Reason* of 1787, where Kant mentions Stahl, Toricelli, and Galilei, as "merely" referring to empirical "non-proper" science would put the emphasis wrongly. For Kant, these scientists illustrate the "Copernican revolution" of actively posing questions to nature and Kant praised Stahl in particular for devising experiments to answer theoretically significant questions.[36]

The first thing to note is that Kant's aim is not reductive. His view is not that to be "a proper science," it must be reducible to mathematical physics. The problem we are addressing is situated between two three partitions. On the one hand to be a proper science, mathematical, metaphysical, and principles specific to that science should be clearly separable (kept in their own domain); on the other hand, their relations should be laid out. Against the background of this tripartite scheme, there are three possible prospects for chemistry, following Lequan's (2000) reconstruction:

(i) to become a proper science in the strict sense; constituted by purely philosophical principles, i.e., completely *a priori*;
(ii) to become a proper science in the weak sense (containing a pure and applied part);
(iii) to remain an improper science, i.e., systematic and rational, but merely experimental or descriptive in the sense of offering experimental generalizations, not necessary laws of nature.

Here, "improper science" is still science: it is systematic; it may use mathematics, but it is fully dependent on empirical generalizations—to be distinguished from *mere* description (which may be an "art," governed by reflexive judgment, but without much of "grounds and consequences" and no prospect of using mathematical tools).

On balance it seems that Kant would favor the middle option for both chemistry and physics (in the modern sense of these terms). For chemistry, to achieve the status of proper science in the weak sense, laws in mathematical form have to be found for the affinities between substances,[37] and the universal laws for the attraction and repulsion between substances (Lequan 2000, 19).[38] For this, it will be necessary to find the proper "simple quantities" and the relevant extensive and intensive magnitudes. One mark of the imperfection of chemistry is that it cannot properly define its subject matter: it has no neat metaphysical picture of what makes possible the variety of substances.[39] In order to achieve the status of proper science in the weak sense, not only should chemistry use mathematics, it should also resolve the metaphysical issue of the possibility of *a priori* knowledge concerning the variety of substances.

Though chemistry can only achieve the status of proper science in the weak sense, it has other characteristics that give it a high status for Kant. I already hinted at this when I mentioned the analogies Kant sees between the methods of chemistry and philosophy.[40] One might speculate that if chemistry would rise to the status of proper science in the weak sense, it would still have a double or even dialectical role. On the one hand, through its connection with the *a priori* of mathematics and of the metaphysics of providing an *a priori* account of the variety of substances via the transcendental side of his dynamic theory of matter (see below), it would become part of physics in the sense of natural science proper (having a pure and an applied part). On the other hand, it would keep (or even strengthen) its position of becoming an analogical model for all inquiry that is not "dominated" by mathematics; thus having a unifying function—a theme later developed by Hegel.

Therefore, the notorious passages in the preface of the *MAdN* should not be read excluding chemistry from ever becoming a proper science. If a universal law of "affinity" could be formulated capable of explaining *a priori* the attractions and repulsions between substances, i.e., the relative distances among parts of matter, then chemistry would become a proper science.[41] This should give, perhaps not an *a priori* way of calculating the densities of all materials, but it should be able, in principle, to describe all possible chemical reactions.

THE OPUS POSTUMUM OR ÜBERGANG

At the end of his life, Kant came to realize that something was missing in his system. In the years 1796–1803, he was working on a draft of a work he had entitled *Transition from the Metaphysical Foundations of Natural Science to Physics*.[42] It is usually referred to as the *OP*; although among the hundreds of pages of text, there is at best the draft of one chapter; the rest is "working notes."

For a long time, it was assumed that Kant could not have done any serious philosophy in his old age. Typically, Adickes, charged with editing the *Akademie Ausgabe*, says that in the 1790s Kant could not have been following the chemical literature seriously.[43] Kant's *OP* appeared in German (in the Academy edition) after more than a century of problems delaying its publication. And what was published is generally considered a mess, random texts without editing.[44] The first abridged English edition appeared in 1993.[45] However, it is now generally agreed that Kant's *Transition* project has to be taken seriously.

From correspondence with Kiesewetter,[46] we know that Kant was thinking of his new project before 1795 and perhaps as early as 1790; but he started on it only in 1796 and more intensely from 1797 onwards. Visiting friends testify to the importance this project had to Kant. In reviewing this episode Marty (1986, VI) uses the expression "chef-d'œvre" to emphasize the project's importance. Kant's old age only started to interfere seriously with his project from 1800 onwards.

Kant had become convinced that a new *a priori* science must be added to his 1786 *MAdN*. Without this new *a priori* science, the "pure doctrine of nature" remains incomplete.[47] In a letter to Garver he says:[48]

> The project on which I am now working concerns the "Transition from the metaphysical foundations of natural science to physics." It must be completed, or else a gap will remain in the critical philosophy (1798, 12: 257).

And in the *OP*:

> These two territories (metaphysics of nature and physics) do not immediately come into contact; and, hence, one cannot cross from one to the other simply by putting one foot in front of the other. Rather, there exists a gulf between the two, over which philosophy must build a bridge in order to reach the opposite bank, for, in order for metaphysical foundations to be combined with physical (foundations) (which have heterogeneous principles) mediating concepts are required, which participate in both (1798, 21:475).

Kant here comes back on his famous statement in the *Critique of Judgement* (1790) where he says in the preface: "hereby I bring my entire critical undertaking to a close" (5: 170). Reality (the world of experience) is only partly determined by Newtonian mechanics. The world of substances and their properties remains ununderstood; the variety of substances (as apparent in their properties) remains completely unaccounted for (in terms of a metaphysics of nature). For example, the concept of gold encompasses an indefinite number of properties, which only experience can reveal. But they belong to reality too. So how can the whole world of experience be categorically encapsulated? That is the main question of the *OP* according to Heyse (1927) and more recent interpretations can be considered to be a variation on this theme. Kant's more concrete problem was: How to relate the diversity of substances to the metaphysical concept of matter. The *MAdN* provided only an analysis of "matter in general," and not the "doctrine of body" (which would include the density of bodies) it had claimed and intended.

In the 18th century, there were two competing theories of matter: the corpuscular and the dynamic concept. The first was a form of atomism: all physical objects are composed of minute and discrete parts distributed in the void. According to the dynamical concept, which Kant supported and developed, substances are constituted by the interplay of moving forces. Substance is definable in terms of the moving forces of attraction and repulsion. The well-known distinction between primary and secondary properties does not exist on the dynamic account of matter: *all* properties of matter can be understood in terms of attraction and repulsion. There are no corpuscular primary properties. All "Kantian" properties are relational and equally manifestations of the activity of those fundamental forces that present material reality to the knowing subject (Edwards 2000, 110). In the *OP*, using the concept of ether (see below), the concept of a universal continuum (or plenum) of material forces is made the centerpiece of Kant's dynamical theory of matter. Chemistry can only be transformed into a proper science (in the weak sense) by being grounded in this dynamical theory of matter, which provides the possibility of a mathematization of "secondary" qualities. When focusing on these fundamental dynamic forces in the *OP*, physics and chemistry often seem to merge for Kant (more or less as in modern physical chemistry).[49]

Kant disagrees with the "ridiculous" solution of the atomists, to presuppose one primordial type of homogeneous matter and reduce the qualitative differences between substances to differences in quantity. He argues explicitly against explanations of differences in density along atomist's lines. Both atomists and the appeal to metaphysical monads reduce matter to mathematical points, which denies the great variety of substances.[50] Kant says that the chemist is mistaken if he thinks that through analysis one can obtain elements that are "absolutely simple."[51] Kant is a partisan of a continuum thesis (at least as a regulative idea). His theory of matter is dynamist, plenist, and continualist and stands in sharp contrast to that of Boyle and Newton.[52]

Westphal (1995, 403) has argued convincingly that the central problem that the *MAdN* left for the *OP* to sort out was to explain how equal volumes of different basic matters could differ in density. Density is central to Kant's problem.[53] The failure of Kant's dynamic theory of matter in the *MAdN* was to give a proper account of density, as well as related notions such as cohesion, rigidity, and friction.[54] In addition to density, in particular an account of cohesion has to be given in order for balance scales to have any form and function at all. Only the first part of the *MAdN*, the "Phoronomy" escapes unscathed in the *OP*.[55] In the *OP*, he aims at a new "transcendental dynamics." Kant's "moving forces" are at the bottom of anything that happens, including how matter can affect our sensory organs. This also leads to a new theory of "self-positing" (Förster 1989). We can only see objects if we first identify ourselves as beings, who are centers of active force. Hence, we perceive ourselves and objects through our dynamic interaction.

The relation of the *MAdN* and the *OP* has been the subject of extensive Kant scholarship, which as yet has not brought about consensus.[56] Perhaps, the wisest option is to see the *OP* as not so much different in content from earlier work, but representing a change in perspective. The "gap" Kant speaks of is the need to give a unified account of all systematic disciplines. He has dropped particular worries about this or that science being proper or not; even if not proper, it has to be included somehow. Further, in Kant's later work, mathematics is demoted from a model to be imitated in metaphysics to a mere auxiliary aid.[57] The domain of science is to be identified "topically"; the aim is not to produce empirical laws by simple reflection. The "scientific" problem of how to account for the "hanging together" of bodies (in different states of aggregation),[58] is so central for Kant, because it is the linchpin of how to account for the "hanging together" of systematic knowledge.[59]

Hence, the *Übergang* is supposed to fill the gap in the structure of Kant's science of nature (*Naturlehre*) and thus fill out the architectural plan of his transcendental philosophy. The transition from the metaphysical principle to the empirical part of physics and chemistry hinges on the systematic formulation of a dynamical theory of matter. Some will argue that even more is at stake here, viz. that it is not so much filling a gap, but a complete revamping of the conditions of our experience of objects *in general*.

ON THE CALORIC ETHER AND THE VARIETY OF SUBSTANCES

The "gap" in Kant's philosophy of nature has to be closed by a dynamic theory of matter, which leads to the central role of a "world ether" or "caloric ether."[60] The ether is a universal continuum of dynamic forces (attraction and repulsion) of matter and the material basis for the interaction between all empirically observable bodily entities. Moreover, it is the carrier of all phenomena of heat and light. Kant's dynamic theory of matter is the necessary precondition of all experience. Therefore, the concept of matter is both empirical and *a priori* (21: 289); and similarly for the concepts of caloric (*Wärmestoff*) and ether. I will use the terms "caloric" and "ether" interchangeable,[61] as Kant does most of the time in the *OP*.[62]

In the *MAdN*, the ethereal caloric is nothing more than a hypothetical substance, but in the *OP* it is elevated to the rank of transcendental philosophy.[63] One could see the *MAdN* as giving an account of "the One," i.e., matter *per se*, whereas the ether of the *OP* provides the basis for the multiplicity and heterogeneity of matter.[64] For Kant, space is neither empty nor homogeneously filled, but filled with caloric ether to varying degrees and he advocates an explanation of the variety of substances in terms of a varying distribution of the ether over space and the qualitative variation in the combination of original dynamic forces.

The ether or caloric is a kind of primordial fluid element. It is an elastic fluid that fills the universe. It has both matter- and wave-like characteristics. It is the cause of the difference between heat and cold and explains since the origin of the universe rarefication and condensation (concentration) of matter. It is "fluidity itself," the condition for all fluidity. The fundamental "state of aggregation" is the fluid ether; solidity derives from it. It constitutes all chemical substances and envelops them. It is the matrix of all bodies. It is the "mother of all matter" from which all bodies derive their cohesion.[65] The "internal proportion" of ether (in interaction with the moving forces) determines not only the chemical "type," but also the state of aggregation (solid, liquid, vapor). Kant does not make a big distinction between physical and chemical "mixing"; in fact, his model of chemical reaction is very similar to Aristotle's model of homogeneous "mixts."[66] The only difference is the (ir)reversibility. In a chemical reaction, there is a complete reciprocal penetration, possible because of the infinite divisibility, resulting in a homogeneous combination. A chemical reaction takes place if the attractive forces of chemical affinity outweigh cohesive attraction.

Similarly, the notion of caloric is conceived as an expansive substance, endowed with strongly repulsive and penetrating forces. When caloric permeates material bodies, its vibrations intermix their parts. Kant's notion of the ethereal caloric can be seen, in retrospect, as combining aspects of Lavoisier's caloric theory and the kinetic theory of heat. The properties of the *Wärmestoff* (or *Äther*) determine the sensible properties of matter.[67]

There exists only one "generic" matter (not matter*s*), whose elements are qualitatively different. Kant's dynamical theory of matter entails that his notion of (chemical) element is qualitative (not quantitative as in atomistic theories).[68] The most important of all problems is how to move from the general concept of matter to *a priori* knowledge of the variety of chemical substances [such as nitrogen, carbon, oxygen, hydrogen: (22: 360)] and their differing densities (e.g., that of water and mercury). In order to understand the differences between substances, Kant adds to the dynamic hypothesis of two moving forces the ether hypothesis.

Matter in general is made possible by the inseparable bond between attraction and repulsion. Because the proportion of these forces may vary, the possibility of an infinite variety of chemical substances is given. Hence, the "first causes" are not mechanical (as in atomist theories), but physico-dynamic.[69] The interaction between the omnipresent ether and the two originary forces creates the specific differences between "types" of matter. Each chemical substance is characterized by a quantity of ether and three forces [universal attractive force, "proper" repulsive force, and "proper" attractive force (chemical cohesion)]. Hence, the ether is the ultimate origin of the variety of substances (together with the two moving forces).

Kant distinguishes four "elements," although they are very different from Aristotelian elements.[70] They stand more for classes of chemical reactions. The ether or caloric is a "fifth essence" (replacing phlogiston as "mysterious" fifth essence).[71] The caloric ether is categorically an *a priori* given "stuff," which is the foundation of the variety of substances and their properties. It implies, for example, that the statement that all matter can be weighed is not an empirical observation, nor is it an analytic statement concerning the concept of matter. It expresses the condition that has to be presupposed to make experience of "quantity of matter" possible.[72]

The ether fulfills several functions for Kant.[73] It is necessary because it makes possible all quantification of matter, in particular weighing bodies on a balance.[74] Further, it has both a cosmological significance and has the status of a "special" (chemical) element. From a cosmological point of view, chemistry is superior to the other natural sciences, because cosmology needs the notion of ether.[75] The ether provides the subject-independent "causal" basis for the perception of any and all external objects. It is necessarily real in order to account for the phenomena, i.e., it is a necessary condition for all experience. It fulfills the function of a middle concept that Kant is looking for to "fill the gap." As a necessary condition, it cannot be identified by empirical means.[76] Empirically, the ether is a "pure hypothesis" (its existence is problematic); but from the transcendental point of view, it is a postulate of purely speculative reason (its existence is necessary).It is a categorical given, which therefore requires a new "deduction" of the categories (Lehmann 1963). Whether Kant succeeded in actually "proving" his "solution" of closing the "gap" is an open question.[77]

KANT AND THE CHEMICAL REVOLUTION

In 1786, Kant still seems to doubt whether chemistry is just an art, not even an improper science, but he tends to favor the latter and hopes for more in the future, because of its double role (the analogy between the art of chemistry and philosophy). In the 1786 preface of *MAdN*, Kant denies the status of proper science to chemistry, but this has changed in the later *Reflections* and the *OP*.

Friedman (1992) ascribes Kant's *Transition* from *MAdN* to *OP* primarily to the influence of Lavoisier's "new chemistry." He shows that by 1785 Kant has become aware of the new discoveries in pneumatic chemistry; between 1785 and 1790 he had assimilated the developments in the science of heat; between 1790 and 1795 he had completed the conversion to Lavoisier's system of chemistry.[78] The *Reflections on Physics and Chemistry* of the late 1780s and the early 1790s mention Lavoisier's anti-phlogistic theory of combustion. Around 1790, Kant connected the caloric theory of gases with the problem of solidity and the concept of latent heat.[79] In his 1792–1793 *Lectures on Metaphysics* (*Metaphysik Dohna*), Kant associates the science of chemistry with Lavoisier's doctrine of the composition of water.[80] By 1795 (after some intermediate stages), Kant acknowledges as a "very plausible hypothesis" that water is constituted by hydrogen and oxygen.[81] Moreover, around 1795 Kant develops his own version of the caloric theory of the states of aggregation.[82] In the 1797 preface to the *Doctrine of Right*, Kant has Lavoisier "represent" chemistry (instead of Stahl).[83]

Hence, it may be surmised that from 1795 onwards, under the influence of the work of Lavoisier, Kant considers chemistry to have achieved the status of proper science (in the weak sense). But the change was slow and Stahl was not simply dismissed: he had prepared the way, because he tried to use *a priori* principles to "order" chemistry. Also Kant's transition from phlogiston to oxidation was slow; for some time Kant seems to have used "oxidation" and "dephlogistination" as synonyms.

According to Schulze (1994, 40n), Friedman overlooks that the *OP* is not merely a revision of Kant's theory of matter in the light of new scientific developments, but a revision of fundamental traits of his whole philosophy. He argues (1994, 105 and *passim*) that the *OP* is an extension of Kant's metaphysics of nature, making explicit room for a philosophy of nature in between metaphysics and empirical science, in particular by providing "middle concepts" (such as the caloric ether) to connect the "purely" *a priori* and the "purely" empirical.[84] Also Edwards (2000) accuses Friedman (1992) of serious misreadings and misinterpretations of the *OP*,[85] as does Westphal (1995, 412–413). The latter criticizes Friedman's claim that Kant was motivated to integrate chemistry and the theory of heat with mathematical physics, because Friedman does not address the question why Kant in 1798 "rejected the mathematical model." Nor does he explain why Kant,

> focuses on physics and repeatedly formulates his project in terms of a transition to *physics*... he extensively discusses this... transition to physics without mentioning chemistry or biology... Physics should not be stressed so often or so centrally... if Kant's problem was only to relate physics with the other new physical sciences.

I think Friedman's critics are right when they argue that Kant was deeply dis-satisfied with the *MAdN* (or even his whole critical philosophy), but the other disagreements are easily resolved. Kant uses the term "physics" often in a sense in which it includes all "new physical sciences" (Westphal's term). Kant's view of mathematics as the "ideal" model changed, because he realized that he would not be able to account for the experience of the variety of substance on the basis of a metaphysics modeled on mathematics: hence his tripartite "pact" between philosophy, mathematics, and natural science.

Without entering the exegetical battle and acknowledging that the hermeneutic circle is never closed, in defense of Friedman it can be said, as Dussort (1956) had noted too (see also Lequan 2000), that there are many things about the "chaotic" texts of the *OP* that support a "chemical" biased reading, i.e., taking as central focus the "gap" of not being able to connect the variety of substances (and their phenomenal properties) to the *a priori*; and seeing Lavoisier's work as a clue to closing the gap. First, at no point is there any reference to phlogiston in the *OP*. On the other hand, the notion of caloric occurs on almost every page. Although Lavoisier did not invent the term "caloric," he certainly made it respectable. Second, the caloric ether starts to take on the role of transcendental object, something most interpreters agree on. Third, Kant explicitly uses the term "physics" now such as to include chemistry.[86] Further, there is Kant's constant preoccupation with the use of the workings of a balance for weighing, which is relevant for various reasons.[87] Furthermore, Lavoisier's account of weighing, and explanations of states of aggregation, are similar to those of Kant. Finally, Dussort points to similarities between the philosophical views of Lavoisier and those of Kant, in particular the assumption crucial to Kant that sensation (experience) must be based on the movement of matter.[88]

CONCLUDING REMARKS

There seems little doubt that Kant followed the developments in chemistry of his time carefully (leaving aside the question whether he made any substantial contribution to it). In 1804, an obituary of Kant appeared in the *Neues allgemeines Journal der Chemie* by the editor of this journal, who praised Kant's wide knowledge of chemistry and the attention he devoted to it. There is also no doubt that Kant was the first to give chemistry a central place in his (later) philosophy. This was later followed up by Schelling and Hegel.[89] This development in natural philosophy has been of little relevance to 20th century philosophy of science, but at least those philosophers gave chemistry some serious thought. The same cannot always be said of those who quoted Kant's notorious suggestion that chemistry is not an *eigentliche Wissenschaft*.

Given the dominant view in the past few centuries on the relation of chemistry and physics, greatly stimulated by Kant's comment that chemistry is *not* an *eigentliche Wissenschaft*, there is every reason to be suspicious of Kant scholars who underplay the impact of the "chemical revolution" on Kant's thinking. This suspicion is kept alive by tiny oversights. For example Förster (1993, xxv), the editor of the first (abridged) translation of Kant's *OP* in the "definitive" *Cambridge Edition of the Works of Immanuel Kant* and author of numerous publications on the *OP*, refers to Kant's *Reflections on Physics and Chemistry* as "*Reflexionen* on physics." Of course, he could justify this by saying that Kant himself says explicitly in one of these reflections that chemistry is part of physics.[90] Nevertheless, I find it significant that Förster did not take the trouble of quoting the title in full (or instead use an acronym such as *Refl.* or *RPC*).

I tend to favor the part of Dussort's, Friedman's, and Lequan's interpretation that puts considerable emphasis on the impact the "chemical revolution" had on Kant's thinking, without denying at the same time the "bigger" metaphysical ramifications for Kant(ians). The two views are not incompatible if one assumes that from the beginning Kant was concerned to fit it all neatly together. And as he says in the *Danziger Physik*, before he had started on the *Transition* project, there was a gap in his philosophy. The "chemical revolution" could help to close the gap between (all) sciences of nature and philosophy—basically the gap between giving an account of matter in general and giving an account of the variety of substances.[91]

Perhaps one possible confusion in reading Kant through modern eyes is that for us the change from Stahl to Lavoisier is a "chemical revolution," whereas for Kant, it is precisely that period that brings chemistry into the domain of science, loosing its alchemist roots, and like Kant's *Critiques*, aimed to be rid of dogmatic philosophy.[92] Both chemistry and philosophy operate with regulative ideals on a road of historical progress and proper science. The *preface* of 1786 represents chemistry in its adolescence (Lequan 2000, 41). In the *Reflections* and the *OP* chemistry gets its proper place being tied to the central ether concept, which is empirical, transcendental, and cosmological.

Kant had always been against atomism and his resolution to explain the variety of substances was to introduce an ether or caloric and the principle of two fundamental opposing dynamic forces. His notion of ether not only was an empirical hypothesis, but a transcendental principle; in that sense it was part of philosophy, not of chemistry. Hence, this suggested the need for a second Copernican revolution, or at least "filling a gap" in his philosophy. Given the crucial *philosophical* importance of the problems thrown up by chemistry, perhaps the notion of science had to be rethought such that chemistry could be a proper part of it. Whether it would be called part of physics would then be a terminological matter. Physics, chemistry, philosophy, and mathematics could become one seamless whole, without each losing its autonomy, or at least that is what I suggest Kant would suggest.[93]

NOTES

1. All quotes taken from Kant in this text refer to the *Akademie Ausgabe* (giving volume and page number), Kant's collected works published between 1900 and 2000; only page numbers of the *Critique of Pure Reason* (*KrV*) are given in the well-known A/B format. English translations have been consulted where available. German citations in the notes are always from the *Akademie Ausgabe*.
2. See for the argumentative structure of the *MAdN*, the *Metaphysical Foundations of Natural Science*, for example, Plaass (1994) and Watkins (1998). Quotations from the *MAdN* in English are from Ellington (1970).
3. According to von Weizsäcker (in Plaass 1994, 174) the *MAdN* deals with "the conditions under which—in modern terminology—the assignment of physical meaning to mathematical concepts is possible."
4. One should distinguish here between "the transcendental part of the metaphysics of nature" and "a special metaphysical natural science" (4: 470).
5. All systematic knowledge can be called science (4: 468). If the systematicity is that of a connection of "grounds and consequences," science is rational.

6. "So lange also noch für die chemischen Wirkungen der Materien auf einander kein Begriff ausgefunden wird, der sich konstruieren läßt, d.i. kein Gesetz der Annäherung oder Entfernung der Teile angeben läßt, nach welchem etwa in Proportion ihrer Dichtigkeiten u.d.g. ihre Bewegungen samt ihren Folgen sich im Raume a priori *an*schaulich machen und darstellen lassen (eine Forderung, die schwerlich jemals erfüllt werden wird), so kann Chemie nichts mehr als systematische Kunst oder Experimentallehre, niemals aber eigentliche Wissenschaft werden, weil die Prinzipien derselben bloß empirisch und keine Darstellung a priori in der Anschauung erlauben, folglich die Grundsätze chemischer Erscheinungen ihrer Möglichkeit nach nicht im mindesten begreiflich machen, weil sie der Anwendung der Mathematik unfähig sind."
7. Cf. *Danziger Physik* [1785 (29: 97)]: "Die Mathematik reicht gar nicht zu, den chemischen Erfolg zu erklären oder man hat noch keinen einzigen chemischen Versuch mathematisch erklären können; daher ließ man die Chemie aus der Naturlehre aus, weil sie keine Prinzipien a priori hat."
8. This requirement follows from the fact that chemistry would be (part of) a proper science of *corporeal* nature (B202).
9. Cf. the words "ever" (*jemals*) and "never" (*niemals*) in the citation from (4: 470–471). In the *Critique of Practical Reason*, Kant suggests that chemistry might achieve *a priori* natural laws "if our insight went deeper" (5: 26).
10. According to Miller and Miller (in Plaass 1994, 156) Kant's notion of proper science limits "the everyday objectificational paradigm to only what can be described in terms of spatio-temporal determinations" because this "is necessary to assure the kind of reproducibility and measurability needed for mathematization of nature." Such reproducibility demands the experimental isolatability characteristic of modern physical sciences.
11. The *Gesellschaft deutscher Naturforscher und Ärzte*.
12. In particular, many German chemists seem to have wrestled with Kant's assessment of chemistry.
13. Ellington (1970, 8n), the translator of the *MAdN*, makes the following "excuse" for Kant: "The beginnings of the modern science of chemistry were made by Lavoisier shortly before the *MAdN* appeared in 1786. Kant did not foresee the development of atomic physics, which was to make chemistry a science." I will come back to Kant and Lavoisier below.
14. Dingle (1949) adds: "Reluctant as I am, and as a loyal physicist should be, to say anything good of chemistry, I cannot deny that, quite apart from its necessity for the amenities of life, it has been indispensable in making possible the rapid progress of physics."
15. Bader, Popelier and Keith (1994, 620): "Eine wissenschaftliche Disziplin beginnt mit der empirischen Klassifizierung von Beobachtungen. Sie wird exakt in dem Sinne, daß Vorhersagen möglich sind, sobald die Klassifizierung die Physik widerspiegelt, die einer Beobachtung zugrunde liegt."
16. The *3rd Erlenmeyer-Colloquy for the Philosophy of Chemistry* (Janich and Psarros 1998).
17. For a review of the complexity of the relation between molecular chemistry, quantum chemistry and quantum mechanics see van Brakel (2000, Chapter 5).
18. Later Dirac (1939) contemplated a further Pythagorean reduction: "If we express the present epoch, 2×10^9 years, in terms of a unit of time defined by the atomic constants, we get a number of the order 10^{39}, which characterizes the present in an absolute sense. Might it not be that all present events correspond to properties of this large number, and, more generally, that the whole history of the universe corresponds to properties of the whole sequence of natural numbers?"
19. For the enormous impact of the Dirac citation see van Brakel (2000, 120).
20. An example is Hoffmann (1998, 4) who says in a "viewpoint paper" presented at a meeting honouring pioneers of computational quantum chemistry: "only the wild dreams of theoreticians of the Dirac school make nature simple."
21. "But the empirical doctrine of the soul must always remain yet even further removed than chemistry from the rank of what may be called a natural science proper. This is because mathematics is inapplicable to the phenomena of the internal sense and their laws ... But not even as a systematic art of analysis or as an experimental doctrine can the empirical doctrine of the soul ever approach chemistry, because in it the manifold of internal observation is separated only by mere thought, but cannot be kept separate and be connected again at will; still less does another thinking subject submit to our investigations in such a way as to be conformable to our purposes, and even the observation itself alters and distorts the

state of the object observed. It can, therefore, never become anything more than a historical (and as such, as much as possible) systematic natural doctrine of the internal sense, i.e., a natural description of the soul, but not a science of the soul, nor even a psychological experimental doctrine." (4: 471)

22. "Nachricht von der Einrichtung Vorlesungen Winterhalbenjahren von 1765–66" (2: 316): "metaphysische Erfahrungswissenschaft vom Menschen." Cf. *MAdN* (4: 470): "a pure doctrine of nature concerning determinate natural things (doctrine of body and doctrine of soul) is possible only by means of mathematics."
23. Cf. citation in note 21.
24. According to Makkreel (2001), for Kant, psychology, history, geography, and especially anthropology do not come in the category of science in the sense that physics is a science. They are systematic disciplines that are guided by reflective rather than determinate judgement, guided by a *practical* idea of the "best world". It is important to note that though such disciplines can never claim to become exact sciences, this does not undermine their objectivity for Kant. Similarly, Roqué (1985) argues that Kant relegates self-organization to the realm of reflective judgement, not to Newtonian-type science. Here, self-organization refers both to biological organisms and autocatalytic chemical reactions (which would have, in Kant's terms, an intrinsic physical end).
25. See B207f.
26. See A179/B221, where Kant seems to suggest that the intensity of color experiences is ordinally and cardinally measurable. Cf. Sturm (2001, 168) and for a more detailed account Nayk and Sotnak (1995), who argue that we can ascribe to Kant the view that psychometrics (or psychophysics) could be a proper science, but we can never apply mathematics to the psyche and elevate psychology to the status of a proper science.
27. For Kant, there is a connection between sense perception and chemistry via what is now called neurophysiology, cf. (7: 157; 12: 34).
28. Cf. notes 32 and 33.
29. This is a selection of 63 text fragments (14: 63–537); in the notes referred to as *Refl.*
30. All citations in English from the *OP* are from Förster (1993).
31. According to Lequan (2000, 6n, 81n), Kant was familiar with the work of Priestley, Scheele, Cavendish, Muschenbroek, Crawford, Boerhaave, Beccher, Macquer, De Luc, Hube, Girtanner, Gren, Hagen, Hales (and of course Stahl and Lavoisier), cf. (11: 185). Apart from absorbing and reworking their theories he also asked colleagues at the University of Köningsberg to do experiments for him (Förster 1993, 275n90). This may explain the comment by a chemist of Kant's time that "es sei ihm unbegreiflich, wie man durch bloße Lektüre ohne Hilfe anschaulicher Experimente die ganze Experimentalchemie so vollkommen wissen könne" (Groß 1912, 129).
32. See for a comparison of the philosopher and the chemist using a synthetic method Bxxi; for the method of analysis or separation see B870 and (5: 163); for the method of abstraction (8: 199).
33. Kant, *Critique of Practical Reason* (1787): "a process similar to that of chemistry, i.e., we may, in repeated experiments on common sense, separate the empirical from the rational, exhibit each of them in a pure state, and show what each by itself can accomplish" (5: 163); for a direct comparison of the philosopher and the chemist on this point see (5: 92). Cf. as well the *Vorarbeiten zur Rechtslehre* (23: 284): "der Analogie zwischen dem Lavoisierschen System der chemischen Zersetzung u. Vereinigung und dem moralisch-practischen der gesetzlichen Formen u. der Zwecke der practischen Vernunft."
34. Kant scholars disagree about where exactly the gap should be located (whether it is a methodological gap or more serious), but there is no doubt that the term "gap" is appropriate to indicate the seriousness of what is missing in Kant's system hitherto (apart from the fact that Kant repeatedly uses the term himself). Edwards (2000, 152f), who stresses continuity in Kant's philosophy, also uses it: "(Kant's) transitional science is supposed to fill in a gap in the structure of the Kantian metaphysics of nature (and thus fill out the architectural plan of Kant's transcendental philosophy). The actual passage from metaphysical principles to the empirical part of physics is supposed to take place by means of the systematic formulation of a dynamical theory of matter. This theory of matter is founded on the concept of a cosmic aether."
35. "Die Chemie hat sich in neueren Zeiten zur großen Vollkommenheit emporgehoben; sie verdient auch mit allem Recht den Anspruch auf die gesamte Naturlehre: denn nur die wenigsten Erscheinungen

der Natur lassen sich mathematisch erklären—nur der kleinste Teil der Naturbegebenheiten kann mathematisch erwiesen werden so z.B. kann es zwar nach mathematischen Lehrsätzen erklärt werden: wenn der Schnee auf die Erde fällt; wie sich aber Dünste in Tropfen verwandeln oder auflösen können, hier giebt die Mathematic keinen Aufschluß sondern dies muß aus allgemeinen Erfahrungs Gesetzen der Chymie erklärt werden . . . " (29: 97–98; cf. 29: 100).

36. There are already traces of the influence of the "new chemistry" in the first *Critique*, second edition (1787), where Kant adds a physico-chemical example (state transition from fluidity to solidity) to the transcendental deduction (B162); though he still mentions Stahl's theory of the calcination of metals (Bxiii). Stahl's phlogiston theory is still in full force in the *Danziger Physik* of 1785, e.g. (29: 163).
37. Some will say that "substances" should be understood here as "corporeal substances in motion." I have to leave the question of how to take Kant's notion(s) of substance in the *OP* largely unresolved. This is an issue that would deserve further study to assess the relation of Kant and the philosophy of chemistry.
38. Kant did not seem to be aware of the work of J.B. Richter who obtained his doctorate at Königsberg in 1789 with a dissertation on *De usu matheseos in chymia*, an early attempt at what later came to be called the theory of combining proportions (culminating in the work of Dalton). Cf. Friedman (1992, 339n167). In connection with Kant, Heinig (1975) notes that as early as 1741, Lomonossov had written about "the elements of mathematical chemistry." This is a work in which Lomonossov aims to prove all chemical theorems, modeling his method on that of Euclid, starting from atomistic presuppositions: "Wohl verneinen viele daß man der Chemie die Prinzipien der Mechanik zugrunde legen und sie zu den Wissenschaften zählen kaan; aber er verneinen dies solche, die sich in dem Dunkel verborgener Eigenschaften verirrt haben und nicht wissen, daß man in den Veränderungen der gemischten Körper stets die Gesetze der Mechanik beobachten kaan . . . " (Lomonossov 1961, 72).
39. Kant notes that in studying substances chemistry is limited in three ways: (i) by time, because nature has had aeons to bring about chemical changes; (ii) their products are never as pure as natural products; (iii) only a few of the substances found in nature can be made artificially. For Kant there is no sharp boundary between inorganic, organic, and biochemistry; crystallization for example has many of the characteristics of living processes for him.
40. Lequan (2000, 112–117) gives a useful overview of different places where, for Kant, chemistry provides theoretical models for meteorology (8: 323–324), cosmology (8: 74), psychology (8: 456), medicine (7: 193, 287, 293), and others. Chemistry is also the background for physiological processes that "create" secondary properties of objects (5: 349; 6: 400) For example, in the *Anthropology from a Pragmatic Point of View* (7: 177), Kant draws an analogy between the catalytic function of chemical affinity when discussing the role of the "sensory productive faculty of affinity" in social discourse: "The word affinity (*affinitas*) here reminds one of a catalytic interaction found in chemistry, an interaction analogous to an intellectual combination, which links two elements specifically distinct from each other, but intimately affecting each other and striving for unity with each other, whereby the combinations creates a third entity that has properties which can only be brought about by the union of two heterogeneous elements."
41. See the first part of the citation in note 5. Cf. Carrier (2001), Lequan (2000, 12).
42. *Übergang von den Metaphysischen Anfangsgründen der Naturwissenschaft zur Physik*. For exegesis see contributions in Bad Homburg (1991), Hoppe (1969), Tuschling (1971), Mathieu (1989), Schulze (1994), Lequan (2000). There is by far no consensus how to interpret the *Übergang*. Förster (1991) comments that in the Kant literature "herrscht darüber weitgehend Ratlosigkeit."
43. "Auf jeden Fall kann keine Rede davon sein, daß Kant in den 90er Jahren hinsichtlich der chemische Literatur auch nur einigermaßen auf dem laufenden gewesen wäre" (Adickes 1924, I, 63f). Perhaps Adickes' judgement was colored by his frustration expressed in a note of six pages appended to a sentence at (14: 489), quoted in note 78 below, where he says: "Doch ist es mir, trotz langen Suchens in der chemischen Literatur der letzten und auch der früheren 80 er Jahre des 18. Jahrhunderts, nicht gelungen, einer Stelle habhaft zu werden, die Kant als unmittelbare Vorlage hätte dienen können" (14: 491).
44. "Die Bände XXI und XXII tragen den prätentiösen Titel *Opus postumum*, aber sie bieten keine Ausgabe der Vorarbeiten Kants zu seinem geplanten Werk, sondern drucken die zufällig zusammengeratene

Papiermasse aus dem Besitz der Familie Krause ab. Die Herausgeber erzeugen durch die blinde Wiedergabe der Notizen den Eindruck, der alternde Philosoph habe nicht mehr zwischen einer Ätherdeduktion und seinen Rotweinflaschen unterscheiden können" (Brandt 1991, 8). Regarding the fragmentary character of the manuscripts and the resulting difficulties for any systematic interpretation of the *OP* see Adickes (1924, 36–154), Tuschling (1971, 4–14).

45. There have been several editions in French and other languages. All editions are abridged, partly because Kant started the same project over and over again; hence, there is considerable repetition. As to the specific interest to chemistry the *OP* should be studied in connection with the *Reflections*, in particular 20, 21, 40–45, 54, 63–66, 73–74, 79.
46. Letter from Kiesewetter dated June 8, 1795: "Sie haben schon seit einigen Jahren einige Bogen dem Publiko schenken wollen, die den Übergang von Ihren metaph. Anfangsgründen der Naturwissenschaft zur Physik selbst enthalten sollten u(nd) auf die ich sehr begierig bin." (12: 23)
47. Rothbart and Scherer (1997) refer to Friedman (1992) and mention the requirement that Kant needs "a 'transition', filling the gap on *a priori* grounds between the metaphysical foundations of nature, a pure product of thought, and nature," but they do not seem to have picked up the relevance to Kant's *Übergang* of developments in chemistry. The same applies to many other Kant scholars.
48. See also his letter to Kiesewetter dated October 19, 1978: "with that work the task of the critical philosophy will be completed and a gap that now stands open will be filled. I want to make the 'Transition from the Metaphysical Foundations of Natural Science to Physics' into a special branch of natural philosophy (philosophia naturalis), one that must not be left out of the system." (12: 258) English translations of Kant's correspondence are taken from Zweig (1999).
49. At several places Kant says that chemistry is part of physics or that they are both part of natural science. For example *Refl.* 61: "Chemie ist bloß physisch" (14: 470) and in the *OP*: "Die ganze Chemie gehort zur Physik—in der Topik aber ist vom Übergange zu ihr die Rede" (21: 288); "Die Chemie ist ein Theil der Physik aber nicht ein bloßer Übergang von der Metaph. zur Physik.—Dieser enthält blos die Bedingungen der Möglichkeit Erfahrungen anzustellen." (21: 316). Cf. "This property, however, belongs to physics (chemistry) as a system" (22: 138); and also (7: 177n; 14: 40; 22: 161; 4: 530).
50. Moreover, there is no experience of atoms or monads or mathematical points (21: 218; 22: 555).
51. For example he says in *Refl.* 45 that a drop of salt water put in an ocean of sugar water will give salt throughout.
52. "Matter does not consist of simple parts, but each part is, in turn, composite, and atomism is a false doctrine of nature." (22: 212; cf. 22: 554, 611) First, extension and impenetrability are not primary properties of matter but derive from more fundamental forces; second, matter fills space completely; third, matter is infinitely divisible—there are no atoms (Carrier 2001). For an overview of the different positions held on the divisibility of matter from Kant's *Monadologia physica* to the *MAdN*, see for example Malzkorn (1998).
53. Carrier (2001, 228n14) has suggested [referring to (4: 526)] that Kant might have recognized that the density of macroscopic bodies is affected by a host of causes other than fundamental forces.
54. Hence, the numerous times he discusses theses notions in the different parts of the *OP* and also in the *Reflections*.
55. For Westphal, the issue is important because of the "problem of circularity" in the *MAdN*, which was already pointed out by reviewers during Kant's life. I will not enter into this difficult interpretative issue. Allegedly, Kant would have discovered the circularity in his definition of the quantity of matter around January 1792. The *MAdN* has four chapters: *phoronomy*, in which "motion is considered as pure quantum" (here "matter" only figures at an extremely abstract level); *dynamics*, in which "motion is regarded as belonging to that quality of the matter under the name of an original moving force"; *mechanics*, which is more or less what we might call mechanics: describing the motions of material objects; and *phenomenology*, in which "matter's motion or rest is determined... as an appearance of the external senses" (4: 477). Dynamics studies the quality of substances in terms of the fundamental dynamic forces. More precisely: the quality of matter in so far motion is determined by an originary moving force. See discussion in Lequan (2000, 23–27, 57–62).

56. See Bad Homburg (1991, xii–xiii), Friedman (1992), Tuschling (1971), Mathieu (1989). Kant briefly discusses the dynamic theory and the problem of the specific differences in density in the *MAdN* at (4: 530–535).
57. "For mathematics is the finest instrument for physics and the knowledge which falls therein (for that mode of sense) but it is still always only an instrument for another purpose." (22: 490; cf. 21: 105, 139)
58. The state of aggregation and what substance it is depends on the ratio of attractive and repulsive forces (21: 382).
59. See also Edwards (2000) on the *Grundsatzes der Gemeinschaft*. According to Kant there has to be a "community of substances" because this is needed for the unity of the world, in which all phenomena must be interlaced; "die Erklärung der Möglichkeit der Gemeinschaft verschiedener Substanzen, durch die sie ein Ganzes ausmachen" [Preisschrift über die Fortschritte der Metaphysik, 1791 (20: 283)]. Cf. A218/B265 and (17: 97, 580; 21: 374, 600).
60. That is, experience in the sense of something that can be an object of the external sense; not experience "in general" (as dealt with in the *KrV*).
61. "Caloric" is one of Kant's lifelong "invariants," only adjusted by new developments in chemistry; see for example (2: 184–185; 29: 119; 21: 522).
62. See (14: 287–291), though there are places where Kant distinguishes between ether and caloric (*Wärmestoff*). On the issue of using *ether* and *Wärmestoff* often interchangeably see Schulze (1994, 134–141). See for a more or less complete inventory of occurrences in Kant's text, Schulze (1994, 137); for Kant's own formulations of the properties of caloric, see (21: 605, 610; 22: 550–551) and in particular (22: 214): "This aether (the only *originally* elastic matter; the name of fluid would not, however, apply to it), moving as elastic matter in straight lines, would be called light material; when absorbed by bodies, and expanding them in all three dimensions, it would be called caloric." Other congeners of caloric and ether Kant uses include *Elementarstoff, Weltstoff, Urstoff, Lichtstoff*. For a detailed but concise "definition" of Kant's ether concept, see Edwards (1991, 91f), who suggests that ether, caloric, and light matter designate "the universal field-entity" constituted by the activity of the action of attractive and repulsive forces (ibid. 155).
63. *OP* (21: 571): "Der Übergang der metaph. Anf. Gr. der NW zur Physik geschieht eben durch die Idee von Wärmestoff welcher darum kein blos hypothetischer sondern der allein all Körper in allen Räumen Erfahrungsmäßig leitende und continuirlich in Einer Erfahrung zusammenhängende Stoff sein muß." In the *Reflections*, the ether does not only appear as the foundation of all forms of cohesiveness, but also as the "matrix of all bodies." See in particular *Refl.* 44. Each body is more or less "saturated" with ether. The ether "inside" and "outside" forms a continuum (*Refl.* 52). However, how to take these statements will depend strongly on how one interprets Kant's use(s) of *Wissenschaft* and *ether* in the *OP*.
64. And, crucially, "Die Heterogeneitaet bringt die Perceptibilitaet hervor." (21: 611) Plaass (1994, 293) suggested that already in the *MAdN*: "Kant is just barely able to derive how matter of different densities is necessarily (i.e., *a priori*) possible, namely, by means of the necessary opposing interplay (limitation) of repulsion (reality) and attraction (as its negation)."
65. See *Refl.* 41 and 44.
66. Cf. Carrier (2001, 223) and (4: 530): "Wenn aber zwei Materien und zwar jede derselben ganz einen und denselben Raum erfüllen, so durchdringen sie einander." For Aristotle's "mixts" see *De generatione et corruptione*, 327a, and Needham (1996).
67. This is a somewhat speculative statement with which not all Kant scholars will agree. In support, one could point to (21: 223, 563, 576, 579, 600; 22: 160, 161, 551).
68. See (22: 13, 205–212, 474).
69. "Die erste Ursachen sind nicht mechanisch, sondern dynamico physisch." (14: 211) The idea is worked out further in the *OP* (22: 205, 239–240, 474). Cf. *Refl.* 41, 43 (14: 165, 270).
70. Kant's notion of element would warrant separate attention. Cf. Carrier (2001), B673–681, (29: 161–166, 341–361). Kant followed Stahl in having five elements, which for him were more like regulative ideas of reason and can never be identified empirically. Kant's understanding of these elements changed over time under influence of developments in chemistry. Kant's theory of elements in the *KrV* is transcendental (not empirical or metaphysical).
71. For an early example of phlogiston as fifth essence in Kant's writings, see (1: 212).

72. Förster (1989, 36); Hoppe (1991, 60; 1969, 99); Mathieu (1989, 70); Friedman (1992, 297).
73. Carrier (1991, 223) suggests that Kant's use of the words *Wärmestoff* and *ether* in the *OP* is threefold; it has a transcendental, a chemical, and a cosmological function.
74. As Lequan (2000, 92n) notes, considering the ether as a necessary hypothesis is not an invention of Kant. The same view can already be found in the article "Heat" in Gehler's *Physikalisches Wörterbuch*.
75. For Kant, both cosmology and chemistry had always been closely related to the notion of ether (cf. his *Theory of Heavens* of 1755); also (4: 534).
76. On the question whether it is an empirical physico-chemical hypothesis or a transcendental *a priori* truth see (21: 216–217, 551, 535–536; 22: 217, 550) and discussion in Lequan (2000, 89–98).
77. I leave the difficult interpretative issues concerning Kant's *Ätherbeweise* in the *OP* to the Kant specialists: Edwards (2000, 152–157), Schulze (1994), Friedman (1992, 293–341), Mathieu (1989, 231–271), Förster (1991, 41–45; 1989), Lehmann (1963).
78. Vasconi (1996, 157) suggests "it is reasonable to assume that Kant was already familiar with the new discoveries as early as 1793," referring to a letter from Erhard (11: 408), which refers to Girtanner's *Anfangsgründe der antiphlogistischen Chemie*, of which Kant had been sent a copy. The first mention of Lavoisier is in *Refl.* 66, around 1789–1790 (14: 489): "Nach Lavoisier, wenn etwas (nach Stahl) dephlogistirt wird, so kommt etwas hinzu (reine Luft); wird es phlogisticirt, so wird etwas (reine Luft) weggenommen."
79. Cf. (21: 417, 424).
80. "Ist Wasser Element? Nein denn es läßt sich noch auflösen, es besteht aus Lebensluft und brennbarer Luft, und wir nennen etwas was keine Spezies enthält elementarisch" (28: 664); see also *Refl.* 73.
81. In an enclosure to a letter to S.T. Soemmerring of August 10, 1795: "Das reine, bis vor kurzen noch für chemisches Element gehaltene, gemeine Wasser wird jetzt durch pneumatische Versuche in zwei verschiedene Luftarten geschieden." (12: 33). In the *Physische Geographie* (1802) Kant writes (following Gehler): "das Wasser (besteht) aus Wasserstoff und Sauerstoff, und zwar in einer Mischung die bei einhundert Theilen, 15 des ersten und 85 des letßten enthält" (11: 184).
82. *OP*, around 1795 (21: 453): "What is chemistry? The science of the inner forces of matter." The distinction between fluid and rigid bodies cannot be explained without invoking different intensities of cohesion. Kant borrows from Leibniz the idea that the liquid state is the originary state from which the solid and vapour state derive (he also refers to Thales). Metaphysically this is supported by the chemical and cosmological supremacy of the primordial fluid element, viz. the ether. See further previous section.
83. (6: 207): "there is only one chemistry (that according to Lavoisier)." But Kant still writes in the same book: "Chemists base their most universal laws . . . entirely on experience" (6: 215). In the *Anthropology from a Pragmatic Point of View* of 1798, we can read (7: 326): "What a mass of knowledge, what discoveries of new methods would now be on hand if an Archimedes, a Newton, or a Lavoisier with their industry and talent would have been favored by Nature with hundreds of years of continuous life without the loss of vital power?" The translation is from Dowdell (1978) who informs us (p. 288n119) that Kant substituted "Lavoisier" for "Galilei" in the manuscript stage. Quotations in English from the *Doctrine of Rights* are taken from Gregor (1991).
84. Hence, he also dismisses Tuschling's (1971) view, who suggested that the function of the *OP* was merely to resolve some inconsistencies in the concept of matter as outlined in the *MAdN*.
85. See in particular notes 4, 5, 12 of his Chapter 8.
86. See references in note 49.
87. See for example (21: 294, 299; 22: 136–137, 158, 587). For the significance of the ether in explaining the working of a balance, see Hoppe (1969, 99). One might say that Lavoisier's work, in which the use of a balance played a crucial role, made possible the application of mathematics to chemistry. Cf. *Refl.* 66 (14: 489–494); Adickes dates this text at about 1789 (based on the ink used). This is the year Lavoisier published his *Traité élémentaire de chimie*.
88. Laplace, *Œvres* (II, 645): "En général, nous n'avons de sensation que par le mouvement; en sorte qu'on pourrait poser comme un axiome: point de mouvement, point de sensation . . . Ce principe s'applique au sentiment du froid et du chaud." For Lavoisier's views on this, see in particular his *Réflexions sur le*

phlogistique of 1783 (*Œvres*, II: 623–655). Duhem (1899, 214) quotes Lavoisier as saying: "la science des affinités est à la chimie ordinaire ce que la géometrie transcendante est à la géometrie élémentaire" (without giving a reference). This could be read as a Kantian influence on Laplace.

89. According to Lequan (2000, 122), Hegel completed Kant's project and elevated chemistry fully to the rank of proper science. Moreover, developing the analogical role model of chemistry, with Hegel chemistry became philosophical and philosophy became chemical.
90. The confusion of thinking that Kant only focuses on physics in the *OP* arises, because it is true that Kant often uses the term "physics" to cover both physics and chemistry. For example (22: 501): "As a science of experience, however, physics is naturally divided into two subjects. The one is the subject of the forms in action and reaction of forces in space and time. The other is the complex of the substances which fill space. The one could be called the systematics of nature, the other is called (following Linnaeus) the system of nature." However, other passages make clear that he might also call the former physics and the latter chemistry (see citations in note 49). In the end, the question whether Kant would locate his "gap" or "transition" in connection with physics or chemistry is neither here nor there.
91. Cf. my distinction between "the ontology of *matter in general*, to be dealt with in relation to developments in micro- and astrophysics" and "the ontology of *particular kinds of matter*, i.e., chemical kinds" (van Brakel 1991). Cf. Kant's "matter in general" (21: 307) and "der Gemeinschaft verschiedener Substanzen" (21: 571).
92. Cf. "Critique stands in the same relation to the common metaphysics of the schools as chemistry does to alchemy" (4: 366—*Prolegomena*).
93. I am greatly indebted to Martin Moors who helped me to grapple with the intricacies of Kant's philosophy of nature. This research was supported by a grant from FWO-Vlaanderen (project number 1.5.012.02).

REFERENCES

Adickes, E. 1924. *Kant als Naturforscher*. Berlin: de Gruyter.

Arabatzis, T. 1998. How the Electrons Spend their Leisure Time: Philosophical Reflections on a Controversy between Chemists and Physicists. In: Janich, P. and Psarros, N. (eds.), *The Autonomy of Chemistry: 3rd Erlenmeyer-Colloquy for the Philosophy of Chemistry*. Würzburg: Königshausen & Neumann, 149–159.

Bad Homburg 1991. *Übergang: Untersuchungen zum Spätwerk Immanuel Kants*, Forum für Philosophie Bad Homburg, Frankfurt: Klostermann, 1991.

Bader, R.F.W., Popelier, P.L.A. and Keith, T.A. 1994. Die theoretische Definition einer funktionellen Gruppe und das Paradigma des Molekülorbitals. *Angewandte Chemie* 106: 647–659.

Brandt, R. 1991. Kants Vorarbeiten zum Übergang von der Metaphysik der Natur zur Physik: Probleme der Edition. In: Forum für Philosophie Bad Homburg, *Übergang: Untersuchungen zum Spätwerk Immanuel Kants*. Frankfurt: Klostermann, 1–27.

Carrier, M. 1990. Kants Theorie der Materie und ihre Wirkung auf die zeitgenössische. Chemie. *Kantstudien* 81: 170–210.

Carrier, M. 1991. Kraft und Wirklichkeit. Kants späte Theorie der Materie. In: Forum für Philosophie Bad Homburg, *Übergang: Untersuchungen zum Spätwerk Immanuel Kants*. Frankfurt: Klostermann, 208–230.

Carrier, M. 2001. Kant's Theory of Matter and his Views on Chemistry. In: Watkins, E. (ed.), *Kant and the Sciences*. Oxford: Oxford University Press, 205–230.

Dingle, H. 1949. The Nature of Scientific Philosophy. *Proceedings of the Royal Society of Edinburgh* 4: 409.

Dirac, P.A.M. 1929. Quantum Mechanics of Many-Electron Systems. *Proceedings of the Royal Society of London* A123: 714–733.

Dirac, P.A.M. 1938/1939. The Relation between Mathematics and Physics. *Proceedings of the Royal Society of Edinburgh* 59: 122–129.

Dowdell, V.L. (transl.). 1978. *Immanuel Kant: Anthropology from a Pragmatic Point of View*. Carbondale: Southern Illinois University Press.

Drews, A. 1894. *Kants Naturphilosophie als Grundlage seines Systems*. Berlin: Mitscher & Röstell.

Duhem, P. 1899. Une science nouvelle, la chimie-physique. *Revue philomatique de Bordeaux et sud-ouest*. Paris: Hermann, 205–219; 260–280.

Dussort, H. 1956. Kant et la chimie. *Revue philosophique* 81: 392–397.

Edwards, B.J. 1991. Der Aetherbeweis des Opus postumum und Kants 3. Analogie der Erfahrung. In: *Übergang: Untersuchungen zum Spätwerk Immanuel Kants*, Forum für Philosophie Bad Homburg, Frankfurt: Klostermann, pp. 77–104.

Edwards, J. 2000. *Substance, Force, and the Possibility of Knowledge: On Kant's Philosophy of Material Nature*. Berkeley: University of California Press.

Ellington, J. (transl.). 1970. *Immanuel Kant: Metaphysical Foundations of Natural Science*. Indianapolis: Bobbs-Merrill.

Förster, E. 1989. Kant's Selbstzetsungslehre, In: Förster, E. (ed.), *Kant's Transcendental Deductions*. Stanford: Stanford University Press, 217–238.

Förster, E. 1991. Die Idee des Übergangs. Überlegungen zum Elementarsystem der bewegenden Kräfte. In: Forum für Philosophie Bad Homburg, *Übergang: Untersuchungen zum Spätwerk Immanuel Kants*. Frankfurt: Klostermann, 28–48.

Förster, E. (transl. and ed.). 1993. *Kant's Opus postumum*. Cambridge: Cambridge University Press.

Frenking, G. 1998. Heretical Thoughts of a Theoretical Chemist About the Autonomy of Chemistry as a Science in the Past and the Present. In: Janich, P. and Psarros, N. (eds.), *The Autonomy of Chemistry: 3rd Erlenmeyer-Colloquy for the Philosophy of Chemistry*. Würzburg: Königshausen & Neumann, 103–108.

Friedman, M. 1992. *Kant and the Exact Sciences*. Cambridge, MA: Harvard University Press.

Gregor, M. (transl. and ed.). 1991. *Immanuel Kant: The Metaphysics of Morals*. Cambridge: Cambridge University Press.

Groß, F. (ed.). 1912. *Immanuel Kant: Sein Leben in Darstellungen von Zeitgenossen: Die Biographien von L. E. Borowski, R.B. Jachtmann und A.Ch. Wasianski*, Berlin: Deutsche Bibliothek.

Hartmann, M. 1948. *Die philosophische Grundlagen der Naturwissenschaften*. Jena: Fischer.

Heinig, K. 1975. Immanuel Kant und die Chemie des 18. Jahrhunderts (in der Darstellung der Wissenschaftsgeschichte). *Wissenschaftliche Zeitschrift der Humboldt-Universität zu Berlin, gesellschafts- und sprachwissenschaftliche Reihe* 24(2): 191–194.

Heyse, H. 1927. *Der Begriff der Ganzheit und die Kantische Philosophie*. München: Reinhardt.

Hoffmann, R. 1998. Qualitative Thinking in the Age of Modern Computational Chemistry—Or What Lionel Salem Knows. *Journal of Molecular Structure* 424: 1–6.

Hoppe, H.G. 1969. *Kants Theorie der Physik: Eine Untersuchung über das Opus postumum von Kant*. Frankfurt a/M: Klostermann.

Hoppe, H.G. 1991. Forma dat esse rei. Inwiefern heben wir in der Erkenntinis das aus der Erfahrung nur heraus, was wir zuvor in sie hineingelegt haben? In: *Übergang: Untersuchungen zum Spätwerk Immanuel Kants*, Forum für Philosophie Bad Homburg, Frankfurt: Klostermann, pp. 49–64.

Janich, P. and Psarros, N. (eds.). 1998. *The Autonomy of Chemistry: 3rd Erlenmeyer-Colloquy for the Philosophy of Chemistry*. Würzburg: Königshausen & Neumann.

Kant, I. 1900–2000. *Akademie Ausgabe*.

Körner, S. 1991. On Kant's Conception of Science and the *Critique of Practical Reason*. *Kantstudien* 82: 173–178.

Lehmann, G. 1963. Zur Frage der Spätentwicklung Kant. *Kantstudien* 54: 491–507.

Lequan, M. 2000. *La chimie selon Kant*. Paris: Presses Universitaires de France.

Lomonossov, M.W. Ausgewählte Schriften in Zwei Bänden. Band I Naturwissenschaften. Berlin: Akademie Verlag. 1961.

Mainzer, K. 1998. Computational and Mathematical Models in Chemistry: Epistemic Foundations and New Perspectives of Research. In: Janich, P. and Psarros, N. (eds.), *The Autonomy of Chemistry:*

3rd Erlenmeyer-Colloquy for the Philosophy of Chemistry. Würzburg: Königshausen & Neumann, 33–50.

Makkreel, R.A. 2001. Kant on the Scientific Status of Psychology, Anthropology, and History. In: Watkins, E. (ed.), *Kant and the Sciences*. Oxford: Oxford University Press, 185–204.

Malzkorn, W. (1998) Kant über die Teilbarkeit der Materie, *Kant Studien*, 89: 385–409.

Marty, F. (ed.), 1986. *Emmanuel Kant. Opus postumum. Passage des principes métaphysiques de la science de la nature à la physique*. Paris: Presses Universitaires de France.

Mathieu, V. 1989. *Kants Opus postumum*. Frankfurt am Main: Klostermann.

Meyer, V. (1889) The chemical problems of to-day, *Journal of the American Chemistry Society*, 11: 101–120.

Nayak, A.C. and E. Sotnak (1995) Kant on the impossibility of the "soft sciences", *Philosophy and Phenomenological Research*, 55: 133–151.

Needham, P. 1996. Aristotelian Chemistry: A Prelude to Duhemian Metaphysics. *Studies in History and Philosophy of Science* 27: 251–270.

Nye, M.J. 1993. *From Chemical Philosophy to Theoretical Chemistry: Dynamics of Matter and Dynamics of Disciplines 1800–1950*. Berkeley: University of California Press.

Paneth, F.A. 1962. The Epistemological Status of the Chemical Concept of Element. *The British Journal for the Philosophy of Science* 13: 1–14; 144–160 [originally published in German as Über die erkenntnistheoretische Stellung des Elementbegriffs. *Schriften der Königsberger Gelehrten Gesellschaft, naturwissenschaftliche Klasse* 8:4 (1931) 101–125.]

Plaass, P. 1994. *Kant's Theory of Natural Science*. Dordrecht: Kluwer (translation, analytic introduction and commentary by A.E. Miller and M.G. Miller; introductory essay by C.F. von Weiszsäcker).

Roqué, A.J. 1985. Self-organization: Kant's concept of teleology and modern chemistry, *Review of Metaphysics*, 39: 107–135.

Rothbart, D. and I. Scherer (1997) Kant's *Critique of Judgement* and the scientific investigation of matter, *Hyle*, 3: 65–80.

Schulze, S. 1994. *Kants Verteidigung der Metaphysik: eine Untersuchung zur Problemgeschichte des Opus Postumum*. Marburg: Tectum.

Sturm, Th. 2001. Kant on empirical psychology: How not to investigate the human mind. In: Watkins E. (ed.) *Kant and the Sciences*, Oxford University Press, pp. 163–176.

Tuschling, B. 1971. *Metaphysische und transzendentale Dynamik in Kants Opus Postumum*. Berlin: de Gruyter.

van Brakel, J. 1991. Chemistry. *Handbook of Metaphysics and Ontology*, Vol. 1. München: Philosophia Verlag, 146–147.

van Brakel, J. 2000. *Philosophy of Chemistry: Between the Scientific and the Manifest Image*. Leuven: Leuven University Press.

Vasconi, P. 1996. Kant and Lavoisier's Chemistry. In: Mosini, V. (ed.), *Philosophers in the Laboratory*. Rome: Euroma, 155–162.

Watkins, E. 1998. The Argumentative Structure of Kant's Metaphysical Foundations of Natural Science. *Journal of the History of Philosophy* 36: 567–593.

Westphal, K.R. 1995. Kant's Dynamic Constructions. *Journal of Philosophical Research* 20: 382–429.

Zweig, A. (transl. and ed.). 1999. *Immanuel Kant: Correspondence*. Cambridge: Cambridge University Press.

SECTION III

CHEMISTRY AND CURRENT PHILOSOPHY OF SCIENCE

CHAPTER 5

THE CONCEPTUAL STRUCTURE OF THE SCIENCES

Reemergence of the Human Dimension

OTTO THEODOR BENFEY

Guilford College, Greensboro NC 27410, USA; Home address: 909 Woodbrook Drive, Greensboro NC 27410, USA; E-mail: benfeyot@infionline.net

INTRODUCTION

Half a century ago during a period of professional transition, I became aware of a common conceptual pattern underlying certain areas of subject matter that I had been teaching. It became apparent first in two fields and then in other areas of science. The pattern for long seemed to be ignored by others, but some years later the ideas were picked up and developed by Jensen (1988).

In my freshman course, I dealt with the ideal gas laws and the kinetic molecular theory that accounted for them. My organic chemistry classes explored the structural theory of organic chemistry developed in 1858 by Kekulé and Couper. In each of those fields, I had to point out that the initial simple theory had to be amplified. Gases in fact were not ideal; real gases did not exactly obey the simple laws of Boyle, Charles, and Gay-Lussac; and the beautifully simple structural theory of 1858 could not account for all cases of isomerism.

What became apparent was that the failures of the two theories and the ways they were amended had something in common. The simple theories did not initially incorporate all the dimensions that were part of our real world. Especially, noteworthy was the fact that neither had incorporated the evolutionary perspective, time's arrow, by which we who live in the 21st century make sense of large areas of the sciences and their interconnections. Structural theory in fact ignored time completely. Also neither theory in its classic form took account of the fact that atoms, molecules, and chemical bonds required space to exist, they were not points. Those were two crucial omissions.

THE KINETIC THEORY OF GASES

Looking first at the kinetic theory of gases, we begin with the familiar equation $PV = RT$ expressing the relation of pressure, volume, and temperature on the Kelvin scale for one mole of gas. But we also know that real gases diverge from this ideal gas law at high pressures and low temperatures. Those divergences also follow patterns,

D. Baird et al. (eds.), Philosophy of Chemistry, 95–117.

one of which was proposed by Van der Waals in his equation

$$(P + a/V^2)(V - b) = RT$$

Here the term b accounts for the fact that the original ideal gas law ignored the fact that gas molecules occupy space. Under high pressure, when there is little empty space left, doubling the pressure will no longer halve the volume. The term b introduced the internal space dimension hereafter designated as 3d. The ideal gas equation did not ignore space completely. The gas molecules move in three-dimensional external space, a dimension designated 3D. The simple kinetic molecular theory that was developed to account for the equation $PV = RT$ also recognizes number, it counts the molecules, the dimension N.

But the dimension N becomes problematic at low temperatures. At low temperatures or high pressures, molecules aggregate, the number of particles decreases, and before that happens there is attraction between the particles so that the pressure exerted by the gas is less than expected for ideal behavior. For obedience to the ideal gas law, the pressure has to be augmented by a second term expressed by Van der Waals as a/V^2 where a recognizes the non-numerical coalescing aspect of nature, the continuity characteristic which is as real in nature as numerical discontinuity. The gas units coalesce forming larger units. This aggregating, continuity aspect we designate as A.

Finally, ideal gas theory required the inclusion of time. Molecules move randomly and if there are enough of them in enough space, the gas would look uniform. This time dimension we call t.

Now suppose a chamber of slow and hence cold molecules is connected with a chamber of warm, fast moving molecules. Soon, the two chambers will have the same mix of fast and slow molecules. But if you begin with two chambers containing the same mixture and connect them, you will never find one chamber collecting all the fast molecules and the other the slow ones. Newtonian physics was reversible, but the world he was describing was not. To cope with this fact, enshrined of course in the second law of thermodynamics and the entropy concept, we need to introduce directional time designated here as $\nearrow t$.

In summary, we can say that ideal gas theory required only N, 3D, and t, molecular units, space within which to move, and time to permit movement. The Van der Waals equation added 3d and A, while the one-way behavior encountered in mixing added $\nearrow t$.

These six dimensions are here focused on and they rather naturally divide into three pairs, two involving time, two involving space, and two recognizing the continuity/discontinuity dichotomy. For ease in visualization, the pairs can be presented as opposite faces of a cube (Figure 1) on which ideal gas theory requires only half the faces, one face from each pair, the three meeting at one vertex of the cube.

Before we go more deeply into the philosophical implications of this sextet, we need to test whether the exercise is worth it, whether it has relevance beyond the behavior of gases. We therefore look at the organic chemist's quest—the development of a language that will unambiguously predict the number of isomers of a given molecular formula, and will depict each isomer as distinct from all others, and preferably in a form that will suggest some of its properties.

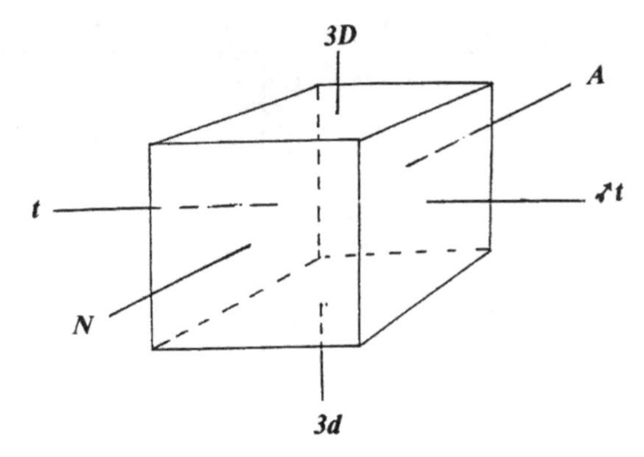

Figure 1: A three-dimensional representation of the six concepts focused on in this paper. The six appear in three pairs, one pair relating to time (*t* symmetrical time, ↗*t* directional time), the second to space (3D external space relations, 3d internal structure), and the third to the continuity–discontinuity dichotomy (*N* number, discreteness, *A* association, continuity).

THE STRUCTURAL THEORY OF ORGANIC CHEMISTRY

The structural theory of organic chemistry, usually dated from 1858, refers to the original two tenets of that theory, the tetravalence of carbon and the peculiar ability of carbon atoms to link with each other. This theory is associated with the names of Friedrich August Kekulé and Archibald Scot Couper, who independently enunciated the theory within a few weeks of each other. It should be noted, as Senior (1935) and Wheland (1949) elegantly pointed out, that the structural theory as it is almost universally called had nothing to do with geometric structure. It was in fact a topological theory, merely specifying which atom was linked to which others and by how many bonds. Topology is rubber-sheet geometry. As long as X is connected to Y that in turn is connected to Z, the theory says nothing as to whether X–Y–Z is linear or angular or shows any other contortion. To emphasize this, Wheland wrote some formulas as connectivity tables merely listing the number of connections or bonds (1, 2, or 3) from each atom to each of the others (Figure 2).

The dimensional structure of the structural theory was simply *N*. And it is to be noted that the number *N* refers not to atoms but to bonds.

The 3D space dimension was introduced—again simultaneously—by two chemists, Van't Hoff and Le Bel, in order to distinguish between pairs or even larger numbers of isomers depicted by the same connectivity table or classical structural formula but lowering each other's melting point and therefore not identical. The two optically active tartaric acids and their inactive *meso* isomer are examples. The new organic subfield of stereochemistry specified the direction in space of the four bonds radiating from carbon.

We can now use our dimensional sextet to search for areas of the structural theory not covered by *N* or 3D. What about 3d, isomers that exist only because atoms occupy finite volumes? Such a case of isomerism surfaced in the chemistry of the biphenyls, two benzene rings linked by a single bond. Whereas 2,2′-difluorobiphenyl exists as

	C	C	H	H	H	H	O
C	-	1	1	1	1	0	0
C		-	0	0	0	1	2
H			-	0	0	0	0
H				-	0	0	0
H					-	0	0
H						-	0
O							-

	C	C	H	H	H	H	O
C	-	1	1	1	0	0	1
C		-	0	0	1	1	1
H			-	0	0	0	0
H				-	0	0	0
H					-	0	0
H						-	0
O							-

(a) (b)

Figure 2: Topological, non-spatial representation of two isomers of formula C_2H_4O, acetaldehyde (left) and ethylene oxide (right). Each number represents the number of covalent bonds linking the atom at the left of the row to the atom at the top of the column. All atoms of the molecule are listed both at the top and on the left-hand side. Each bond is only listed once, accounting for the half-filled appearance of the tables. These tables express all that was originally intended by the more conventional structure diagrams.

only a single isomer, when the fluorine atoms are replaced by larger atoms or groups, optically active isomers can be isolated. We say the latter molecules exhibit steric hindrance, the size of the substituents preventing planarity or free rotation about the single bond linking the two benzene rings (Figure 3). This discovery is associated with the names of Christie and Kenner (1922); see Wheland (1949, 202–214).

The time dimensions have a longer history. The pivotal substance always exhibited as example is acetoacetic ester (ethyl acetoacetate), which Frankland insisted was a ketone while Geuther equally convincingly showed to be enolic. The former published ever more positive keto tests only to be countered by more tests proving enolic properties.

$$CH_3COCH_2CO_2Et \text{ (keto)} \quad \text{versus} \quad CH_3C(OH){=}CHCO_2Et \text{ (enol)}$$

E.C.C. Bailey even suggested, citing spectroscopic data, that the substance might be something between the two. The controversy was finally settled by Knorr in 1911, who cooled the material to −78°C where one of the two forms crystallized while the other remained a liquid. Furthermore, at that temperature each could be preserved unchanged exhibiting its own properties.

The puzzling phenomena at room temperature were due to rapid equilibration. If a keto reagent is added, the ketone will be removed and enol will convert to ketone

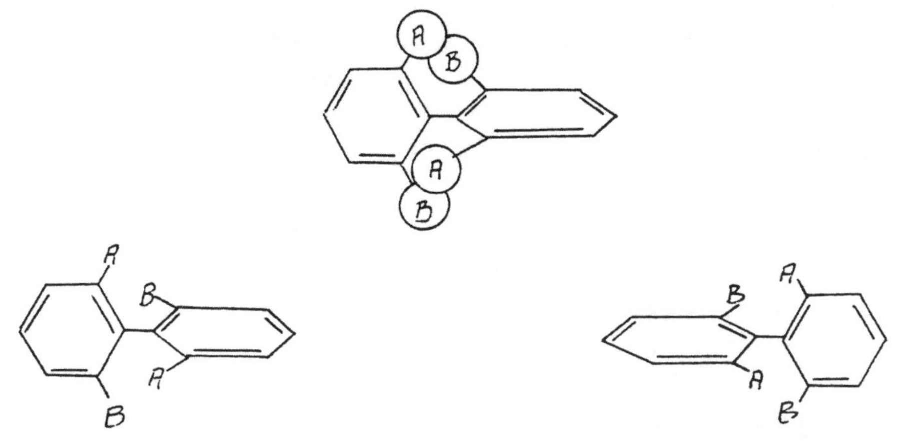

Figure 3: Isomerism due to finite size of atoms. If in substituted biphenyls, atoms A and B are large enough, the molecule cannot become planar. Hence, two mirror images result.

until all the material has reacted. The analogous situation would occur if an enol reagent is used. Here, clearly the dimensional characteristic is t, symmetrical time, since equilibrium can be disturbed in either direction and changes will occur in either direction in order to re-establish the equilibrium (Ihde 1959).

Are there isomer problems that demand introduction of $\nearrow t$, isomers or absence of isomers accountable only if we introduce time's arrow? The answer is yes—in the case of predicted isomers that have never been isolated. We usually account for their absence by the statement that if prepared they immediately convert into a more stable form. Vinyl alcohol $CH_2{=}CHOH$ was for long taken as the key example, but recently it was isolated at very low temperatures. There are, however, numerous others that only exist in the chemist's imagination, never in a test tube. We usually say that they do not exist either because they too easily convert into a more stable substance, that is, the energy barrier to change is too low, or else the valence requirements are satisfied, but there is no way atom X can link up with atom Y because of the complexity of the structure. Time here goes one way, a substance only being recognized if it can exist long enough to be isolated or at least to give a spectroscopic signal of its existence. Those that cannot, have changed too quickly in unidirectional time to a different structure.

The above refers to non-existent structures. There is another class of structures where the number isolated is less than the number predicted, but not zero. This is the class subsumed under the theory of resonance. The classic text on the subject is Wheland's (1955) and the classic example is that of 1,2-dichlorobenzene (Figure 4) which should exist in two forms, one with a single and the other with a double bond between the chlorine-substituted carbons.

But only one 1,2-dichlorobenzene has ever been isolated and the bond between the substituted carbons has properties intermediate between those of a single and a double bond. Instead of a low energy barrier between two substances, no barrier exists

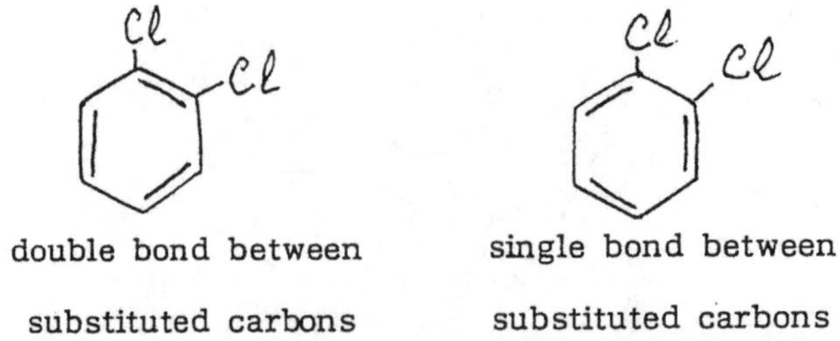

double bond between substituted carbons

single bond between substituted carbons

Figure 4: Two classic representations for 1,2-dichlorobenzene, one with a single bond the other with a double bond between the substituted carbons. The actual molecule has a bond intermediate between single and double.

at all. In fact, the actual structure has a lower energy than the hypothetical valence bond structures depicted in Figure 4. Here, the structural theory confronts the limit to the claim that it involves the dimension N, that bonds must occur in units, 1, 2, or 3 and cannot be intermediate or fractional. In fact, the "bond order" of the ring bonds in benzene is uniform throughout, each bond having a bond order of 1.5. We confront here the dimension A.

Interestingly, Westheimer (1965) at Harvard developed a summer short course in organic chemistry structured much like the development above. He began with the simple structural theory and then looked at cases where it failed:

a) where there were more substances than formulas: stereochemistry;
b) fewer substances than predicted but not zero: resonance theory; or else tautomeric systems such as acetoacetic ester where interconversion was too facile;
c) structures but no substances: absence due to too easy conversion or steric impossibility of bonding.

We have then encountered the same six dimensions, N, A, 3D, 3d, t, $\nearrow t$ in an area of science totally different from kinetic gas theory, and it gives us courage to use the dimensional sextet in further explorations.

THE THREE PAIRS OF CENTRAL CONCEPTS

We divided the six concepts into three pairs, the first dealing with space, the second with time, and the third with the classic discrete/continuous dichotomy, already evident in the distinction between arithmetic and geometry. Placing the members of each pair of concepts on opposing faces of the cube in Figure 1, each face is in contact with each of the others except for the face directly opposite. Adjacent faces of the cube represent phenomena which require both concepts, can be explained by either concept, or lead to a new conceptual development subsuming both as in the space–time continuum of relativity theory. The concept of velocity requires both direction in space (3D) and duration in time (t). As seen in the previous section, certain failures encountered in the structural theory of organic chemistry can be explained either in spatial terms

(3D) or in terms of insufficient detection time ($\nearrow t$). Similarly, geologic strata can be described either as temporal or as spatial sequences.

On examining the cubic representation, we notice that the three concepts central to the classic mechanical description of particles—*N*, 3D, *t* (particles, directional space, time)—appear as three faces sharing a vertex. The remaining three faces—directional time $\nearrow t$, inner structure 3d, and continuity or associative character *A*—sharing the opposite vertex, are characteristics of the organicist tradition as we will see below.

Opposing faces, in contrast to those sharing an edge, cannot be easily incorporated in a single logical presentation. It should therefore come as no surprise that the three pairs of opposite faces correspond to contrasting but complementary "themata" using Gerald Holton's term (1973). In fact, they are at the center of the great riddles and the controversies that have marked the growth and development of the sciences in the last two centuries: the problem of how the concept of irreversibility can be incorporated in classical physics; the particle/wave duality that precipitated the quantum mechanical controversies; and the inner/outer duality lying at the heart of discussions regarding levels of organization, emergent levels and reductionism, the question whether the whole is more that the sum of its parts. Attempts have repeatedly been made to overcome these dualities, to subsume one concept under its opposite, to reduce one to the other. But the difficulties of such tasks have prompted others to suggest that both concepts may be necessary though not logically compatible. These three pairs of concepts need to be discussed in more detail.

Reversibility/irreversibility (t / $\nearrow$t)

The direction of time, "time's arrow," is part of all evolutionary and organic growth processes as well as of order–disorder transitions. One of the earliest moments when the distinction between reversible and irreversible phenomena had to be grappled with in the sciences occurred in the establishment of the second law of thermodynamics. Based largely on the work of Carnot who introduced the idea of the reversible cycle for the attainment of maximum work from a heat engine, Clausius added to thermodynamics the concept of entropy. Gibbs and Helmholtz then developed a statistical interpretation of the second law according to which systems tend to change in the direction of states of greater probability and greater randomness. Ever since, attempts have been made to derive the conclusions of probability predictions from classical laws or else to suggest that all laws of physics are of a statistical nature. Only slowly did the recognition develop that both types of conception are required for any adequate description of physical systems.

Cassirer summarized his discussion of the history and current understanding of probability (and hence irreversibility) statements in the comment: "From all this we can conclude that 'causality' and 'probability', 'order according to law' and 'accident', not only can but must exist side by side when we want to determine an event as completely as possible. In classical physics causality refers essentially to the knowledge of the course of the event, and probability to the knowledge of its initial conditions. From the two combined there arise the theorems of statistical mechanics" (1956, 104).

Particle/wave; discrete/continuous (A/N)

A scientific description often begins with positing a set of totally distinct countable entities (N). However, phenomena are usually encountered which lay open to question the validity of the initial definition of objects as separate entities. The objects exhibit behavior suggesting that they have an affinity for other members of their class or else are part of something larger. The criteria for counting the number of individual entities sometimes lead to the discovery of fractional rather than integral units. In response, field or wave approaches are attempted, treating the entities earlier considered distinct as parts of a continuum. The original objects are now viewed as being associated (A) with other objects of similar kind in a larger whole, which is not a simple aggregate or collection of its parts.

Before such radical transformations are proposed, attempts are usually made to retain the original paradigmatic structure by adding attractive forces or associative tendencies operating among the separate entities. Thus, valence forces are invoked in chemistry, the force of gravity in Newtonian physics. In contrast, the continuum view in chemistry looks at a molecule as held together by orbitals encompassing the molecule as a whole, while in relativity physics material particles are seen as local disturbances in a four-dimensional space–time continuum.

The best known case of the discrete/continuum duality is of course the crisis precipitated by phenomena now dealt with by quantum theory. After the controversies concerning the wave or particle nature of light appeared to have been solved in favor of the wave viewpoint, Einstein's explanation of the photoelectric effect suggested that the question was not whether light was wave *or* particle, that an adequate answer might require the acceptance of both particle and wave characteristics. De Broglie then proposed a similar duality of properties for matter, soon confirmed by electron diffraction studies. Again attempts were made to explain both types of behavior from a single viewpoint. Schrödinger's wave mechanics represented electrons as infinite wave disturbances, whereas statistical quantum mechanics interpreted the square of the wave amplitude as a measure of the probability of finding an electron at a given point. Bohr's complementarity principle then suggested the necessity of accepting the two types of observations as logically irreducible yet necessary ingredients of any adequate world picture. This development is ably described by Cassirer (1956, Chapter 9).

Holton (1973, 29, 100) has pointed to the discrete/continuum duality as an instance of pairs of themata constantly reappearing in the thought and the theoretical elaborations of scientists. Mendelsohn (1980) has surveyed the periodic appearance of this pair of concepts in the development of the various sciences and has suggested correlations with religious, social, and political developments. From the point of view of the present chapter, much of the history of the pendulum swings between discrete and continuum conceptions is based on the widely held conviction that the truth must lie on one side or the other. Instead, insights from both sides are certainly needed for a full understanding of any episode in science. Mendelsohn points out that the two sides have lived in tension with each other, a tension that contributed to the healthy growth of the sciences. Their coexistence, however, now seems also to be epistemologically necessary.

In pure mathematics, the discrete/continuum dichotomy plays a central role. Justus Grassmann according to Lewis (1977) had already in 1844 pointed to the fundamental distinction between the algebraic approach based on the discontinuous unit, and geometric conceptions which originate from the idea of a point from which a continuous line is generated. His son Hermann Grassmann developed his classification of mathematical fields based on this distinction.

*External/internal, outer/inner (*3D/3d*)*

This pair of concepts appears at first as simple and straightforward and not likely to cause philosophical problems. Remarkably often, however, a research field begins with the atomistic illusion that its building blocks are unanalyzable units. The Daltonian (and Newtonian) atom, uncuttable and eternal, the point molecule of the kinetic theory of gases, the "fundamental" particles of physics, the unitary valence bond, the biological cell, the cell nucleus, the gene, all were thought of at one time as the units of discourse only having external relations with other entities. But each was later seen to have internal complexity whose nature had to be understood in order to explain external relations. The attempt was then made to account for external behavior in terms of internal composition and structure, a venture never fully successful because of the vast increase in complexity usually occurring in the transition from any level of the parts to the next higher "whole" (Benfey 1977). The problem is central to the controversies surrounding reductionism. By analogy with the other pairs of concepts here discussed, it would seem plausible to suggest that the outer/inner dichotomy is also in principle unbridgeable.

EXPLORATIONS IN OTHER AREAS OF SCIENCE

In this section, a number of additional research areas will be examined by the procedure previously used. They suggest an analogous conceptual structure for numerous fields.

The chemical elements

N/*A*: Although Dalton's conception of the chemical atom arose from complex considerations, quite limited properties were needed to explain the quantitative data of chemistry. Dalton summarized his views of chemical reactions in 1808:

> Chemical analysis and synthesis go no farther than to the separation of particles from one another, and to their reunion. No new creation or destruction of matter is within the reach of chemical agency. We might as well attempt to introduce a new planet into the solar system, or to annihilate one already in existence, as to create or destroy a particle of hydrogen. All the changes we can produce, consist in separating particles that are in a state of cohesion or combination, and joining those that were previously at a distance (1808, 162–3).

Dalton then proceeded to develop a relative atomic weight scale, taking the atomic weight of hydrogen as his unit. He assumed that all atoms of the same element had identical atomic weights and that the weights of atoms of different elements differed. Chemical combination occurred between small numbers of atoms of two or more elements. Finally, he believed that in most circumstances the most common compound of two elements involved combination of one atom from each element to form the smallest particle of the compound.

These propositions are essentially limited to the integer concept N except that one of its main purposes was to explain the combination of atoms and hence to their ability to associate (A). The early discussions regarding these associations include the electrochemical view of Berzelius and the non-electrical and quite mysterious proposal of Laurent and Dumas. The realization that association can totally erase the identity of an element through nuclear fusion had its origin in the nuclear bombardment experiments of Rutherford in 1917 who succeeded in converting nitrogen (mass 14) when bombarded with helium nuclei (mass 4) into oxygen (mass 17) and hydrogen (mass 1). This line of inquiry culminated in the recognition of hydrogen fusion in the sun and its simulation in the hydrogen fusion bomb.

3D/3d: The spatial relations among atoms (3D) were dealt with in the organic structural theory discussion (The Structural Theory of Organic Chemistry, see also Benfey 1963, 1964.). That atoms had an internal composition (3d) and were not the ultimate uncuttable units of matter was proposed within seven years of Dalton's publication. William Prout, on the basis of Daltonian atomic weight tables, noticed that most weights were close to being integral multiples of the atomic weight of hydrogen. He therefore suggested that all atoms were hydrogen aggregates. He may have been motivated by philosophical concerns about the multiplicity of elements since he suggested that hydrogen might be the "protyle of the ancients" (Prout 1815, 1816; Benfey 1952; Brock 1985). Prout's proposal led to much fruitful research throughout the 19th century and later to some ingenious speculations as to why atomic weights deviated from integral values. Mendeleev (1872), for instance, suggested a possible mass–energy conversion.

Parallel to the concern for more reliable atomic weights, attention was focused by Döbereiner and Dumas on numerical relations among the atomic weights of chemically similar elements, while Newlands, Mendeleev, Lothar Meyer, and others pointed to the periodic recurrence of elements of similar properties when elements are ordered by increasing atomic weight. Meyer and after him Mendeleev drew attention to the Pythagorean integral step-wise variation in valence or combining capacity (1, 2, 3, 4, 3, 2, 1) along the atomic weight scale, suggesting some type of building block construction. (Benfey 1992–1993).

In 1891, G. Johnstone Stoney pointed out that Faraday's laws of electricity implied a particulate character of electricity. He proposed the name electron for the smallest electrical charge. The study of cathode rays by Crookes and others, and J.J. Thomson's quantitative measurements of the electron's properties led the latter to surmise that the electron was a constituent of all atoms. Rutherford in 1911 discovered evidence for the atom's nucleus. The arrangement of electrons within the atom on the basis of spectroscopic and chemical evidence has been explored extensively during the past century. The 3d character of atoms has become the basis of a large section of theoretical chemistry concerned with the explanation of chemical behavior.

$t/{\nearrow}t$: As long as atoms were seen as eternal and indivisible, the time concept was that of reversible time. Atomic spectra are explained as the reversible absorption of energy by an atom raising it to a higher energy level, the excess energy being readmitted as radiation.

With the prominence of theories of biological evolution in the second half of the 19th century, some chemists speculated on the process by which atoms may have been constructed from simpler units. Crookes (1886) presented a view of element evolution before the British Association for the Advancement of Science in which as temperature decreased, hydrogen slowly congealed into the larger elements. Lockyer (1900) proposed that element building was still occurring in stars.

The fact that atoms were not eternally stable became evident in the discovery of radioactivity by Becquerel, followed by the extensive investigations of nuclear transformations by Marie Curie, Rutherford, Soddy, and others. Atoms thus are not indivisible eternal building blocks, but rather entities having a life history from their birth from simpler atomic nuclei by nuclear fusion to their decay or fusion to form other atoms. Radioactive decay is the conversion in time of some atoms into others and thus makes clear the inadequacy of using the atoms now known as starting points for the construction of chemistry.

The biological cell

That cells are the building blocks of plants and animals is a view usually associated with the names of Schwann and Schleiden in the 1830s though many others contributed to the theory's early formulation. R. Brown in 1831 supplied the 3d generalization that all cells contain nuclei, while the structural relations of cells (3D) in multi-cellular systems are studied via microscopy and other approaches. Schleiden pointed to the cell's associative characteristics *A*: "Plants, developed in any higher degree [than unicellular organisms], are aggregates of fully individualized, independent, separate beings, namely the cells themselves. . . . Each cell leads a double life, one independent, pertaining to its own development alone, the other incidental, as an integral part of a plant" (quoted in Singer 1959, 335; originally from Schleiden 1838). However, the delineation of cell boundaries is sometimes difficult, leading Weiss (1940) to propose that the cell be defined by its nucleus.

The fusion of cells and of cell nuclei (*A*) occurs in sexual fertilization. Hybrid cells containing several nuclei, including examples of union from different species, have been prepared by fusion of the separate cells using killed virus to effect the fusion. In some cases, fusion of nuclei was observed also (Harris and Watkins 1965).

The reversibility (*t*) of cellular association has been observed in organisms such as certain sponges, where cells at times operate as individuals while at others they merge into larger systems (Weiss 1940, 34).

The growth, development, and reproduction of organisms (${\nearrow}t$) has always been an integral part of biological studies. In the history of cell theory, the major conceptual generalization with regard to cell generation was Virchow's aphorism "Omnis cellula e cellula":

> Where a cell arises, there a cell must have been before, even as an animal can come from nothing but an animal, a plant from nothing but a plant. Thus in the whole series of living things there rules an eternal law of continuous development. There is no discontinuity nor can any developed tissue be traced back to anything but a cell (quoted in Singer 1959, 344; originally from Virchow 1858).

The evolutionary construction of the earliest cells from non-cellular components, the attempt at understanding and hopefully reconstructing the conditions for the "origin of life" occupies a significant segment of biochemistry today.

Genetics

Mendel's 1866 paper on plant hybridization proposed a numerical, integral pattern (*N*) for the distribution of hereditary qualities, a view contrasting strongly with Darwin's non-quantitative, infinitesimal variations. Mendel, in his paper, did not explore the nature of possible particle units that might account for his laws. Others before him, Maupertuis, Buffon, Diderot, and Herbert Spencer in his *Principles of Biology* of 1864 had propounded ideas of particulate inheritance.

Mendel added a second concept, the idea of dominance. Inheritable characteristics (tall, short; smooth, wrinkled; green, yellow; etc.) appeared in pairs, one member being dominant establishing its observable character when present, while the other, the recessive gene, would not be noticeable in the presence of the dominant opposite. These pairs of genes were thus not totally independent units, since some genes were capable of strongly influencing the expression of others. Therefore, some deviation from the totally atomistic *N* concept was already implicit in Mendel's laws. However, as far as his statistical rules were concerned, Mendel assumed a complete independence of genetic units.

It soon became apparent that the rules were not always followed, that when a certain gene was present, a second gene was also, and the concept of gene pairing or association (*A*) developed. That this pairing was not absolute but was sometimes more, sometimes less effective, suggested that gene linking was a matter of degree. It was at this point that a topological view of gene association developed, commonly known as the beads-on-a-string model. Gene linkage could be strong or weak depending on the closeness of two genes to each other, a phenomenon later known as the position effect. The recent realization that in humans a mere 35,000 genes supply instructions for coding 100,000–200,000 proteins (Lewis and Palevitz 2001) underscores the significance of gene interaction. Dobzhansky already suggested that genes do not simply aggregate mechanically that the chromosome is more than the mere sum of its parts:

> A chromosome is not merely a mechanical aggregation of genes, but a unit of a higher order... The properties of a chromosome are determined by those of the genes that are its structural units, and yet a chromosome is a harmonious system which reflects the history of the organism and is itself a determining factor in this history (1936, 163).

But gene interaction and gene shuffling, no matter how extended in time, could never produce anything essentially new. The Mendelian innovation was incapable of yielding insight into species evolution. New variations, differing significantly from earlier generations, were obtained experimentally by DeVries in 1900 and led to the concept of mutation as an internal change within the gene which thereby could no longer be a single undifferentiated unit but had to have an internal structure (3d). T.H. Morgan achieved artificial mutations in fruit flies with chemical agents, and H.J. Muller with x-rays in 1927, the latter developing a picture of a particulate gene of finite size (3d) because of his ability to bombard it and thereby to change it.

The gene-shuffling process was an irreversible process in time. Two parents, with genes RR and rr, respectively, produced in the second generation only Rr. If the latter are crossed among themselves we do not revert only to the pure parental types but to three types of offspring RR, Rr, and rr in the ratio 1:2:1. The ($\nearrow t$) character of the theory was therefore present from Mendel's first paper.

From the beads-on-a-string linear topology, research moved to the mapping of genes in a chromosome, and later to the discovery that genes were nucleic acids. Their chemical constitution was then determined, first topologically according to the classic structural theory procedures by Todd, and finally in the three-dimensional structural pattern (3D) of the DNA double helix proposed by Watson and Crick in 1953.

From there, emphasis has become more and more focused on the function of DNA in heredity, the process in space and time by which DNA duplicates itself (3D$-\nearrow t$) and controls the protein syntheses and metabolic processes that determine the inheritance of characteristics from parents to offspring.

The earth's continents and the theory of plate tectonics

The theory of plate tectonics provides a succinct modern example of the transformation of a scientific field toward a conceptual structure that includes the six concepts emphasized in this chapter.

> During the 1960's a conceptual revolution swept the earth sciences. The new worldview fundamentally altered long established notions about the permanence [t] of the continents and ocean basins and provided fresh perceptions of the underlying causes and significance of many major features of the earth's mantle and crust. As a consequence of this revolution it is now generally accepted that the continents have greatly altered [$\nearrow t$] their geographical position and pattern of dispersal [3D], and even their size [3d] and number [N]. These processes of continental drift [$\nearrow t$], fragmentation [3d] and assembly [A] have been going on for at least 700 million and perhaps for more than two billion years (Valentine and Moores 1978).

Classically, the earth's continents were viewed as distinct entities (N), whose major features would suffer no fundamental alteration so that any observed changes would be reversible (t). Biological and geological characteristics within the continents (3d) as well as the geographic relations of the continents to each other (3D) were noted. The approximate complementarity of shape of West Africa and the Eastern coast of South

America suggested a former land bridge, strengthened by the close similarity of plants in those two regions, noted by Darwin among others. The land-bridge theory had to be abandoned when no evidence for a submerged land link could be found. Then Wegener in 1915 presented evidence suggesting the non-permanence of continents, their rising and subsiding, drifting, fragmentation, and fusion. The fact that mammals on the two sides of the Atlantic are less similar to each other than plants suggests that Africa split from South America before the onset of major mammalian evolution. Thus, continents also have an evolutionary past (➶t) and cannot be considered ultimate unit entities since they are now believed to have fused and fragmented and were once part of a single land mass (A). (Wegener 1966, Takeuchi et al. 1970)

What has happened to our conception of the earth's continents is rather similar to our view of stars. Beginning as unchanging and independent, they are now known to be in process of evolution and to be part of the larger systems, the galaxies, and the galactic systems.

Comparative studies—the periodic table and the great chain of being

Perhaps the examples given are sufficient to show the wide applicability of the six-concept analysis to particular areas of study. These analyses should be useful for comparison purposes also, an example being the parallel analysis of the periodic classification of the chemical elements and the biological concept of the Great Chain of Being. The comparison was suggested by Lovejoy:

> The theory of the Great Chain of Being, purely speculative and traditional though it was, had upon natural history in this period an effect somewhat similar to that which the table of the elements and their atomic weights has had upon chemical research in the past half-century (1936, 232; cf. Benfey 1965).

The element classifications of Newlands, Meyer, and Mendeleev, and the chain of being of plant and animal species, both pointed to missing members with rather clearly defined characteristics, allowing for a more focused rather than a random search. It became possible to look in particularly likely places for elements or species to fill the gaps with a good chance of success.

The chain of being and the table of the elements acquire a timescale (➶t)

The chain of being originally was a timeless scheme, all species capable of creation having been created at the same moment in time. Only in this way was it thought that God's love and omnipotence could have been manifested. Yet the scheme when analyzed more minutely led to philosophical and ethical difficulties and at the same time encountered the growing weight of evidence that some species though potentially capable of existence, never appeared while others, though once present, are now extinct, thus breaking the chain. The chain was then "temporalized" and later branched, being transformed in the 19th century into an evolving process.

Atoms, the building blocks of the chemical elements, held on to their timeless character longer than species, yet, as indicated above, radioactivity and nuclear transformation studies led to the conviction of the evolutionary origin of the elements. The periodic table of the elements has been temporalized, the heavier atoms being seen as having been produced from the lighter ones early in the solar system's history while element building continues to proceed in the stars.

Continuity in the atomic weight scale and discontinuity in the chain of beings (A/N)

For Dalton, an element was characterized by the mass of its particular atoms. The atomic weight scale was discontinuous and, with the development of the periodic table, no new elements between already known ones in adjacent places could be expected. However, early in the 20th century isotopes, that is atoms of the same element yet of different mass, were discovered, and uranium samples of widely varying atomic weights became known. Isotopes of an element were then considered aggregates of a given number of protons with varying numbers of neutrons, suggesting an integral pattern of isotope masses (the masses of protons and neutrons are almost identical, while electrons, the third constituent of atoms, are much lighter). This rebirth of Pythagoreanism, however, was short-lived because atomic weights reflect the intensity of internal cohesion, some mass being lost in particle synthesis, the magnitude of the loss being given by Einstein's equation $E = mc^2$. Thus, the atomic mass scale became less Pythagorean, more continuous.

On the other hand, the continuous gradations assumed within the Great Chain of Being and in the view of spontaneous variability postulated by Darwin gave way to discontinuous change in Mendel's genetic laws, and the later discovery of biological mutations. Thus, the two classifications have encountered both the discrete and the continuous parameters N/A in their development.

External relations and internal composition (3D/3d)

Spatial relations among organisms are central to ecological studies and to the concept of territoriality. The chemical counterpart is the study of chemical interactions, stereochemistry (The Structural Theory of Organic Chemistry, and valence forces between atoms.

As has been mentioned, the internal structure of atoms has become an essential basis for the explanation of their external behavior. In the case of an organism, internal pattern and structure determine its place in the chain of being or in the evolutionary scheme. The internal pattern, once limited to anatomical details, now has reached down to the amino acid sequences in the plant or animal protein and to DNA.

Single concept studies

It should clearly be possible to carry out cross-disciplinary studies of particular concepts. Toulmin and his co-author Goodfield (1962–1963; 1965) have looked at the incorporation of the concept of directional time ➶t in the physical and biological sciences while Mendelsohn (1980), as mentioned, has explored the concepts of the discrete and the continuous.

Many histories of science still write their scientific chronology as if the mechanical tradition, symbolized by Galileo's approaches to natural phenomena, triumphed during the renaissance over all contenders. Normally, only its apparently decisive conquest over the "organicist" viewpoint is even mentioned, ignoring completely a third strand in the intellectual world of the Renaissance, sometimes known as the magical tradition. An analysis of the six concepts demonstrates that the three traditions are still with us.

THE THREE SCIENTIFIC TRADITIONS

Kearney (1971) and Kirsch (1981) have pointed to three major ways in which nature was viewed in the Renaissance, the organicist, the mechanical, and what they call the magical tradition and which Debus (1977) called the "Chemical Philosophy."

The six concepts identified in this chapter and depicted on the cube in Figure 1 correlate impressively with these three traditions. The mechanical tradition as we saw in discussing the kinetic theory of gases relates closely to the three concepts N, t, and 3D, number, time, and spatial extension. These meet at an apex of the cube. The opposite apex unites A, $\nearrow t$, and 3d, continuity, time's arrow, and inner complexity. These we meet in all discussions of organisms. And the magical tradition prepares us to accept apparently irreconcilable opposites, wave/particle, reversible and directional time, the whole and its inner complexity. (Debus 1977, 1980) These pairs appear on opposite faces of the cube.

In the library of the University of Pennsylvania a display case for several years contained original copies of classic books of this third tradition together with a commentary by Arnold Thackray:

> The world of the sixteenth and early seventeenth century natural philosopher was one in which astrology and astronomy were hardly differentiated—both Brahe and Kepler cast horoscopes and indulged in general prognostication—and chemistry was as much an occult philosophy as an empirical science. Cosmic harmonies, astral influences, neo-Platonic number mysticism, and the philosopher's stone were integral parts of the Weltanschauung of Paracelsus, Dee, Porta, Fludd, and their contemporaries and they perceived the universe as a "great chain of being." Elements of this tradition of alchemy, natural magic, and Hermeticism included the search for sympathies and antipathies in nature and a stress on the likeness of man the microcosm to the macrocosm. Experience was emphasized as the path to the proper understanding of nature and through such a grasp of phenomena the magus gained control. (1981).

The Hermetic tradition, as curious as it seems from the modern viewpoint, represents another thread which was woven into early modern science. Until recently neglected in the history of science, it was an inseparable portion of the intellectual climate of the times, with an appeal that was both widespread and long-lived.

The Hermetic tradition arose in part because in medieval Europe the Aristotelian and other Greek views, though incorporated extensively into Christian thought, were seen as of pagan origin. The Judaeo-Christian tradition was felt to have new insights to contribute to the understanding of the material world. The sacraments spoke of the transformation of matter to more than matter, the creation stories claimed a common origin of living and non-living entities, and the doctrine of the holy spirit encouraged meditation and trust in intuition as ways other than reason for gaining knowledge and understanding. Furthermore, the belief in continuing revelation encouraged reliance on experimentation as important for the discovery of unexpected truths.

The magical tradition had little chance of flourishing during the period of early successes of the mechanical viewpoint of Galileo, Newton, and Descartes. Chemical phenomena were resistant to the approach, and were made sense of for a while by the iatrochemical and phlogiston generalizations. Newton's physics never could explain either selective or limited affinity exhibited by one chemical element when exposed to another. If two hydrogen atoms had an affinity for each other why not three or four or thirty-four? Only quantum theory a century later could offer an explanation.

The mechanical viewpoint faced other problems. Faraday introduced the field concept—probably coming via Leibniz from the Chinese—to explain magnetic phenomena, followed later by the concept of the ether to carry light waves. And the ether was not banished until Einstein showed no need for it. The second law of thermodynamics of course was thoroughly un-Newtonian.

In the 20th century, a number of thinkers began suggesting that we must look once more at the organicist tradition of Aristotelian and medieval physics for insight and direction. Needham (1962) pointed out that the remarkable successes of Chinese science prior to Western influence, that is, until the 17th century, were based on an organicist viewpoint. When the mechanical approach was brought to China—and this occurred almost as soon as it was developed in Europe—it met with suspicion and faced much resistance. Needham suspects that the mechanical tradition needed to be fully explored but now needs to be seen as a necessary detour on the way to a more mature science.

More than half a century ago, Whitehead (1948), looking at the new physics, suggested that the basic particles, atoms, and molecules were becoming more like biological organisms far removed from Greek atomistic ideas. Physics he suggested studied small organisms, chemistry intermediate ones, and biology the large ones.

The Chinese also had insights reminiscent of the magical tradition in their yin–yang symbolism. They believed that embedded in nature and in human experience are numerous polar opposites that necessarily coexist, such as light and dark, male and female, growth and decay.

Holton (1973) has discussed at length opposing concepts such as those that we find on opposite faces of the cube of Figure 1 and that relate to the magical tradition. He points out that in addition to data and their rational, analytical manipulation, the scientist brings to his work basic pre-suppositions, an intuitive element, a thematic dimension. Newton thought of light as particles, Huyghens as waves. In the 20th century, both the continuum approach of wave mechanics and statistical quantum mechanics seek to explain the behavior of electrons in atoms. The Greeks, to cope with the unpleasant fact of change, tended to follow the atomistic ideas of Empedocles, Democritus, and Leucippus, change being seen as merely the rearrangement of the same unchanging elements. The Chinese on the other hand accepted change as fundamental, explaining everything in terms of change. For them change was not a problem. Problems were explained in terms of change.

The themas and antithemas (Holton's terms) were argued vehemently for decades and only recently have we come to realize that perhaps both of the contrasting approaches are often needed for a full understanding of phenomena. Niels Bohr above all emphasized the frequent need to accept seemingly opposing concepts in his emphasis on complementarity both in the natural sciences and in the other fields such as psychology. When in 1947, he was honored with the Danish Order of the Elephant, he chose as his coat of arms the yin–yang symbol with the words above it *Contraria sunt Complementa*, contraries are complementary (Holton 1973, Chapter 4 "The Roots of Complementarity").

With the recognition that opposite faces of the cube point to conceptual pairs of the kind emphasized in the magical tradition, we come to the startling conclusion that contemporary science has by no means discarded any of the earlier three traditions. Although the general opinion still holds that the mechanical tradition decisively conquered and dispensed with the organic viewpoint while the magical tradition is not even considered worthy of mention, an analysis of the six concepts demonstrates that the three traditions are very much alive (Benfey 1982).

CONCLUSION

During the last two centuries, in spite of the enormous proliferation of scientific research and speculation, the sciences have grown closer together in the sense that they are conceptually less diverse. One need only contrast the extent of presently attained unity with the diversity of central concepts that characterized the sciences of the late 18th and early 19th centuries: the phlogiston theory, Daltonian atomism, caloric fluid, Newtonian physics, the great chain of being, catastrophism, uniformitarianism. Conceptual gulfs separated and isolated many of these from each other preventing their amalgamation into a larger synthesis.

Today, no sharp dividing lines separate the various sciences. Research frontiers overlap whatever boundaries still nominally separate them. The conceptual framework underlying the various fields of science is now largely unified, and many key concepts are used identically throughout the sciences. Other concepts, more or less confined within a particular field, nonetheless are compatible with each other and with the larger framework. How did each field or subfield become fit conceptually to take its place in this contemporary rapprochement? It is unlikely that many scientists foresaw this convergence of their science with others as an outcome of their labors. They encountered conceptual problems within their fields of expertise and used what insights and ingenuity they could muster to overcome them. In order to introduce new concepts to be added to or replace the old, they borrowed from other fields of science, from non-scientific disciplines or from their general cultural experience. Couper found the clue to order in the bewildering mass of chemical formulas by analyzing them as if they were words of an unknown language, while Kekulé viewed them from the viewpoint of his earlier exposure to architecture. This chapter suggests a methodology for demonstrating the path of concept accretion, the growth of those conceptual clusters that are part of the modern synthesis.

A useful approach appears to be to focus on certain subfields, choosing initially those that began with a particularly simple conceptual structure. The subfields especially conducive to the tracing of concept accretion are those developed in what might be called the Galilean mode.

Galileo carried out his inclined plane experiments with balls as spherically perfect as possible rolling down a plane as smooth as possible. Thereby he hoped to avoid complications due to air resistance and friction. These he felt could be incorporated later, once the basic "ideal" law was established. Francis Bacon objected bitterly to this procedure, demanding that the phenomena of nature be looked at in all their complexity, that hypotheses were to be developed by recognizing what all the phenomena had in common. By that path he claimed he had found that heat was a form of motion.

Bacon probably saw in Galileo's procedure the continuation of the medieval astronomers' epicyclic method, which described a planet's path first by the best circle, the deferent, to which are then added epicycles to account for deviations from pure circular motion. Galileo used an analogous procedure in his study of dynamic processes on earth. Perhaps, he did this as a direct consequence of the fact that Copernicus had placed the earth among the planets, thus suggesting that the same methodological procedures are applicable to both.

This chapter therefore began with the analysis of two subfields of the physical sciences. Each began with an extremely simple conceptual structure and then incorporated new concepts until the conceptual structure of the two became the same. This leads to the question as to whether this was a trivial coincidence or had a broader significance. Neither the sciences nor the study of the history of science can rest content merely with the accumulation of research findings. They must seek organizing patterns, and the history of science looks for insights into the paths by which the sciences develop. This chapter suggests one such trend, the inner conceptual transformation of every field, subfield, and paradigm, until it reaches a form assimilable to the modern schema. It is my hope that other historians will take on this task in the fields of their special interest. When detailed analyses are available from a number of fields, comparisons will become possible as to parallelisms or individuality among them and suggestions can be made as to the reasons for parallelisms where found. Furthermore, where a conceptual set in a research field is found to lack one or two concepts common to most others, it might suggest the likelihood that significant discoveries may result if scientists look in a particular direction. This is how new elements were found when the periodic table was developed.

The set of concepts identified in this chapter as common to most areas of modern science is a set far removed from one describing an eternal, Newtonian or Daltonian particle because its concepts are far richer. The set contains, as we have seen, directional time, fitting far more comfortably a description of organisms in an evolutionary framework.

A previous integration of the sciences at all approaching the magnitude of the contemporary one can only be discerned in the medieval-Aristotelian synthesis, similar to the present viewpoint in some of its organismic aspects but differing from it in its timeless, hierarchical, as against the present evolutionary approach. The modern view places everything from fundamental particles, through, molecules, cells, organisms, and the events of human history in one continuous time-frame. The Newtonian approach, which was heralded with so much hope and confidence, and under whose sway we still largely operate, was never able to subsume chemistry, let alone the even less mechanical sciences.

The Newtonian attempt at synthesizing the sciences was an approach by way of domination and reduction. The success of Newtonian physics raised the possibility that other areas of science could be freed from non-operational notions and could become analytical-prescriptive fields, by limiting themselves to the underlying paradigm of interacting eternal particles obeying Newtonian-type laws. The present chapter suggests a synthesis by an essentially opposite route—not by reduction, simplification, and domination, but by growth, enrichment, and mutual adaptation of the various areas of science. The Newtonian synthesis was a program and a hope. The unity achieved under the contemporary framework by contrast is a unity we have stumbled upon largely unawares. Scientists were not working toward some unifying goal, rather they were solving their own particular problems.

However, I cannot agree with the view that the sciences have developed their current conceptual form almost at random by acquiring, discarding, and periodically alternating certain concepts or themata based on individual predilection or local cultural experience. Far larger trends seem to be operating in the world's intellectual community turning us from a hierarchical fixed view of nature to a developing, historical, evolutionary view in which real significant novelty occurs and in which "magical" complementary yin–yang aspects can no longer be ignored. Perhaps much of what has happened in the last two centuries has been the adjustment of every aspect of science to this major intellectual transformation.

We have identified a set of six concepts more descriptive of organisms than of building blocks but differing significantly from the organismic view of medieval physics. That view sought to understand the behavior of material bodies by endowing them with intentions, abilities, and feelings related to those of humans. The set of six concepts discussed in the present chapter is, however, also a set closely linked to human experience. As a person, I am aware of myself as an isolated entity capable of performing actions and being acted upon. I recognize other individuals similar to myself. Seeing them as separate atomic selves I treat them as integers, I count them and group them, involving the numerical concept N. I order my experience within a matrix of space (3D) and time (t), seeing space as symmetrical and time as reversible in the sense that in much of life what is done can also be undone. Reflection and experience introduce time's arrow $\nearrow t$ such as the finality of death, the irreversible consequences of certain actions, the experience of growth, maturing, and aging. I also become aware of disordering processes, the impossibility of unscrambling a scrambled egg. Physical pain and exercise make me aware of internal structure, *3d*, when I notice that there are parts of my body that may malfunction, influencing the whole. I also become aware of an internal structure to my mental processes, as I grapple with my tangled thoughts and feelings. Finally, we learn that we are political animals, that we are nothing if not part of society, that we are also inextricably part of our physical and biological environment. We begin to question the validity of considering ourselves as isolated, atomic selves. The religious consciousness of our being part of something larger relates to the associative concept A.

Whereas some conceptual incorporations have now become so commonplace that they seem self-evident, others have not significantly influenced the way we think of the entities of nature. In organic chemistry, the idea of a geometric structure for molecules was fought bitterly when first put forward, but it has become central to the way we now visualize them. Although time (in two senses) is equally important to the definition of molecules, chemists have done little to change their concept of molecules as timeless structures.

The path that modern science has trodden from its sublime optimism in the universal applicability of Newtonian-type laws to its present recognition of profound philosophical hurdles slowing our way to further understanding, may make us less surprised at the marked success of the Chinese approach to science and technology during the first 15 centuries of our era (Needham 1959). Their conception of the complementary yin and yang phases, and of a universal organism composed of component organisms, led to much insight that helped order human experience, and to much discovery, some of it of critical importance to the meteoric growth of modern science since the Renaissance. That China did not go further was no doubt due to the one-sidedness of its own scientific development. It now appears possible that the characteristic rational analytic, quantitative approach of modern science may also be one-sided and only through transcending that limitation can the sciences move forward to a new level of comprehension.

ACKNOWLEDGMENTS

My thanks go to Kenneth Caneva of the University of North Carolina at Greensboro with whom I had many discussions about an earlier version of this chapter during a sabbatical 1979–1980; and to Gerald Holton of Harvard University for helpful comments.

REFERENCES

Benfey, O.T. 1952. Prout's Hypothesis. *Journal of Chemical Education* 29: 78–81.

Benfey, O.T. (ed.). 1963. *Classics in the Theory of Chemical Combination.* New York: Dover Publications; Benfey, O.T. (ed.). 1981. *Classics in the Theory of Chemical Combination.* Malabar, Florida: Robert E. Krieger Publishing Co.

Benfey, O.T. 1964. *From Vital Force to Structural Formulas*. Boston: Houghton Mifflin; Benfey, O.T. 1975. *From Vital Force to Structural Formulas.* Washington, DC: American Chemical Society; Benfey, O.T. 1992. *From Vital Force to Structural Formulas*. Philadelphia, PA: Chemical Heritage Foundation.

Benfey, O.T. 1965. The Great Chain of Being and the Periodic Table. *Journal of Chemical Education* 42: 39–41.

Benfey, O.T. March 1977. The Limits of Knowledge. *Chemistry* 50: 2–3.

Benfey, O.T. 1982. The Concepts of Chemistry—Mechanical, Organicist, Magical, or What? *Journal of Chemical Education* 59: 395–398.

Benfey, O.T. Winter/Spring 1992–1993. Precursors and Cocursors of the Mendeleev Table: The Pythagorean Spirit in Element Classification. *Journal for the History of Chemistry* 13/14: 60–66.

Brock, W.H. 1985. *From Protyle to Proton: William Prout and the Nature of Matter 1785–1985*. Boston: Adam Hilger.

Cassirer, E. 1956. *Determinism and Indeterminism in Modern Physics*. Benfey, O.T. (trans.), Chapters 7 and 8. New Haven, CT: Yale University Press.

Christie, G.H. and Kenner, J. 1922. The Molecular Configurations of Polynuclear Aromatic Compounds. Part I. The Resolution of 6:6′-Dinitro- and 4:6:4′:6′-Tetranitro-diphenic Acids into Optically Active Components. *Journal of the Chemical Society* 121: 614–620.

Crookes, W. 1886. *Reports of the British Association for the Advancement of Science*, 567. Excerpts, and his diagram showing the "giant pendulum swings" in element evolution are reprinted in Farber, E. 1963. *Chymia* 9: 193.

Dalton, J. 1808. *A New System of Chemical Philosophy*, reprint edition. New York: The Citadel Press, 1964, 162–163.

Debus, A.G. 1977. *The Chemical Philosophy*. New York: Science History Publications.

Debus, A.G. 1980. Mysticism and the Rise of Modern Science. *Journal of Central Asia* 3: 46–61.

Dobzhansky, T. 1936. Position Effects on Genes. *Biological Reviews* 11: 382; quoted in Dunn, L.C. 1965. *A Short History of Genetics*. New York: McGrawHill, 163.

Harris, H. and Watkins, J.F. 1965. Hybrid Cells from Mouse and Man: Artificial Heterokaryons of Mammalian Cells of Different Species. *Nature* 205: 640–646.

Holton, G. 1973. *Thematic Origins of Scientific Thought*. Cambridge: Harvard University Press, 107–109.

Ihde, A. 1959. The Unravelling of Geometric Isomerism and Tautomerism. *Journal of Chemical Education* 36: 333–336.

Jensen, W.B. 1988. Logic, History, and the Chemistry Textbook: I. Does Chemistry Have a Logical Structure? *Journal of Chemical Education* 75: 679ff., fn.1.

Kearney, H. 1971. *Science and Change 1500–1700*, Chapters 1 and 4. New York: McGraw-Hill.

Kirsch, A.S. 1981. Introducing 'Chemical Bonds.' *Journal of Chemical Education* 58: 200–201.

Lewis, A.C. 1977. H. Grassmann's 1844 *Ausdehnungslehre* and Schleiermacher's Dialektik. *Annals of Science* 34: 103–162.

Lewis, R. and Palevitz, B.A. 2001. Genome Economy. *The Scientist* 15(12): 19–21.

Lockyer, N. 1900. *Inorganic Evolution as Studied by Spectrum Analysis*. London: Macmillan, 166; quoted in Farber, E. 1963. *Chymia* 9:196.

Lovejoy, A.O. 1936. *The Great Chain of Being: A Study in the History of an Idea*. Cambridge: Harvard University Press; reprinted by New York: Harper and Row, 1960, 232.

Mendeleev, D. 1872. Das periodische System der Elemente: Prognose neuer Elemente. *Liebig's Annalen* (Suppl. 8), 206–207; English translation in Farber, E. 1963. *Chymia* 9: 187.

Mendelsohn, E. 1980. The Continuous and the Discrete in the History of Science. In: Brim, O.G., Jr. and Kagan, J. (eds.), *Constancy and Change in Human Development*. Cambridge, MA: Harvard University Press.

Needham, J. 1959 ff. *Science and Civilization in China*. 7 Vols. Cambridge: Cambridge University Press (many published in multiple sections with more in press).

Needham, J. 1962. *Science and Civilization in China*, Vol. 2. Cambridge: Cambridge Univeristy Press, 291–294, 493–505.

Prout, W. 1815. On the Relations Between the Specific Gravities of Bodies in the Gaseous State and the Weights of Their Atoms. *Annals of Philosophy* 6: 321–330 (published anonymously but identified in the next article).

Prout, W. 1816. Correction of a Mistake in the Essay on the Relations Between the Specific Gravities. *Annals of Philosophy* 7: 111–113.

Schleiden, M.J. 1838. *Müller's Archiv für Anatomie und Physiologie* 137; quoted in Singer, C. 1959. *A History of Biology*, 3rd ed. New York: Abelard-Schuman, 335.

Senior, J.K. 1935. An Evaluation of the Structural Theory of Organic Chemistry. *Journal of Chemical Education* 12: (I) 409–414, (II) 465–472.

Takeuchi, H., Uyeda, S., and Kanamori, H. 1970. *Debate About the Earth*. San Francisco: Freeman, Cooper and Co.

Thackray, A. 1981. Statement of the Hermetic Tradition. University of Pennsylvania, Courtesy of Arnold Thackray.

Toulmin, S. 1962–1963. The Discovery of Time. *Memoirs and Proceedings of the Manchester Literary and Philosophical Society*, 105(8): 1–13.

Toulmin, S. and Goodfield, J. 1965. *The Discovery of Time.* New York: Hutchinson.

Valentine, J.W. and Moores, E.M. 1978. Plate Tectonics and the History of Life in the Oceans. In: Laporte, L.F. (ed.), *Evolution and the Fossil Record.* San Francisco: W. H. Freeman and Co., 193–202. Reprinted from *Scientific American* 230, April 1974, 80–89.

Virchow, R. 1858. *Cellularpathologie*; quoted in Singer. 1959. *A History of Biology,* 3rd ed. New York: Abelard-Schuman.

Wegener, A. 1966. *The Origins of Continents and Oceans.* New York: Dover Publications, translated from the 1929 4th edition of *Die Entstehung der Kontinente und Ozeane*, 1915.

Weiss, P. 1940. The Problem of Cell Individuality in Development. *American Naturalist* 74: 37–46.

Westheimer, F.H. 1965. A Summer Short Course in Carbon Chemistry: The Structural Theory of Organic Chemistry. *Chemistry* 38(6, 7): 12–18, 10–16.

Wheland, G.W. 1949. *Advanced Organic Chemistry*, 2nd ed. New York: John Wiley and Sons, 87.

Wheland, G.W. 1955. *Resonance in Organic Chemistry.* New York: John Wiley and Sons.

Whitehead, A.N. 1948. *Science and the Modern World.* New York: New American Library.

CHAPTER 6

NORMATIVE AND DESCRIPTIVE PHILOSOPHY OF SCIENCE AND THE ROLE OF CHEMISTRY

ERIC R. SCERRI

Department of Chemistry and Biochemistry, University of California at Los Angeles, 405 Hilgard Avenue, Los Angeles, CA 90095-1569; E-mail: scerri@chem.ucla.edu

INTRODUCTION

Since the demise of Logical Positivism the purely normative approach to philosophy of science has been increasingly challenged. Many philosophers of science now consider themselves as naturalists and it becomes a matter of which particular variety they are willing to support. As is well known, one of the central issues in the debate over naturalism concerns whether philosophy offers a privileged standpoint from which to study the nature of science, or whether science is best studied by studying science itself.[1]

The situation is often summarized by the aphorism that one cannot derive an "ought from an is." It does not appear as though we should be entitled to draw normative conclusions about how science should ideally be conducted by merely observing the manner in which it is conducted at present, or was conducted in the past, since this appears to be a circular argument. Indeed the only circumstance in which the die-hard normative epistemologist would contemplate any input from naturalism would be if it could be shown conclusively that rational agents were literally incapable of thinking in the manner which the normative scheme requires that they should.[2]

I am speaking as though the normative approach is somehow more respectable and that naturalism represents a recent intruder onto the philosophical scene. I am suggesting that the burden of proof lies with the naturalists to show that their position can have some influence on normative philosophy and not vice versa. This may indeed be the case at the present time but things were not always this way. Normative philosophy and philosophy of science did not always hold center stage. It arose as a result of the work of Gottlob Frege in Jena at the end of the 19th century. Frege's development of modern logic rapidly led to attempts to base all of philosophy on the analysis of language, a program that was furthered by Russell, Wittgenstein, Carnap, and many others. Frege's program led to the overthrow of the then current naturalistic philosophy, which looked to scientific discoveries and scientific practice for philosophical enlightenment.

Before Frege, the study of psychology and evolutionary biology *were* indeed highly regarded among philosophers of the day. In particular, psychologism was not

D. Baird et al. (eds.), Philosophy of Chemistry, 119–128.

considered to be the 'mortal sin' that Frege later convinced analytical philosophers that it might be.

Today confidence in the analysis of the logical structure of scientific theories and the approach involving an *a priori* or a "first philosophy" has waned. Philosophers *do* now explore ideas from psychology, biology, physics, political science, economics, computation, and as it would appear are even beginning to consider the field of chemistry (van Brakel 2000; Bhushan and Rosenfeld 2000). These excursions may all be said to fall under the generic label of naturalism. The proponents of these approaches share an opposition to the Frege–Wittgenstein line whereby philosophy is somehow more fundamental than all these other branches of learning.

I want to spend a few more moments in exploring how the stranglehold that Frege had around analytical philosophy has been loosened. Frege's position may be stated in simple terms as requiring that:

(i) Logic and not psychology is the proper medium for philosophy.
(ii) Philosophical reflection is to be regarded as *a priori*.

Modern naturalists, not surprisingly, dispute both (i) and (ii). One of the main reasons why (i) has been reconsidered, and why there has been a return to psychologism, has come from work in epistemology. In 1963, Edmund Gettier presented some examples to show that there are cases when one might possess true and justified belief and yet fail to have knowledge (Gettier 1963). Philosophers like Dretske and Goldman have led the way in formulating responses to these puzzles that invoke the psychological states of knowing agents (Dretske 1981; Goldman 1992).

Meanwhile (ii) has been challenged on various fronts. First of all, Quine produced his famous articles in which he argued that the distinction between analytic and synthetic statements could not be clearly drawn (Quine 1951). As a result, any hopes of a completely *a priori* approach to philosophy would appear to be discredited. And yet in spite of other well-known contributions from Quine such as his slogan that "philosophy of science is philosophy enough," and his general support for a naturalized epistemology, his own work remained steeped in logical analysis and did not shown much sign of engaging in an examination of actual science.

Then came the historical turn in philosophy of science due to Kuhn, Lakatos, Toulmin, Feyerabend, and Laudan among others. These authors have done much to show the importance of the history of science to the study of scientific methodology. In addition, for better or for worse, the writings of Kuhn, in particular, have spawned the sociology of science and science studies industries which claim to study science as it is actually practiced and stress the need to go beyond an analysis of theories. I say for better or for worse because, as I see it, the problem with sociology of science has been an obsession with the context of scientific developments and the demotion of the actual science involved.[3] What many of these authors have done is to take naturalism a little too far such that they end up with relativism. This is of course one of the well-known dangers of naturalism in general.

Two leading current proponents of naturalism include Ronald Giere and Larry Laudan (Giere 1985, 1989; Laudan 1987, 1990). Perhaps, the main distinction between them seems to be that Laudan insists that his brand of naturalism is also normative whereas the claim to normativity is rather weaker in Giere's version, although not altogether absent as some commentators seem to believe.

Another author to champion naturalism has been Philip Kitcher. His 1992 article in the centenary issue of *Philosophical Reviews* presents one of the clearest accounts of the history of naturalism and the various currently available positions (Kitcher 1992). Kitcher followed this with a book called *The Advancement of Science* in which he claimed to be doing naturalistic philosophy of a moderate kind. But as some critics have suggested Kitcher seemingly fails to deliver the goods (Solomon 1995). The book is full of formal analysis of scientific episodes such as the Chemical Revolution and one cannot help wondering, as in the case of Quine's writings, where the real science might be lurking.

Of course there are still many philosophers, including Siegel and Doppelt for example, who continue to dispute the very notion of a naturalized philosophy of science (Siegel 1985, 1989; Doppelt 1990).

CHEMISTRY?

So much by way of introduction but what is the relevance of chemistry to these issues? First of all, if by naturalism one means looking at science itself, rather than purely operating within a logical analysis of concepts, then here in chemistry is a whole field of science that has never been seriously considered. The apparent disdain for chemistry is all the more surprising when one considers that chemists are by far the largest group of scientists among people studying and working in any of the sciences.[4]

But the main point I want to make in this article is that in a sense one *can* indeed derive "an ought from an is," although I intend this claim in a more restricted sense than usual. I will argue that by starting with a naturalistic approach to the nature of chemistry, we can make normative recommendations to chemists and chemical educators. The larger question of whether one may make normative claims about science in general is a more difficult one, which I will avoid for the time being. But if what I say about chemistry is correct then perhaps it could easily be generalized to other branches of science (Siegel 1985, 1989, 1990; Doppelt 1990).

FROM NORMATIVE TO NATURALISTIC

I have begun to realize that my own research in philosophy of chemistry has consisted in a gradual shift from initially advocating a normative view about such issues as atomic orbitals and electronic configurations, to more recently adopting a naturalized position. I now find myself advocating the critical use of such concepts rather than highlighting the fact that they strictly fail to refer to any entities in the physical world. Please allow me to now re-trace some of these steps in my own research.

In the case of a many-electron atom or molecule the commutator involving the Hamiltonian operator and the operator corresponding to the angular momentum of an individual electron, $[H, \ell_x]$, is non-zero. This implies that eigenvalues corresponding to the angular momentum operator for any individual electron is not a constant of the motion, or to use the jargon, is not a good quantum number, and cannot be said to characterize the motion with any exactness. For example, in the absence of spin–orbit coupling only the vectorial sum of all the individual angular momentum operators or L, rather than individual angular momenta, or ℓ, represent good quantum numbers. Another way of stating this result is to say that the individual electrons in a many-electron atom are not in stationary states but that only the atom a whole possesses stationary states. Similar arguments can be made for other quantum numbers like m_ℓ and m_s and these likewise imply the strict breakdown of the notion of assigning four quantum numbers to each electron in many-electron systems.

But more important than the approximate nature of the orbital concept is the categorical fact that an orbital does not refer to any physical entity. Here we are fortunate in having a clear-cut case where it is not left to one's philosophical prejudices as to whether we should adopt a realistic or anti-realistic interpretation. The mathematical analysis dictates quite categorically that the much beloved, and much pictured, concept of atomic orbitals does not have a "real" or independent physical existence. Indeed the use of the term "orbital" rather than orbit does not really convey the radical break with the notion of a continuous path for elementary particles, which took place as a result of the advent of quantum mechanics.

Of course it is still possible to use the complex mathematical expressions, corresponding to the different type of orbital solutions to the hydrogen atom problem, in order to build up a wavefunction that approximates that of a many-electron atom or molecule. In such cases, we are using orbitals in a purely instrumental fashion to model the wavefunction of the atom or molecule and there is no pretense, at least by experts in the field, that the constituent orbitals used in this modeling procedure possess any independent existence. Contrary to the recent claims which appeared in Nature magazine, as well as many other publications, orbitals have not been observed (Scerri 2000b, 2001).

Now what I have been arguing concerning orbitals, for several years, is a normative claim which emerges from paying strict attention to quantum mechanics. But soon after I started to publish these ideas, Robin Hendry, pointed out that according to some philosophical analyses of scientific theories my position was somewhat passé.

Hendry's criticism of my article of 1991 went something like this.[5] He claimed that the issue I was raising did not have quite the philosophical significance that I was attaching to it. Whereas I was pointing out that the realistic interpretation of one-electron orbitals was strictly inconsistent with quantum mechanics, Hendry claimed that in making this assessment I was working within the covering-law model of explanation. This model appeals to fundamental laws for explanation or, in the case I am considering, the Schrödinger equation. Hendry's view, following Nancy Cartwright, was that it is not fundamental theories or laws that explain but scientific models. Of course this represents another version of naturalism. Nancy Cartwright has championed the view that it is models, rather than scientific theories, that are used in scientific practice to give explanations for natural phenomena (Cartwright 1983).

To connect this with my main theme, I am saying that my own normative approach has been to claim that the fundamental theory explains everything. Hendry countered this claim by appealing to Cartwright-style naturalism which required that we look carefully at the way that models are used. As Cartwright correctly points out, it is often models that scientists appeal to rather than high-level theories. From the normative perspective based on the theory there is indeed something wrong with the way in which the orbital approximation is used in many areas of chemistry and even applied physics. But from the perspective of what I shall call Cartwright–Hendry naturalism, scientists regarding orbitals realistically cannot be faulted given the central role of models in modern science. I did not readily accept this criticism at the time when Hendry first circulated his manuscript but recent developments in my work have now shown me that he was making a very valuable point (Hendry 1994).

In addition, some chemical educators have reacted to my work by pointing out that orbitals are here to stay and that no amount of nit-picking about whether or not they exist physically will have the slightest impact on their use in teaching chemistry (Richman 1999a,b; Emerson 1999). Of course my thought had never been that we should do away with orbitals but that we should point out more carefully their limitations in the course of teaching chemistry.[6] From a philosophical point of view the aim was to examine objectively the theoretical status of electronic orbitals and configurations from the perspective of quantum mechanics.

Then one day some years ago while giving a lecture on philosophy of chemistry a thought came to me. The thought was that it is quite appropriate for chemists and chemical educators to not only use orbitals but to do so in a realistic fashion regardless of their status according to the fundamental theory of quantum mechanics. I think, I fully realized at this moment the truly paradoxical situation in that chemistry is an autonomous science while at the same time resting on fundamental physics. These two positions need not be seen as being contradictory just as the normative and the naturalized position need not be seen as contradictory.

In 1997, I met Bernadette Bensaude, the well-known French philosopher of science, whose work is motivated by the history of chemistry. We struck up a conversation about Paneth and his view of the elements which is the subject of one of the classic papers in philosophy of chemistry written in German and translated by his son Heinz Post (Paneth 1962)[7]. The gist of Paneth's paper is that the chemist must adopt an intermediate position between the fully reductive view afforded by quantum mechanics and a naively realistic view that dwells on colors, smells, and such-like properties of macroscopic chemistry. In that paper, Paneth is concerned with how elements are to be regarded and he upholds a dual view of elements as unobservable "basic substances" on one hand and observable "simple substances" on the other. This he claims resolves a major puzzle in the philosophical understanding of substance, namely how it is that an element can survive in its compounds although the properties of the compound appear to bear very little resemblance to those of the element.

I suggest that this is a way of seeing the relationship between a normative view, or what quantum mechanics says about chemistry, and the more naturalistic view which tries to consider how chemists and chemical educators actually regard chemical concepts and models. Although it may seem paradoxical to embrace both positions at once, perhaps one might rest easier with Paneth's view of an intermediate position. But my emphasis, contrary to Paneth's conciliatory stance, is that both positions should be adopted simultaneously. Admittedly the history of paradox has a long history in Western thought. Usually a paradox is regarded as a serious problem that must be struggled against and overcome at all costs. Interestingly in Eastern thought, especially Chinese philosophy, paradoxes are not serious dilemmas to be resolved. Instead a paradox is to be embraced for what it is. It is in this rather esoteric sense that I am proposing to regard the reduction of chemistry, on one hand, and the continued use of reduced concepts as being paradoxical.[8]

In the 1960s and 1970s, Kuhn and others showed philosophers of science that it was futile to insist on a normative view of scientific theories which did not bear much relationship to the historical development of real science. Similarly, the case of atomic orbitals, which I continue to concentrate upon, shows us that it is somewhat unhelpful to insist only on the normative view from quantum mechanics. One needs to also consider what is actually done in chemistry and the fact that chemists get by very well by thinking of orbitals are real objects. In fact we need both views, the normative and the descriptive. Without the normative recommendation orbitals are used a little too naively as in the case of many chemical educators who do so without the slightest idea that orbitals are strictly no more than mathematical fictions. Hopefully my previous work was not in vain if I have managed to convince some people in chemical education to be a little more careful about how far an explanation based on orbitals can be taken.

To make a general point now, I think it is still of great value to question the status of the orbital approximation even if one eventually returns to using it in a realistic manner in chemistry. This is because the eventual use of the orbital approximation is greatly improved by such questioning.[9]

FROM NATURALISTIC TO NORMATIVE: THE REDUCTION OF CHEMISTRY

There is another area in philosophy of chemistry where I have been urging a naturalistic approach ever since the beginning of my work. This concerns the question of the reduction of chemistry to physics or more specifically quantum mechanics or relativistic quantum mechanics if one insists on being very precise.

As far back as the 1994 PSA meeting in New Orleans I suggested that the reduction of chemistry should not be regarded in the classic fashion of trying to relate the theories of chemistry to the theories of physics or in the formal sense of seeking bridge principles (Scerri 1994). The problem with such a Nagelian approach is that it requires axiomatized versions of the theory to be reduced as well as the reducing theory. Now although a case can be made for the existence of axiomatic quantum mechanics, clearly the same cannot be said for chemistry since there is no axiomatic theory of chemistry.[10]

But my main reason for advocating a naturalistic approach was that scientists themselves have an entirely different approach to the question of whether chemistry has been reduced to physics. Instead of considering a formal approach linking the theories in both domains a scientist would try to examine the extent to which chemically important quantities such as bond angles, dipole moments and the like can be derived from first principles from the reducing theory or quantum mechanics. A more challenging question for the theory is whether the liquidity of water, for example, can be predicted from first principles. One suspects that this will remain unattainable for some time to come or even than it may never be achieved. Rather than seeking a relationship between theories, a naturalistic approach to the question of the reduction of chemistry should examine the relationship between chemical properties on one hand and the reducing theory on the other hand.

Of course the hope is to go beyond data such bond lengths or dipole moments and to be able to calculate the feasibility or rate of any particular reaction from first principles without even needing to conduct an experiment. It should be mentioned that considerable progress has been made in the quantum theoretical approach to reaction rates although very accurate treatments have, not surprisingly, been confined to such reactions as that of hydrogen atoms with hydrogen molecules.

Since making this proposal I have found many instances of philosophers who have hinted at precisely this more naturalistic, more pragmatic, approach to reduction which consists in following what computational chemists and physicists do (Suppes 1978; Popper 1982; Hacking 1996). To repeat, whereas philosophers have previously viewed reduction as a relationship between theories, the naturalistic approach I advocate consists in accepting the scientific approach to the reduction of one branch of science to another but without sacrificing any philosophical rigor in analyzing the procedures used. Indeed what begins as naturalism can, I claim, turn to making specific normative recommendations to practitioners in the fields such as computational quantum chemistry (Scerri 1992a, b; 1998a, b). What I discovered was that in many instances in computational quantum chemistry chemists were claiming strict deduction of chemical facts from the Schrödinger equation. But such treatments frequently involved semi-empirical procedures such as using a set of orbitals that are found to work in other calculations or the use of calculations that fail to estimate rigorous error bounds (Scerri 1997). I will not enter into further technicalities here. But I want to suggest is that this is an example of moving in the opposite direction. What begins as a naturalistic or descriptive project ends up by making normative suggestions.[11]

CONCLUSION

I am suggesting that both the normative view about the status or orbitals according to quantum mechanics, and the adoption of a realistic view of orbitals can happily coexist. This I claim is an example of the coexistence of the normative and naturalistic approaches to philosophy of science. This claim will only appear to be a contradiction if one maintains the usual static view about holding philosophical positions. There have recently been signs that philosophers have grown tired of the never-ending arguments regarding realism and anti-realism since they realize that both positions show many virtues. Arthur Fine has coined the phrase "Natural Ontological Attitude"

(Fine 1986). On my reading the important part of this notion resides in the choice of the word "attitude" rather than position and although Fine has not really developed a distinction between attitude and position his articles hint in this direction. What I believe Fine is getting at is that we should no longer think of the discussion as an either/or situation between realism and anti-realism but rather that scientists adopt both positions at different times and sometimes even both at once. These are not so much positions held by scientists and philosophers but the more temporary, more dynamic, attitudes.

I think that this suggests an even further stage in the kind of bootstrapping I have been urging in this paper. The nature of chemistry, more than physics or biology perhaps, could serve as a model for the kind of philosophical approach based on "attitudes" rather than hard and fast positions which one is typically supposed to maintain and defend at all costs. This seems especially appropriate since chemistry, as its practitioners, as well as chemical educators know only too well, requires us to operate on many different levels simultaneously. It demands the adoption of several attitudes. To show you what I have in mind let me end by quoting from the work of a South African chemical educator Michael Laing:

> The difficulty arises because we teach chemistry simultaneously on four different levels, The Realities of Water.
>
> Level 1: macroscopic: tactile (touchable, tasteable, wet to the touch).
>
> Level 2: communicative: language, name: (a) oral and aural, the word, the sound assigned to something that can be seen and felt; and (b) the written version.
>
> Level 3: symbolic, representational: elemental formulaic. This came once chemists understood the concepts of elements, atom, equivalence, valence, stoichiometry, beginning n 1800 and understood by 1860 as confirmed at the Karlsruhe conference. Wet water now becomes H_2O with a relative molecular mass of 18 units. Numeric calculations become important.
>
> Level 4: Atomic scale model. This representation depicts the shape and size of the molecule on a scale so small that it s beyond the comprehension of the average person....
>
> To appreciate water as a chemist you must make use of all four conceptual levels and be able to switch from one to the other as appropriate (Laing 1999).

Chemistry is a fertile new area in which philosophy of science could investigate further the question of normative and naturalistic approaches or the question of whether or not scientists actually adopt philosophical positions in the manner in which they use and interpret scientific theories. My own feeling is that chemists, in particular, tend to adopt different "attitudes," to use Fine's phrase, depending on what level they are operating at.

NOTES

1. The term naturalism has a variety of meanings. For example, in the philosophy of the social sciences it is sometimes taken to mean positivism. This is not the sense in which naturalism is being used in the present article. I am not concerned here with an empirical approach to social science but with a form of naturalism that involves concentrating on the apparatus, techniques, and assumptions of the natural sciences.

2. I am grateful to Martin Curd or Purdue University for discussion on this point.
3. An analogous situation exists in chemical education research at present with nearly all efforts being directed toward the learning process and virtually none at the content of chemistry courses (Scerri 2003).
4. Even sociologists of science appear to have overlooked chemistry, with the possible exception of Bruno Latour's book laboratory life, which is based on observations made in a leading biochemical laboratory (Latour and Woolgar 1979).
5. The article in question is Scerri (1991).
6. In addition, contrary to what Hendry claimed in his article I did not advocate the use of more rigorous forms of quantum chemistry for chemistry at large.
7. See Ruthenberg's short biography of Paneth (Ruthenberg 1997).
8. Needless to say I am not an advocate of parallels between modern science and Eastern mysticism as popularized by the likes of Capra and Zukav (Capra 1976; Zukav 1979). For example see my critique of the alleged parallels between Eastern mysticism and modern physics as published in the American Journal of Physics (Scerri 1989).
9. A recent article by Lombardi and LaBarca analyzes my position on the status of atomic orbitals as well as building on my view (Lombardi and LaBarca 2005).
10. This has not prevented philosophers from trying to axiomatize certain parts of chemistry, such as the Periodic system of the elements for example (Hettema and Kuipers 1988). See also the following critique of these attempts (Scerri 1997).
11. Let me also mention that Paul Needham has produced a detailed critique of my view on the reduction of chemistry in a debate between us which has been published in several issues of the International Journal for the Philosophy of Science (Scerri 1998b; Needham 1999; Scerri 1999; Needham 2000; Scerri 2000a).

REFERENCES

Bhushan, N. and Rosenfeld, S. 2000. *Of Minds and Molecules*. New York: Oxford University Press.

Capra, F. 1976. *The Tao of Physics*. London: Fontana.

Cartwright, N. 1983. *How the Laws of Physics Lie*. Oxford: Oxford University Press.

Doppelt, G. 1990. The Naturalist Conception of Methodological Standards in Science. *Philosophy of Science* 57: 1–19.

Dretske, F. 1981. *Knowledge and the Flow of Information*. Cambridge: MIT Press.

Emerson, K. 1999. The Quantum Mechanical Explanation of the Periodic System. *Journal of Chemical Education* 76: 1189–1189.

Fine, A. 1986. *The Shaky Game*. Chicago: Chicago University Press.

Gettier, E. 1963. Is Justified True Belief Knowledge? *Analysis* 23: 121–123.

Giere, R.N. 1985. Philosophy of Science Naturalized. *Philosophy of Science* 52: 331–356.

Giere, R.N. 1989. Scientific Rationality as Instrumental Rationality. *Studies in History and Philosophy of Science* 20: 377–384.

Goldman, A.I. 1992. *Liaisons: Philosophy Meets the Cognitive and Social Sciences*. Cambridge: MIT Press.

Hacking, I. 1996. The Disunity of Science. In: Galison, P. and Stump, D.J. (eds.), *Disunity of Science: Boundaries, Contexts, and Power.* California, Stanford: Stanford University Press.

Hendry, R.F. 1994. Unpublished Manuscript.

Hettema, T. and Kuipers, A.F. 1988. The Periodic Table—Its Formalization, Status and Relationship to Atomic Theory. *Erkenntnis* 28: 841–860.

Kitcher, P. 1992. The Naturalist Returns. *The Philosophical Review* 101: 53–114.

Laing, M. 1999. The Four-Fold Way. *Education in Chemistry* 36(1): 11–13.

Latour, B. and Woolgar, S. 1979. *Laboratory Life: The Social Construction of Scientific Facts*. Los Angeles: Sage Publications.

Laudan, L. 1987. Progress or Rationality? The Prospects for Normative Naturalism. *American Philosophical Quarterly* 24: 19–31.

Laudan, L. 1990. Normative Naturalism. *Philosophy of Science* 57: 44–59.

Lombardi, O., LaBarca, M. 2005. The Autonomy of the Chemical World. *Foundations of Chemistry* 7: 125–148.

Needham, P. 1999. Reduction and Abduction in Chemistry, A Response to Scerri. *International Studies in the Philosophy of Science* 13: 169–184.

Needham, P. 2000. Reduction in Chemistry—A Second Response to Scerri. *International Studies in the Philosophy of Science* 14: 317–323.

Paneth, F.A. 1962. The Epistemological Status of the Concept of Element. *British Journal for the Philosophy of Science* 13: 1–14, 144–160. Reprinted in *Foundations of Chemistry* 5: 113–111, 2003.

Popper, K.R. 1982. Scientific Reduction and the Essential Incompleteness of All Science. In: Ayala, F.L. and Dobzhansky, T. (eds.), *Studies in the Philosophy of Biology*. Berkley: University of California Press, 259–284.

Quine, W.V.O. 1951. Two Dogmas of Empiricism. *Philosophical Review* 60: 20–43.

Richman, R.M. 1999a. In Defence of Quantum Numbers. *Journal of Chemical Education* 76: 608.

Richman, R.M. 1999b. The Use of One-Electron Quantum Numbers to Describe Polyatomic Systems. *Foundations of Chemistry* 1: 175–183.

Ruthenberg, K. 1997. Friedrich Adolf Paneth (1887–1958). *Hyle* 3: 103–106.

Scerri, E.R. 1989. Eastern Mysticism and the Alleged Parallels with Physics. *American Journal of Physics* 57: 687–692.

Scerri, E.R. 1991. Electronic Configurations, Quantum Mechanics and Reduction. *British Journal for the Philosophy of Science* 42: 309–325.

Scerri, E.R. 1992a. Quantum Chemistry Truth. *Chemistry in Britain* 28(4): 326.

Scerri, E.R. 1992b. Quantum Extrapolation. *Chemistry in Britain* 28(9): 781.

Scerri, E.R. 1994. Has Chemistry been at Least Approximately Reduced to Quantum Mechanics? In: Hull, D., Forbes, M., and Burian, R. (eds.), *Proceedings of the Philosophy of Science Association, PSA*, Vol. 1. East Lansing, MI: Philosophy of Science Assosiation, 160–170.

Scerri, E.R. 1997. Has the Periodic Table Been Successfully Axiomatized? *Erkentnnis* 47: 229–243.

Scerri, E.R. 1998a. How Good is the Quantum Mechanical Explanation of the Periodic System? *Journal of Chemical Education* 75: 1384–1385.

Scerri, E.R. 1998b. Popper's Naturalized Approach to the Reduction of Chemistry. *International Studies in the Philosophy of Science* 12: 33–44.

Scerri, E.R. 1999. Response to Needham. *International Studies in Philosophy of Science* 13: 185–192.

Scerri, E.R. 2000a. Second Response to Needham. *International Studies in Philosophy of Science* 14: 307–316.

Scerri, E.R. 2000b. Have Orbitals Really been Observed? *Journal of Chemical Education* 77: 1492–1494.

Scerri, E.R. 2001. The Recently Claimed Observation of Atomic Orbitals and Some Related Philosophical Issues. *Philosophy of Science* 68: S76–S88.

Scerri, E.R. 2003. Philosophical Confusion in Chemical Education. *Journal of Chemical Education* 80: 468–474.

Siegel, H. 1985. What is the Question Concerning the Rationality of Science? *Philosophy of Science* 52: 517–537.

Siegel, H. 1989. Philosophy of Science Naturalized? Some problems with Giere's Naturalism. *Studies in History and Philosophy of Science* 20: 365–375.

Siegel, H. 1990. Laudan's Normative Naturalism. *Studies in History and Philosophy of Science* 21: 295–313.

Solomon, M. 1995. Legend Naturalism and Scientific Progress: An Essay on Philip Kitcher's the Advancement of Science. *Studies in History and Philosophy of Science* 26: 205–218.

Suppes, P. 1978. The Plurality of Science. In: Asquith, P.D. and Hacking, I. (eds.), *PSA 1978*, Vol. II, East Lansing: Philosophy of Science Association, 3–16.

van Brakel, J. 2000. *Philosophy of Chemistry*. Leuven: Leuven University Press.

Zukav, G. 1979. *The Dancing Wu-Li Masters*. London: Fontana/Collins.

CHAPTER 7

HOW CLASSICAL MODELS OF EXPLANATION FAIL TO COPE WITH CHEMISTRY

The Case of Molecular Modeling

JOHANNES HUNGER

The Boston Consulting Group, Chilehaus A, Fischertwiete 2, 20095 Hamburg, Germany

INTRODUCTION

"We say it is 'explanation'; but it is only in 'description' that we are in advance of the older stages of knowledge and science. We describe better, we explain just as little as our predecessors. We have discovered a manifold succession where the naive man and investigator of older cultures saw only two things, 'cause' and 'effect,' as it was said; we have perfected the conception of becoming, but have not got a knowledge of what is above and behind the conception. The series of 'causes' stands before us much more complete in every case; we conclude that this and that must first precede in order that that other may follow—but we have not grasped anything thereby. The peculiarity, for example, in every chemical process seems a 'miracle,' the same as before, just like all locomotion; nobody has 'explained' impulse. How could we ever explain? We operate only with things which do not exist, with lines, surfaces, bodies, atoms, divisible times, divisible spaces—how can explanation ever be possible when we first make everything a conception, our conception?"

How could we ever explain? This is a question that has bothered philosophers of science for at least six decades. As we can see from the quote above taken from the third book of Nietzsche's *The Gay Science*, it is a question that has also been of great metaphysical importance. What do we require for an explanation to be valid? And what distinguishes explanation from mere description? The quote from Nietzsche, strangely enough, could as well stem from the opus of one of the founders of modern philosophy of science and a catholic physicist, Pierre Duhem. Yet, whereas Nietzsche's aim was to discredit metaphysics for its incapability to deliver valid explanations, Duhem's objective was just the opposite: i.e., to exclude explanatory claims from science and to leave them completely to metaphysics. Because "if the aim of physical theories is to explain experimental laws, theoretical physics is not an autonomous science, it is subordinate to metaphysics" (Duhem 1991, 19).

Hence, if we want to talk, after and with Duhem, about explanation within science we properly mean "description" and the corresponding problem to establish criteria of how to rule out inadequate from adequate descriptions. What then could philosophy of science possibly tell us about the concept of scientific explanation? Does it, as Duhem demands, have to retire from science, leaving only the problem of finding

D. Baird et al. (eds.), Philosophy of Chemistry, 129–156.

good descriptions? Or might philosophy of science, as a discipline of philosophy, recalling Nietzsche's challenge to reach "behind the picture of science,"[1] strive to establish an autonomous conception of explanation, a concept that, like a platonic idea, would bring explanation into an eternal distance from its object while transcending the range of "down to earth" scientific-empirical *de facto* questions toward a *de jure* legitimization. In this chapter, I follow the intuition that philosophy of science cannot be freed from the attempt to get right into the middle, to get "within the picture" of what it is trying to grasp. Even though there are enough "prima facie" reasons speaking in favor of this intuition, according to its own standards it can only maintain the position "within the picture" by delivering *adequate* descriptions of how scientists actually use the term "explanation" and of what they actually do when they explain.

It is striking that Nietzsche's diagnosis is still of interest: Whereas philosophers of science in the last decades have densely populated the pictures of physics and biology by analyzing in detail their scientific practices—they sometimes even were to be found in the laboratories—they did not show much interest in drawing an appropriate picture of *chemistry*. In this respect, the purpose of the paper is a negative one: I shall try to convince the reader that philosophical models of scientific explanation, at least the most popular ones, by and large leave chemical explanations *unexplained*. In particular they cannot cope with the explanations modern chemistry offers for what Nietzsche called "modes of chemical becoming," i.e., explanations for why molecules *become* the way they do, why they "*look like*" the way we experimentally observe them, or, to put it more precisely, why they possess a certain *molecular structure*. This is the positive part of the paper: I first present the underlying principles and features of three different models chemists use nowadays to describe molecular structures. Second I examine the way chemists actually estimate the explanatory power adherent in these models.

By no mean is this survey meant to be complete. Rather it intends to cover a wide spectrum of available methods by discussing three representative points: the first point is located on the non-theoretical end of the spectrum. It is the method of using an artificial intelligence algorithm, "Neural Network Simulations," in order to correctly predict the secondary structure of proteins. It is an interesting point because, as it turns out, it is the only model where the result obtained by applying philosophical accounts of scientific explanation matches the way scientists estimate its explanatory virtue. The second point is located more or less in the middle. It is not a mere heuristic, still it does not satisfy fundamental theoretical standards, chemists speak of "empirical" methods when they refer to the "molecular mechanics" or "force-field" model. The third point on the spectrum is at its "theoretical" ending. The *ab initio* model, to the greatest extent possible, tries to implement the most basic theoretical principles of what keeps the atoms in molecules together, i.e., the principles of Quantum chemistry.

The structure of the chapter is somewhat Hegelian: first the three mentioned molecular modeling methods are discussed. Then I introduce two major models of scientific explanation, the "*deductive-nomological*" and the "*causal*" models of explanation. In synthesis, I merge these first two parts together by applying each of the explanation models to each of the described methods of molecular modeling. The result is a negative one: Apart from the neural network model the diagnosis as obtained by applying the philosophical models does not match the estimate of the explanatory power as reported in the first part. Finally, I offer some constructive, if sketchy, remarks on how a better philosophical account of explanation within science—*including* chemistry—could look.

WITHIN THE PICTURE OF CHEMISTRY—EXPLAINING MOLECULAR STRUCTURES

The *structure* of a molecule, however depicted, is one of the very central concepts in chemistry. It generally refers to the three-dimensional arrangement of the molecule's constituents in space, whereas different possible arrangements one and the same molecule can adopt are called conformations. The space spanned by the structural parameters necessary to fully describe the structure of a molecule is called conformational space. Knowledge of which areas in conformational space are occupied and recognition of regularities in the way these areas are populated is decisive for an understanding of the chemical reactivity of a molecule.

The famous "key-lock" principle is just one of many examples: It states that the tremendous selectivity found in enzymatic reactions is explained by a perfect *geometrical* matching between the structural shape of the enzyme and the shape of a receptor molecule. But how, in the first place, is the geometry of a molecule, i.e., its structure explained? Of course the conception of a static structure is oversimplified. In fact the atoms of a molecule, even in the solid state and at temperatures close to 0 K are in constant, vibrational and rotational motion. What we silently assume when we talk about the "structure" of a molecule therefore is a *stable* three-dimensional arrangement of the atoms, i.e., a stable *conformation* of the molecule. This is where thermodynamics comes into place: "*stable*," defined thermodynamically, means an arrangement that corresponds to minimal free energy. Neglecting the entropy term, this free energy refers to the energy resulting from the interactions of the atoms and their constituents. Assuming we could describe these interactions correctly, the way to find the stable conformations is to look for atom arrangements that are a minimum on the energy hypersurface—a multi-dimensional surface that describes the free energy of a molecule as a function of the three-dimensional coordinates of its atoms. The concept of an energy hypersurface is at the heart of most of the molecular modeling methods that chemists nowadays use to model the structure of a molecule.

The ab initio *model*

Ab initio is a euphemistic term used to characterize a method that is not intended to start at the very beginning, but as close to the "beginning" as possible. But where is the beginning? As far as quantum physics is concerned, in the beginning is the Schrödinger equation. Quantum chemistry has to be more modest. It cannot even *start* without fundamental assumptions, or rather, *simplifications*. Here is another famous euphemism:

> The underlying physical laws necessary for the mathematical theory of a large part of physics and the whole of chemistry are thus completely known, and the difficulty *is only* that the exact application of these laws lead to equations much too complicated to be soluble (Dirac 1929, 714; emphasis added).

The whole enterprise of Quantum chemistry, myriads of Gigaflops of computational resources for decades have been absorbed with Dirac's "is only." Not only do physicists and chemists start at different points, they start in different races. For a theoretical physicist it simply does not matter that there is not a single molecule in the whole universe to which the quantum physical formalism is applicable. Yet for chemistry, at its core a highly taxonomic discipline, a model that cannot be applied anywhere is not a viable model. "It is only" with a massive set of drastic approximations and assumptions that we can get Quantum chemistry on its way.

But this does not make chemistry a simpler or more unrealistic discipline than physics. On the contrary, as it turns out the more realistic a scientific discipline is the more complex it is. It is a "fallacy of misplaced concreteness," as Whitehead (1929) put it, to think that the *first* things are elementary particles. They may be or may be not, in an ontological sense. Historically and epistemologically, they are not.

So what does it take to solve the Schrödinger equation for a many body system like a molecule? First there is the Born–Oppenheimer–Huang approximation: The mass of an atom's nucleus is about 2000 times higher than the mass of an electron. It seems reasonable to assume that, relative to the electrons, the nuclei could be considered fixed. Now, we are only interested in solving the Schrödinger equation for the electrons located in an electric field stemming from a set of fixed nuclei. As Woolley et al. have emphasized (Woolley 1985; Woolley and Sutcliffe 1977) it is only under this assumption that the concept of a molecular structure starts to get intelligible. One of the starting points of modern chemistry hence is not a deduced "concept," but one that is "fitted" into the quantum chemical framework "by hand" (Woolley 1985). A stable molecular structure corresponds to a minimum on the Born–Oppenheimer hypersurface that gives the ground level energy of the molecule's electrons as a function of the nuclei positions. Still the problem is highly complex. Even under the Born–Oppenheimer approximation, the only *molecule* for which an exact solution can be found is the simplest molecule, H_2^+. Hence, further simplifications have to be made. Roughly speaking, the strategy is to split up the wavefunction for the whole system, Ψ, into a product of single-electron wavefunctions, Ψ_i, called "molecular orbitals." Since electrons are fermion particles, and thus anti-symmetric, this product is described as a determinant called *Slater* determinant. It has the desired property of changing sign if electrons, i.e., molecular orbitals, are interchanged:

$$\Psi = \det(\Psi_1 \otimes \Psi_2 \otimes \cdots \otimes \Psi_n)$$

Each of these molecular orbitals, Ψ_i, in turn is described as a linear combination of basis functions, ϕ_i. This is the "linear combination of atom orbitals" (LCAO) approximation:

$$\Psi_i = \sum_{\mu=1}^{N} c_{\mu i} \cdot \phi_i$$

How these atom orbitals look like and how many of them are used in a calculation is an empirical question, and is limited by the size of the problem and the computational power available. In practice, "slater-orbitals" possessing an exponential radial part and being similar to the s, p, d, and f-orbitals of hydrogen, and Gauss-orbitals, having a Gaussian radial part, have become popular. In a manner similar to the splitting of the wavefunction into one-electron wavefunctions, Ψ_i, the Hamiltonian also splits up into a single-electron operator, the Hartree–Fock operator. The corresponding Schrödinger equations are as follows:

$$\widehat{h}^{\mathrm{HF}} \cdot \Psi_i = \varepsilon_i \cdot \Psi_i \quad 1, 2, 3, \ldots, N$$

Where h^{HF} is the Hartree–Fock operator and ε_i is the energy of the ground state of electron i.

This operator treats every electron described by the molecular orbital wavefunction, as if it were alone in the nuclear field and "sees" only an effective field of the other electrons (composed of simple Coulomb and exchange interactions):

$$\widehat{h}^{\mathrm{HF}} = \widehat{h}_0 + \widehat{j} - \widehat{k}$$

Here h_0 is the Hamilton operator for a single electron in a fixed nuclei field, and j and k are the Fock- and Hartree-operators accounting for Coulomb and exchange interactions between electrons. This simplification is called the "independent particle" approximation. According to the variational theorem, every

$$\frac{\partial E}{\partial c_i} = 0$$

possible solution to the Hartree–Fock problem has an energy value that is higher than the energy of the exact solution. The coefficient c in the above equation thus has to be chosen according to the condition

$$\frac{\partial E}{\partial c_1} = 0$$

where E is the ground-state eigenvalue of the simplified Schrödinger equation. The problem of such an approach is that in order to apply the electron-interaction operators j and k, the wavefunctions ψ_i of the other electrons already have to be known. Yet to get these wavefunctions the corresponding single-electron Schrödinger equations

$$\widehat{h}^{\mathrm{HF}} \times \Psi_i = \varepsilon_i \times \Psi_i$$

have to be solved: The Hartree–Fock operator thus is *circular*. Its electron-interaction parts depend on its own results. To overcome this circularity it is necessary to start with some initial guesses for the wavefunctions and to solve the corresponding equations in an iterative way. Thus, the result for the wavefunctions, given by specific values of the coefficients c, at a certain stage t_0, is used to solve the equations, using the coefficients $c(t_0)$ to get new coefficients $c(t_1)$. These coefficients, $c(t_1)$, in turn are used to get a further set of coefficients $c(t_2)$. The whole procedure is repeated until

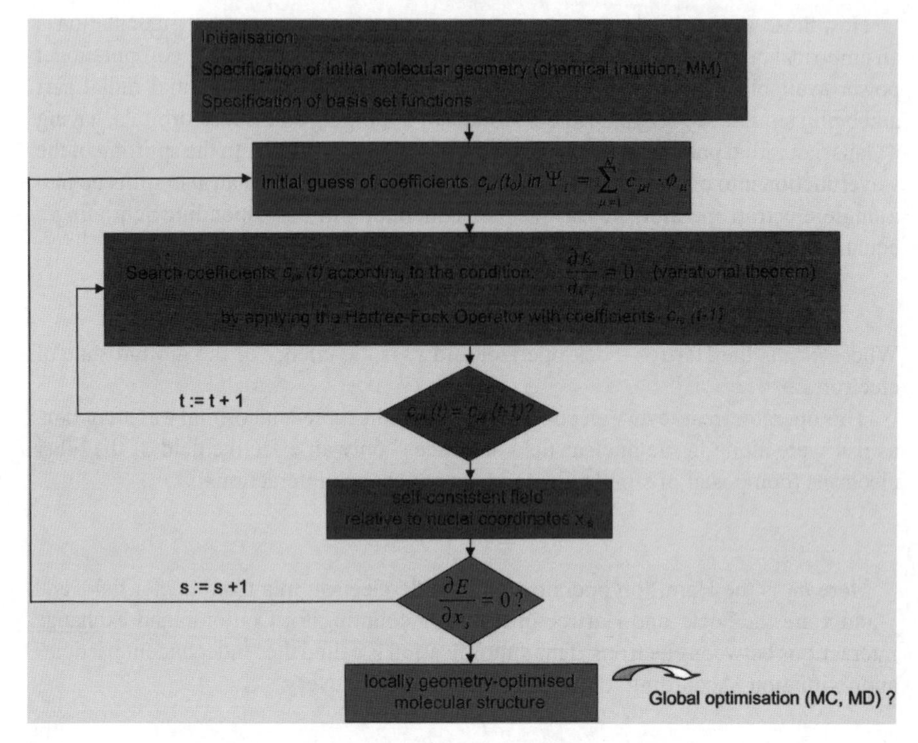

Figure 1: Flowchart of the *ab initio* algorithm.

no significant changes occur. The field stemming from the electrons then is *self-consistent* (this is why this procedure is called the *self-consistent field* method). But we still are not finished. The energy eigenvalue that is obtained as a result of the whole procedure corresponds to only one-single point on the Born–Oppenheimer surface. This point is given by the three-dimensional arrangement of the nuclei which had to be specified in advance and which was considered to be fixed during the self-consistent field calculation. In order to find a *stable* nuclei arrangement these coordinates now have to be changed in an overlaid steepest descent iteration. That is, they have to be changed in such a way that, relative to the starting point, the next local minimum on the Born–Oppenheimer surface is reached.

Figure 1 summarizes this procedure of geometry optimization based on a Hartree–Fock calculation. Note that for each steepest descent step the whole Hartree–Fock routine has to be repeated. The required computational effort, consequently, is huge and such local geometry optimization is the best that can be done by *ab initio* calculations. At the same time this implies that the result of an *ab initio* calculation depends strongly on the geometry specified at the outset. An appropriate guess for these starting positions commonly is taken from experiment.[2] Even though it might run afoul the sense of the term *ab initio*, this fact is readily admitted even by proponents of the method. This is how Hehre, Radom, Schleyer, and Pople—authors of the *ab initio* "bible" "*Ab initio Molecular Orbital Theory*" put it:

> "The simplest and least-involved approach to obtain suitable geometrical structures on which to perform calculations is to assume values for all bond lengths and bond angles. One recourse is to use experimental structural data . . . " (Hehre et al. 1986, 92).

Whenever an appropriate starting geometry is not available the authors recommend staying away from geometry optimization altogether. Another problem stems from the fact that the computational effort necessary to solve the Hartree–Fock equation increases as the fourth power of the number of basis functions used in the LCAO. This number increases drastically with every new electron we have to account for in the system. Consequently, *ab initio* calculations are confined to modeling relatively small molecules.[3] Most biomolecules, like proteins, but also a large number of inorganic molecules containing heavy atoms cannot be treated with the quantum chemical approach. Within this limited range, however, the *ab initio* model is applied with great success. Not only is it used for the calculation of molecular structures, but also is capable of reproducing and predicting electronic properties (energy eigenvalues of ground and excited states), spectroscopic values, bond formation and bond braking processes, reaction pathways, and much more with high accuracy. As the term *ab initio* implies, the calculation is starting "from first principles." Putting aside starting coordinates and coefficients, the only inputs an *ab initio* calculation requires are the spin, charge and mass of the elementary particles in the molecule under consideration. Theoretical chemists praise this method for its theoretical and explanatory rigor. Primas, usually a quite sceptical quantum chemist and an anti-reductionist, writes:

> Quantum chemistry is *the* fundamental theory of molecules; it allows for an *all embracing* and *correct* description of the structure and properties of single molecules. In principle all experimentally measurable properties could be calculated mathematically by applying the quantum chemical formalism; thereby the only empirical parameters are the Planck constant, the velocity of light, the masses, the spins and the electromagnetic momentum of the electron and of the nuclei (Müller-Herold and Primas 1984, 309).

Accordingly, in all scientific papers where an *ab initio* calculation could be performed that fits the experimentally observed structure of a molecule, the result is taken to theoretically explain this observation. But what does this mean for the overwhelming number of molecules that are intractable within the *ab initio* approach? Do we have to assume that their structure is, and must remain, unexplained? Most chemists would deny this. A look in classical textbooks of chemistry, especially organic chemistry, shows that the most common explanatory principle in the context of the everyday molecular structures does not refer to the Schrödinger equation at all. It is rather a principle that lies at the heart of the "molecular mechanics" model: the principle of *steric strain*.

The molecular mechanics model

If we asked a student of chemistry why the ethane molecule (CH_3–CH_3) adopts a staggered conformation and not an eclipsed one (see Figure 2), we would hear

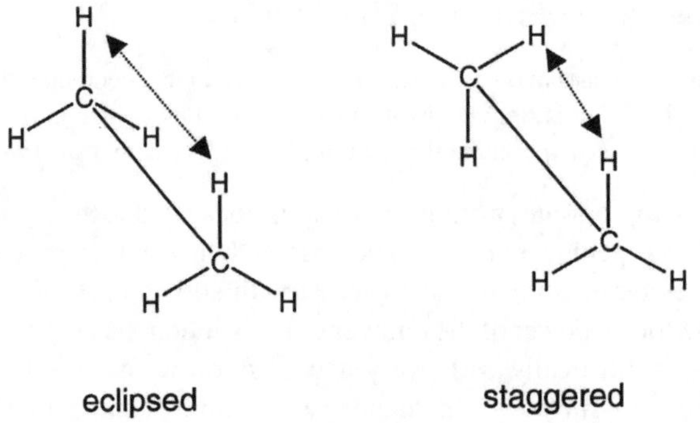

eclipsed staggered

Figure 2: Eclipsed and staggered conformation of ethane.

something like this: In the staggered conformation the hydrogen atoms connected to the two carbon atoms are positioned as far from each other as possible. In the case of eclipsed conformation the opposite is the case. Therefore, the steric strain stemming from the repulsion between the hydrogen atoms is minimal in the staggered case and this is why this conformation is favored [see for instance (**Christen and Vögtle 1988, 69)].

The molecular mechanics, or force-field approach implements this line of reasoning into a computational model. It is based on the simple "*ball and stick*" conception of a molecule. On this conception molecules are composed of atoms connected by elastic sticks. If we knew the equilibrium length and the force of the specific atom-type sticks we could build up a model of the molecule and in its equilibrium state. Simplified, this is the procedure of a force-field calculation. To transpose this procedure into a computer program, however, we need to assume certain force potentials describing the deformation of the "sticks" and we need to assume "numbers," i.e., numeric values for the parameters of the force potentials. As the term "molecular mechanics" already hints, force potentials are akin to simple mechanical force potentials, and describe the force needed for bond stretching, angle bending, torsion around a bond, van der Waals and Coulomb interactions (see Figure 3, upper row).

The total steric strain energy of a molecule is the sum of all the individual mechanical potentials (Figure 3, middle). What a parameterized force field describes is the total strain energy of a molecule as a function of the three-dimensional arrangement of the molecule's atoms. This far, we have assumed that we already knew appropriate values for the force-potential parameters. Yet the *parameterization* of a force field, which is crucial to this method, is by no means a well-defined procedure. Apart from calculating the parameters by *ab initio* calculations, what all parameterization procedures have in common is that they try to generate parameters that best fit available experimental data. This is why this model has been labeled "*empirical.*" Depending both on the parameterization method applied,[4] and on the force-field potentials assumed, different "force fields" with different force-field parameters exist. The question which of them to chose is subject to pragmatic considerations. But again, as in the case of the *ab initio* approach, the possibility of calculating arbitrary points on the

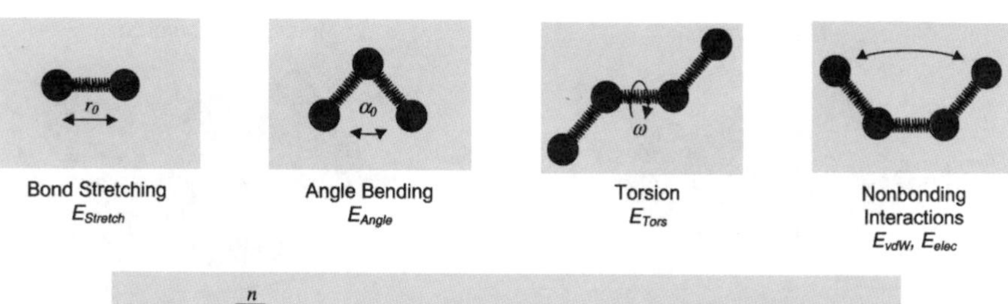

$$E_{Strain} = \sum_{i=1}^{n} (E_{Stretch} + E_{Angle} + E_{Torsion} + E_{vdW} + E_{elec} + E_{cross})$$

Example MM2-Force Field:

$$E_{Stretch} = \frac{1}{2} \cdot k_b \cdot \left[(r - r_0)^2 - 2 \cdot (r - r_0)^3 \right]$$

$$E_{Angle} = \frac{1}{2} \cdot k_a \cdot \left[(\alpha - \alpha_0)^2 + const \cdot (\alpha - \alpha_0)^6 \right]$$

$$E_{Torsion} = k_{t1} \cdot (1 + \cos\omega) + k_{t2} \cdot (1 - \cos 2\omega) + k_{t3} \cdot (1 + \cos 3\omega)$$

Molecular Energy Hypersurface:

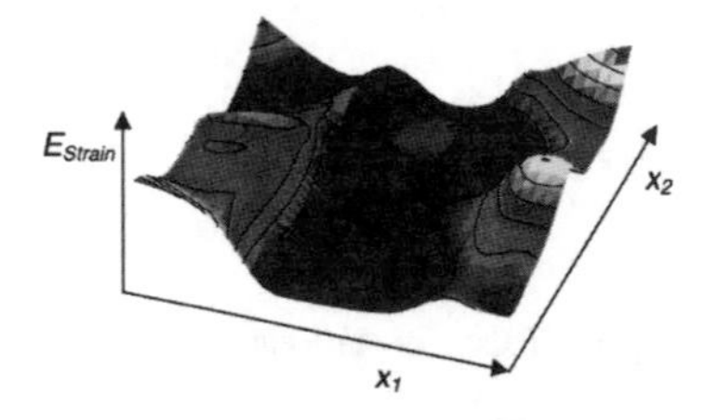

Figure 3: Illustration of a force-field model. Upper row: different interaction types; middle: expression for the sum of the steric strain energy (i: atom i, n: number of all atoms in the molecule); lower left: the MM2-force field as an example (r_0: equilibrium length, α_0: equilibrium angle, ω: torsion angle, k_a, k_b, k_{ti}: force constants); lower right: illustration of an molecular energy hypersurface with two conformational parameters x_1 and x_2.

energy hypersurface does not *eo ipso* allow us to deduce those points on the surface that correspond to *stable* conformations of the molecule observed experimentally. Yet, in comparison with the *ab initio* method, the reduced computational effort of a force-field calculation makes it possible to, in addition to local geometry optimization procedures, perform *global* searches on the hypersurface. I illustrate the fact that, in most cases, the energy hypersurface cannot systematically be screened for energeric minima by a simple example. It was only for the sake of illustration that the energy hypersurface shown in Figure 3 (lower right) was based on only two conformational parameters. Since the number of conformational degrees of freedom of a molecule is $3N-6$ (with N being the total number of atoms in the molecule) the energy hypersurface even of a simple molecule like ethane is already 18-dimensional. A screening on the basis of a 10-point resolution per dimension thus would already require the calculation of 10^{18} different conformations. The only way to tackle such a complex search problem is to apply global optimization algorithms that, in one way or the other, rely on stochastic elements. Among the manifold of methods applied only the two most popular shall be mentioned here.

The *Monte Carlo* method is a stochastic search algorithm that performs a random walk on the energy hypersurface (Figure 4, left). In its simplest form, starting from an arbitrary starting point, $x\vec{x}(t_i)$, new points, $\vec{x}(t_{i+1})$, on the hypersurface are reached by

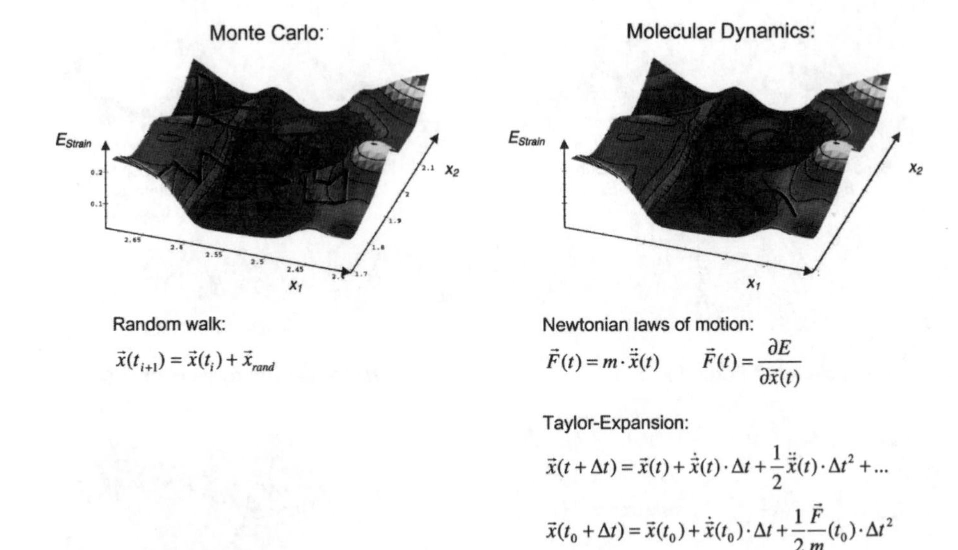

Figure 4: Conformational search; *Monte Carlo* method (left) and *Molecular Dynamics* method (right).

performing random steps, $\vec{x}_{\text{rand}}$. After as many steps as possible have been performed the trajectory through the conformational space of the molecule is screened for its energetic minima that are proposed as the most stable structures. The *molecular dynamics* method physically is more appealing because it tries to simulate the space-time trajectory of the molecule's atoms by integrating the Newtonian laws of motion. Since there is no analytical solution available for this, it has to be done numerically, for instance by using a Taylor series approach and truncating the series after the second member (Figure 4, lower right). Thus, when starting with an initial conformation, $\vec{x}(t_0)$, and initial velocities, $\vec{x}(t_0)$, the conformation after a time step Δt can be calculated if the forces acting upon the atoms are known.

The problem however lies in the fact that these time steps have to be very small in order to make the Taylor series approach valid and typically lie in the order of 10^{-15} s. The full time for which the trajectory of a molecule could be simulated hence is quite short, and, after finishing the simulation, there almost always has to remain some doubt if the energetically most favorable regions of the hypersurface have been passed through.

Especially in the case of large molecules like proteins this can lead to severe problems. For instance, the real time for a protein to fold from its denaturated into its globular state lies between a second and several minutes. To simulate this process within a molecular dynamics simulation would take about 10^8 s, a time interval that exceeds the estimated age of the universe. Still, what can be done is to use cases where the solid-state structure of the protein is known in order to validate the corresponding force field. If the force field is adequate it should reproduce the experimentally observed protein structure as an energetic minimum of the corresponding energy

Crambin x-ray structure (unit cell):

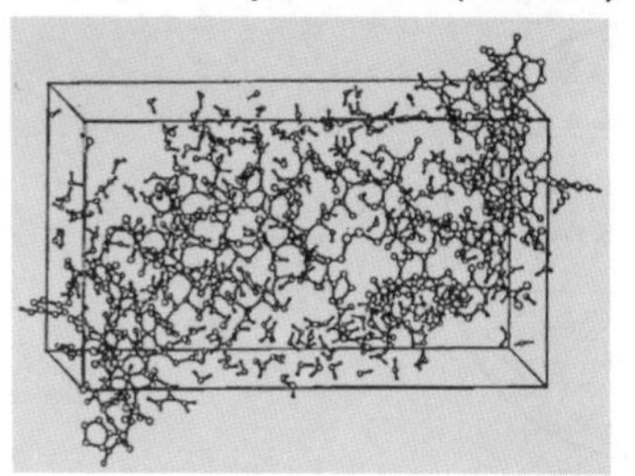

Steepest descent:

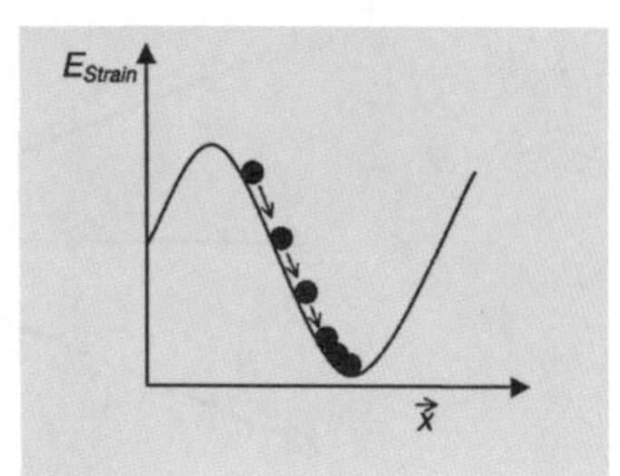

Results:

Force Field	rms [Å]
Amber	0.22
Amber/OPLS	0.17

rms: root-mean-square deviation of all non hydrogen atoms

Computation time: several months

from:
*W. L. Jorgensen, J. Tirado-Rives, J. Am. Chem. Soc. (1988), **110**, 1657.*

Figure 5: Energetic minimization of the crystal structure of crambin. The minimization takes into consideration 1356 atoms. Elementary cell of the crystal structure illustration of the steepest descent minimization procedure result of the minimization with the force fields Amber and Amber/OPLS (rms: root mean square deviation between observed and calculated structure); taken from Jorgensen and Tirado-Rives (1988).

hypersurface. This means that, starting from the observed protein structure, the locally geometrically optimized model structure should be not too far removed from the starting point. Figure 5 illustrates such a validation by local optimization (Figure 5, upper right) in the case of the two force fields Amber and Amber/OPLS and the protein *crambin* (Figure 5, upper left).

The molecular mechanics model is extremely popular among chemists and there is an overwhelming number of articles reporting the application of this method. Their broad application also is considered to raise our understanding and our capability to explain the structural features of the treated molecules.[5] But still, as the last example shows, there exist upper limits concerning the size of the molecules for which a proper prediction of structure can be made. Especially in the case of proteins, such predictions can have tremendous practical importance. The last model, I discuss is a method used to predict the secondary structure of a protein, i.e., its folding mode, starting with only information on its primary structure, i.e., its amino acid sequence.

The artificial intelligent model—Neural network simulations

Derived from cognitive scientists' attempts to model the structure and organization of the human brain, neural network simulations are powerful tools to deal with

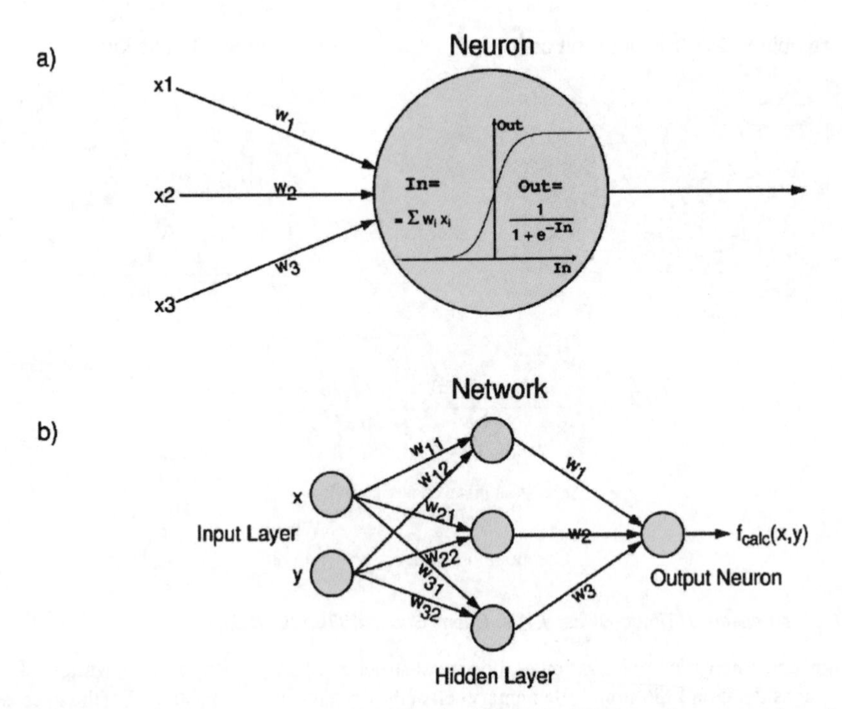

Figure 6: Principal structure of a backpropagation network a) Information processing of a single neuron b) Exemplaric structure of a network with two input neurons, one hidden layer of three neurons and an output layer consisting of one output neuron.

classification and extrapolation problems of nearly all kinds. Especially in chemistry they have found widespread application to such different tasks as the interpretation of spectra, process control and the prediction of protein secondary structure (Gasteiger and Zupan 1993). The type of network architecture that is used in this latter case is called "backpropagation" network algorithm. This network is composed of algorithmic units, called neurons, which are arranged in layers (Figure 6b).

As suggested in Figure 6b each neuron in one layer is connected to all the neurons in the subsequent layer. Data are presented to the neurons in the first layer, the input layer. The data are transformed and passed on to the neurons in the next layer and so on until the output layer is reached. Figure 6a illustrates how this information transformation looks. Each of the incoming signals, x_i, to a specific neuron is multiplied by a specific weight, w_i, attached to the link through which they arrive. These products are added and the sum, "In," is passed on to a signal transformation function which most commonly is a sigmoidal type function (see Figure 6a). This generates the neuron's output. The output thus depends on a given input array of data, the weights of the input links, and the kind of signal transformation function that is used. This means that if we want to "train" the network to produce a desired output when fed with a specific input, we have to adapt the weight matrix of the network. This is done during an iterative adaptation process.

In the chemical case considered here, the training is provided by pairs of data with the amino acid sequence of a part of a protein chain as input and the adherent

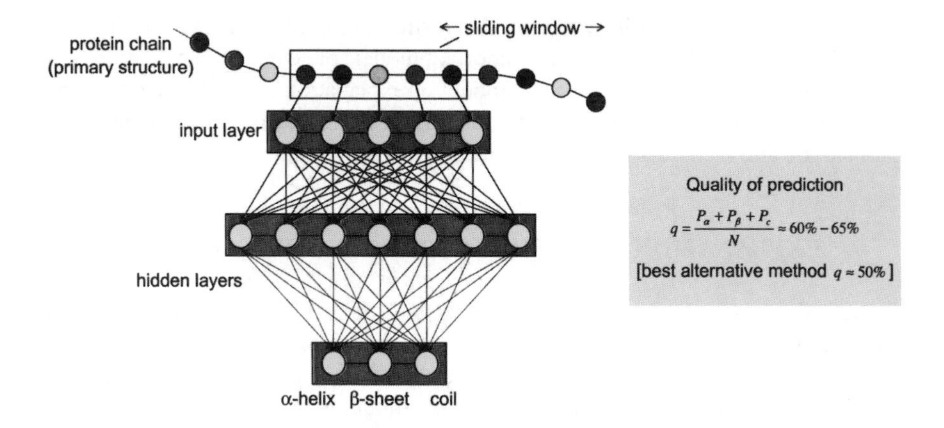

see for instance:

- *N. Quian, T. J. Sejnowski, J. Mol. Biol. (1988),* ***202****, 865 - 88*
- *J. M. Chandonia, M. Karplus, Protein Science (1995),* ***4****, 275 - 285*

Figure 7: Schematic illustration of the use of a backpropagation network for the prediction of protein secondary structures.

folding pattern of the chain as output. Figure 7 illustrates how a network taking into account three different folding patterns (α-helix, β-sheet and coil) typically looks. Of course the training data are only a minor fraction of all the data available and once the training is finished the quality of the trained network will be tested with remaining protein data. In their seminal paper on the neural network approach Quian and Sejnowski reported results remarkably better than results from any alternative method (see Figure 7). Today it is the unrivalled method when it comes to predicting the folding modes of proteins. In contrast to the *ab initio* and the molecular mechanics models it is commonly agreed that this method has no explanatory value. This is how Gasteiger, one of the proponents of neural networks in chemistry, judges the situation:

> Satisfying as we might find the results obtained by the neural network, they cannot give an *explanation* ... (Gasteiger and Zupan 1993, 216; emphasis in the original quote).

There is a similar vote in the prediction of the secondary structure of protein. One of the standard textbooks on the prediction of protein structure summarizes the negative features of neural network models:

(i) *"the method has very poor explanatory power—a Hinton diagram*[6]*—means nothing to a protein chemist"*
(ii) *"there is little use of chemical or physical theory"* (King 1996).

The Molecular Mechanics Model illustrates how chemists proceed when they try to explain molecular structures. It turns out that explanation remains an issue here and that mere prediction, as in the case of the neural network approach, is not, *eo ipso*, explanatorily relevant. On the other hand it is remarkable that not a single, but rather two conceptually quite different models were both praised for their explanatory virtues. In the following, I investigate if this result of scientific practice is mirrored in an analysis that applies the two most important models of scientific explanation—the deductive nomological and the causal model of explanation—to the three modeling methods discussed above. To this end, I first give a short sketch of the two models of explanation and then deal with their application to our "real world" example.

EXPLANATION—THE PHILOSOPHICAL POINT OF VIEW

According to a classification going back to Coffa (1974) there are two major *families* of models of scientific explanation: *epistemic* and *ontic* models. The representative of epistemic model is the Hempelian *Deductive-Nomological*, abbreviated *DN-model* and the main representative of ontic model is the *causal model* of explanation. I discuss each in the following.

For Hempel, explaining an event or a fact—in general that what there is to be explained, the *explanandum*—means to deduce the *explanandum* from a set of laws that have the logical form of a general conditional, and certain *antecedent* conditions, which together are the *explaining* part of an explanation or the "*explanans*" (Hempel and Oppenheim 1948; Hempel 1977). An explanation thus, in the simplest case of only one law and one boundary condition an explanation has the logical form:

Given the law that "*All humans are mortal*" (with Fx: "*x is human*" and Gx: "*x is mortal*") and given the antecedent that "*Socrates is*

$$\forall x Fx \subset Gx \qquad \frac{Fx}{Gx}$$

human," the old Aristotelian syllogism thus turns out

$$\forall x \quad Fx \subset Gx$$
$$\frac{Fx}{Gx}$$

to be an explanation for the fact, that Socrates is mortal and, as we know and as Aristotle knew well, Socrates in fact died.

Of course, certain restrictions apply, for instance it is necessary that the antecedent bears some empirical content. Further, and more important, expressions with the logical form of a law, but which remain mere *accidental* generalizations are not valid components of the explanans.[7] Hence what counts for an explanation is that it raises the *nomic expectability* of the explanandum; explanations provide good reasons that a certain phenomenon has been observed, but they make no further restrictions on the kind of regularities the laws quoted in a scientific explanation have to refer to. This is why *epistemic* models of explanation are, one might say, less dogmatic than ontic models of explanation. In ontic models certain assumptions about the constitution of the world enter the scene. One must assume that the world "*works*" in a certain way that nature, as Salmon puts it, is organized after specific patterns. The answer to the question of how the world is organized and what holds the world together commonly is found in what is called its *causal nexus*. Apart from the question of what causality really consists in, what the different schools of causality-theorist unifies is their conviction, that, in order to explain a certain event or fact, one has to reveal its causal embedding. Salmon put it:

> To give scientific explanations is to show how events and statistical regularities fit into the causal network of the world (1977, 166).

The difference between the epistemic model of explanation and the ontic one becomes clear at this point: "*Cause . . . because*," as Philip Kitcher elegantly puts it, is the slogan of the epistemic school. Explanations frequently refer to causes simply *because* they explain much, but there is no *conditio sine qua non* (1989). If it turned out that other non-causal regularities were better explanatory vehicles the causal explanation would be dropped. The adherent of the ontic model of explanation, on the other hand, maintains that any valid explanation has to cite causes of the explanandum because explanation owes its only possible sense to a thorough assumption about the ontic constitution of our world. Those who want to capture the validity of explanations in their dependency on causes have to give an answer to the question of what causality consists in. Here, I sketch the two major accounts of causality.

According to the process theory of causality recently put forward by Salmon and Dowe, a chain of causal events is not, as Hume defined it, a discrete series of events or objects but rather a continuous non-dividable *process* (Salmon 1981; Dowe 1992). The aim here is to reconcile the empiricist tradition—the tradition of Hume—with the concept of causality by first giving an appropriate account of causal processes and then giving empirically meaningful criteria allowing for a distinction between real causal processes and mere pseudo processes. One of the primordial scenarios in the history of philosophy may serve to illustrate this distinction: It was Plato who described prisoners in the cave, put in chains, as his famous cave allegory. The scenes they observed on the cave's wall are not real, but mere shadows, or to put it in Salmon's terms, mere pseudo processes. How could Plato's prisoners discern this? By attacking the enemies on the wall with a spear, for instance, they could have observed that any injury supposedly caused by hitting an enemy at time t would not have influenced the shadow-enemy's behavior at any time later than t. According to Salmon an essential criterion of any real causal process is its manipulatability or, to be precise, its capability of *transferring a mark*.[8] A shadow-enemy would not carry the mark of being injured at any time later than t because the processes on the wall are not real causal processes but mere pseudo processes. This criterion of transferring a mark is of special appeal to the chemist's account of causality.

Another account of causality that, in contrast to Salmon's, is explicitly based on the Humean assumption of discrete events, comes with Lewis' theory of counterfactuals put forward in 1973. Here is Lewis definition of a causal series of events:

> Let $c_1, c_2, \ldots$ and $e_1, e_2, \ldots$ be distinct possible events such that no two of the c's and no two of the e's are compossible. Then I say that the family $e_1, e_2, \ldots$ of events *depends causally* on the family $c_1, c_2, \ldots$ iff the family $O(e_1), O(e_2), \ldots$ ["O(e)" is the proposition that e occurs] of propositions depends counterfactually on the family $O(c_1), O(c_2), \ldots$. As we say it: whether e_1 or e_2 or . . . occurs depends on whether c_1 or c_2 or . . . occurs (Lewis 1973, 556).

"*Counterfactual dependence*" means that *if events* $c_1, c_2, \ldots$ *had occurred, the events* $e_1, e_2, \ldots$ *would have occurred as well.* The problem of verifying a causal dependency thus is shifted to the problem of verifying a *counterfactual dependency*. In turn, Lewis tries to capture this kind of dependency with a semantics of possible worlds. Accordingly an event *e* depends counterfactually on another event *c*, if, in the actual world W_a, the closest possible world W_c in which *c* is the case, *e* is also the case. Let's consider our example above. The counterfactual: *If the enemy had been hit by a spear, he would have been injured* would be true iff, in the closest possible world to the one in which the counterfactual has been uttered where the enemy has been hit it is also the case that he carries an injury. Of course it is decisive to have a precise *similarity* measure at one's disposal to determine the distance between two possible worlds. As the topic of this paper is not the concept of causality itself but rather its use in theories of scientific explanations, I shall say nothing more about this point and about the delicate problem of how to weigh similarity between two worlds.

Typical of both concepts of causality, different though they are, is the fact that they link their concept of causality to the concept of scientific explanation. Thus on Lewis' view a valid explanation must elucidate the causal background of the event being explained. Salmon also requires that an explanation shows how the explanandum fits into the causal processes of our world.

"HOW COULD WE EVER EXPLAIN?"

I now apply the two concepts of scientific explanation sketched above—the DN-model and the causal model—to the three methods for deriving molecular structure described in the section on Within the Picture of Chemistry—Explaining Molecular Structures. Figure 8 surveys the three molecular modeling approaches.

Is there an artificial intelligent explanation?

The method of neural network simulations regresses only a brief discussion. Let's start with the causal model of explanation. It is obvious that in the process of deriving a prediction based on a neural network simulation no reference is made to a causal fact or event whatsoever. This becomes especially clear from the fact that a neural network could easily reproduce regularities that are classic examples of causal asymmetry. For instance it would be easy to train a neural network to predict the length of a flagpole from the length of its shadow. To be precise, unless no restrictions apply, such that input and output data used in the training process are referring to a causal process, a relation embodied by a trained network cannot possess causal explanation capacities.

The DN-model also does not apply. As described in the section on The Artifical Intelligent Model—Neural Network Simulations the regularities that a trained network has "learned" are represented in its weight matrix. Even though knowing every detail of the network, such a regularity cannot be expressed by a mathematical function but

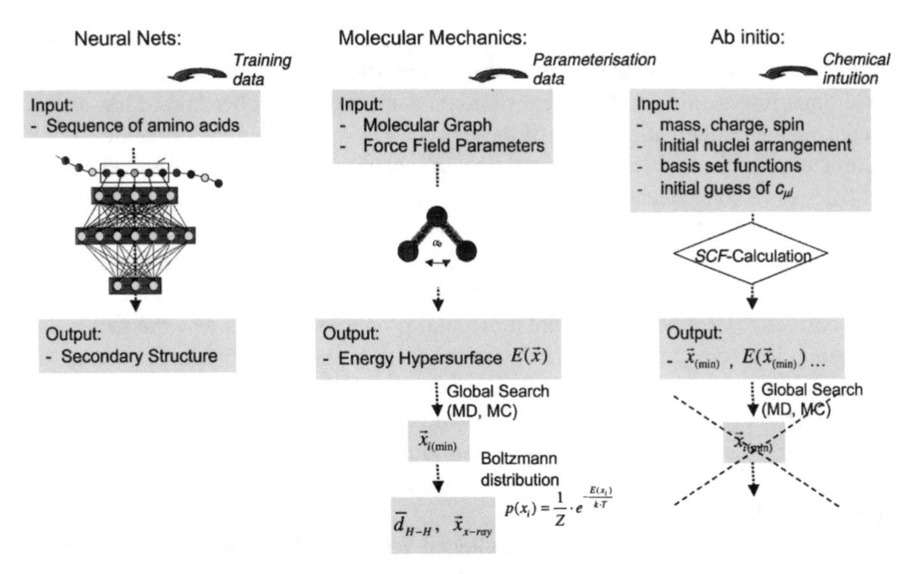

Figure 8: Survey on the three models for the description of molecular structures discussed in the text: left: Neural Network, middle: Molecular Mechanics, right: Ab initio.

only in the following algorithmical way:

> *Add the incoming signals of each neuron, multiplied by the weights of the corresponding links and take the product as an input for the neuron's signal transformation function. Take the output as an input for all the connected neurons in the subsequent layer do the same . . . until the output neurons are reached.*

It is quite obvious that this could hardly be considered a scientific law. Thus, there is no way to describe the process of deriving output values as a nomic deduction or as an argument referring to a certain law or a number of laws. And this is exactly what the deductive nomological model of explanation requires.

Both, the DN-model and the causal model of explanation fail to describe the predictions made by neural network simulations. But since, as mentioned above, scientists working with these methods themselves do not claim to deliver explanations, this result seems satisfying and could count as a confirmation, though a negative one, of both models of explanation. At least for this rather trivial case philosophy of science and scientific practice agree.[9]

Is there an empirical explanation?

This case seems to be more promising. Let's consider the DN-model first. As described above, the molecular mechanics model is based on a number of force-field potentials that refer to proper scientific laws. Given the molecular graph—the different types of atoms contained in a molecule and their connections—and a set of force-field parameters the total strain energy of the molecule can be deduced for every possible atomic arrangement. Is this a DN-explanation? The answer is no, and for two reasons. The first reason concerns the nomic character of derivations made by molecular mechanics models. The single force-field potentials chosen to describe the atomic interactions, to be sure, bear some similarity to proper laws. Still, as has been said, the mathematical form of these potentials and the numerical values of the parameters contained in these potentials vary considerably from force field to force field. This is a consequence of the fact that the potentials used in a force-field model are taken to reproduce the underlying experimental data to the best extent possible when applied to the data *as a whole*. The physical meaning of a force field therefore is *holistic*. Only the sum of all the potentials, not the singular potentials themselves, must correctly describe the relevant molecular properties. This is why the molecular mechanics literature typically warn the reader not to use single force-field potentials or parameters to describe isolated atomic interactions within a molecule.[10] Consequently the sum-expression of all these potentials, which is considered to be the only physically relevant part of a force field (see Figure 3, middle), cannot count as a proper physical law. It is more akin to an accidental generalization, though a generalization for a large number of different molecules.

There is a second problem with the alleged *deductive* character of the derivations in the molecular modeling approach. It is true that the calculation of a single point on the energy hypersurface is a deductive step. But this is only half of the story. In addition, to find the stable conformations one must do a global search on the energy hypersurface for energetic minima. But, as the name "search" already implies, this not a problem that could be solved analytically. Global search algorithms, one way or the other, always contain stochastic elements. Consequently, a Monte Carlo simulation would not necessarily result in exactly the same stable conformation on different runs, but could produce conformations that differ from run to run.

An example illustrates the point: Suppose a blind person stands before an undulated model surface. She must find the deepest basins on this surface by throwing balls on it. She detects height acoustically. Even if this person has a very sophisticated technique of throwing balls on the surface, different runs will lead to different results. She does not *deduce* the deepest basin on the surface.[11] In the same way, the *actual* process of deriving stable conformations by molecular mechanics is not entirely deductive.[12]

Hence the attempt to subsume molecular mechanics under the DN-model of scientific explanation fails. It fails twice: First, the sum of the force-field potentials used in molecular mechanics calculations does not have the status of a proper scientific law and the derivation of the strain energy of a molecule by molecular mechanics is not *nomological*. Second, the conformational search required to find stable conformations is stochastic and not *deductive*.

Consider the causal model of explanation. Is not the molecular dynamics simulation of a molecule a classic example of a simulation of a *causal process*, and is not the causal model of explanation, hence, a perfect candidate to describe these simulations as explanations? To address the first question, I consider the two conceptions of causality discussed above to see if either serves to capture the molecular dynamics method as picturing a causal process. Keep in mind that usually when asking whether a process is causal one cannot look at the process itself, but must rely on a model that simulates it. This is not simply a practical necessity—in our example, for instance, it is not possible to observe the trajectory of a molecule—but is a consequence of the concept of causality itself. Except perhaps for Salmon's early mark-transfer account, the question of what causality consists is deeply intertwined with the causal model proliferated by science itself.

The molecular dynamics example is especially apt to illustrate this point. Is the process described by a molecular dynamics simulation? That is, is the space-time trajectory of all the atoms of a molecule, a *causal process* in the sense required by Salmon? In the first place there are problems with the term *process* itself. On Salmon's account processes are ontologically *continuous*. They cannot be divided into discrete parts or stages (1981). This requirement collides with any attempt to simulate a process on the computer. For instance (see the section on The Molecular Mechanics Model) to describe the time evolution of a molecule the computer calculates its state in discrete time steps Δt. Even though these steps may be very small this remains a problem.[13] One may object that this is just an epistemological problem, produced by the attempt of numerically finding solutions for differential equations, whereas Salmon's requirement is meant to be ontological. However, there is nothing in our differential equations that could guarantee that the process pictured by a molecular dynamics simulation is in fact continuous. We are left to assume that for practical reasons our model fails to describe this trait of the system correctly.[14]

Putting aside this general difficulty, Salmon's account can be applied to our example. We can mark an atom of a molecule in the model by giving it another color, say. The atom will still have the same color after a number of time steps. A Monte Carlo simulation, however, fails. Here the process described by the model, a random walk in conformational space, is not intended to be *in time*, and hence Salmon's criterion cannot even be applied. Salmon's early process account could, and in fact has been used by chemists to determine which, of a number of processes that might have occurred during a chemical reaction, is the *real* one. I refer to the technique of marking molecules with isotopes to find the underlying *reaction mechanism*. Such tracer experiments are a standard technique in chemistry.

It is strange, however, to cite this case in support of Salmon's theory. The aim of a tracer experiment is not to distinguish causal processes from pseudo processes but rather to single-out one causal process from a number of possible causal processes. All these processes are assumed to be causal in the first place. Such experiments conform to Salmon's account. They do not confirm it. Even this matching is doubtful. One of Salmon's essential conditions was that, after marking the process at time t_1 and detecting the mark at time t_2, every possible interaction between t_1 and t_2 must be excluded to use the observation as an indicator of a real causal process (1984). This clearly is not true in the case of a chemical reaction. From the moment where a tracer atom is introduced to the moment where it is detected there is a manifold of interactions occurring. Some of them are essential for the reaction to take place. Even neglecting interactions between the system and its environment in form of energy absorption and emission, there are a large number of "necessary" collisions with other reactant molecules, some of them leading to the formation a "transition complex" out of which the product can emerge. Hence, strictly speaking, Salmon's theory is not applicable to this prominent example from the scientific practice of chemistry.[15] To come back to our original concerns the molecular dynamics simulation, at least in most cases, treats the molecule as isolated. Apart from the problem of modeling time discretely Salmon's account of causality allows us to classify the trajectory of a molecule as a causal process.

I turn now to the second concept of causality. Lewis' account of counterfactuals, intended to be of purely metaphysical import, is very difficult to turn into an empirically applicable theory. This is because in order to verify the counterfactual "If A was the case, B would be the case" we have to find out whether, in the closest possible world in which A, B also is the case. One way to do this is to simulate worlds in a computer. For example, assume a molecular dynamics model has simulated the trajectory of a single ethane molecule. Consider this one-molecule world, "W1," as our reference world. Does a specific conformation, c_1, of ethane at time t_1 *cause* another conformation, c_2, at time t_2? According to Lewis, the answer to this question, depends on the truth of the counterfactual "If ethane had the conformation c_1 at t_1 (A), it also would have a conformation c_2 at t_2 (B)." The closest possible world to our reference world where A is true is a molecular dynamics model where ethane has the conformation c_1, but where we exchanged one hydrogen atom by a Deuterium atom.[16] We change the atom type in the model accordingly. By running the simulation again we may find out that the slightly changed ethane molecule also adopts conformation c_2 at time t_2. We can conclude that, within a molecular dynamics simulation, the conformation of ethane c_1 at t_1 is *Lewis-causing* the conformation of ethane c_2 at time t_2.

Hence, in the light of our both accounts of causality, the molecular dynamics model represents causal processes or chains of events. But is the derivation of a molecule's structure by a molecular dynamics simulation a causal *explanation*? Here the answer is no. The molecular dynamics model alone is not used to explain a causal story elucidating the time evolution of the molecule's conformations. It is used to find the equilibrium conformation situation that comes about a theoretically infinite time interval. The calculation of a molecule's trajectory is only the first step in deriving any *observable* structural property of this molecule. After a molecular dynamics search we have to screen its trajectory for the energetic minima. We apply the Boltzmann distribution principle to infer the most probable conformation of this molecule.[17] It is not a causal principle at work here. This principle is derived from thermodynamics, and hence is statistical. For example, to derive the expression for the Boltzmann distribution, one crucial step is to determine the number of possible realizations there are for each specific distribution of items over a number of energy levels. There is no existing explanation for something like the *molecular partition function* for a system in thermodynamic equilibrium solely by means of causal processes or causal stories based on considerations on closest possible worlds.

In this context, Salmon offers a distinction that seems to deal with this problem. He distinguishes *etiological* and *constitutive* causal explanations. Etiological causal explanations are cases where a causal story can be told leading to the fact that is to be explained. *Constitutive* causal explanations would appear to answer the question posed by causes in a thermodynamic setting. Salmon argues, "We explain the pressure exerted by a gas on the walls of a container in terms of momentum exchanges between the molecules and the walls" (1998, 324). Nobody will deny that there are indeed momentum exchanges between molecules and the wall which are of causal nature. My claim, however, is, that it is an indispensable part of the *explanation* of a specific pressure, that certain distributions of the velocities of the molecules be assumed. This is necessary to derive the fundamental underlying expression of the kinetic theory of gases. These distributions are not explainable in purely causal terms. For instance the fundamental equation that describes the relation between the pressure of a gas and the velocity of the particles refers to the root-mean-square of these velocities c. This depends on the temperature T of the system:

$$p \cdot V = \frac{1}{3} \cdot N \cdot m \cdot c^2$$

while N is the Avogadro constant, V the volume and m is the mass of the molecules. This average velocity is not derived from a causal story or process. It assumed that a great many such causal processes occur, all of which obey certain statistical regularities. This assumption would not be necessary if an explanation in *purely* causal terms were available. This objection applies to nearly all kind of explanations that involve the *constitution* of a system, i.e., the structure of a system that most probably will be found during a certain time interval t. Lawrence Sklar makes a similar point. He puts it, "In any case it is on top of an underlying causal picture of the system that the statistical or probabilistic account of phenomena, essential to the statistical mechanical attempt to account for the thermodynamic features of the world, is superimposed" (1993, 149).

He further points out that this is especially true for the description of systems in thermodynamic equilibrium—as was assumed throughout in all of the modeling methods described above. My point here is simply that in fact in the process of "*explaining*" the structural features of a molecule by molecular mechanics methods, more is needed than the simulation of its causal behavior.[18] While the inner structure of a system may be causal, the capability to correctly describe single causal processes alone is not sufficient to derive a meaningful explanation of any of the systems properties observable by experiment.[19]

Hence, the molecular mechanics approach does not fit either of the models of explanation, Salmon's or Lewis', we have considered. In the case of the neural network simulation model this seemed to be satisfying. Now it conflicts with the chemist's point of view. The main principle used in the molecular mechanics approach, the principle of steric strain, is used as the major explanatory principle among chemists. But perhaps this use is just erroneous, due to a theoretical simple-mindedness chemists and the *ab initio* method is a more promising candidate for a correct explanatory model.

Is there an ab initio *explanation?*

The last step of an *ab initio* calculation, if computationally tractable is a conformational search procedure, similar to the molecular mechanics case. Consequently a full derivation of structural features by quantum chemical methods cannot be exclusively of causal structure. Even the first part of the *ab initio* method, which consists in solving the Schrödinger equation for a specific nuclei arrangement structure, cannot count as a causal explanation. The best one can do in an *ab initio* calculation is start with the *time-independent* Schrödinger equation. Therefore, the whole model from the beginning does not refer to the evolution of the system in *time.* It only allows us to calculate *stationary* points on the Born–Oppenheimer hypersurface.

Hence, quite apart from the general problem of giving a causal account of quantum mechanical explanations [see for instance (Salmon 1998, 23)], *ab initio* methods do not even appeal to causal processes or causal mechanisms. The causal approach to explanation will not work.

What of DN-explanation? Is not the quantum chemical derivation of a molecule's structure a classical example of a scientific explanation in the sense of the DN-model? After all, in an *ab initio* calculation we start with the Schrödinger equation, undoubtedly the central law in quantum physics add some boundary conditions specifying number, kind, spin and charge of the particles in the system and we obtain by mathematical derivation a wavefunction for the ground state of the electrons in a molecule with a fixed nuclei geometry. The use of a proper scientific law and the strictly mathematical derivation would seem to secure the *nomic* and the *deductive* character of such a procedure. But a closer look reveals difficulties. Even for the solution of the time-independent and non-relativistic Schrödinger equation a tremendous number of simplifications and approximations have to be introduced. These approximations are driven empirically. The success of the whole procedure is dependent on further, empirically determined, decisions. For example, nothing in the Schrödinger equation or the boundary conditions tells us which set of basis functions we should use and what should be the starting coefficients of the self-consistent field procedure. Rules of thumb for these decisions are entirely derived by the success of *foregoing* calculations and an appropriate choice of these quantities is decisive for the success of the whole calculation.

The situation becomes even more problematic because we have to specify an appropriate starting conformation to make sure that the geometry optimization will not get stuck in a local minimum on the Born–Oppenheimer hypersurface. More empirical knowledge, in form of the chemist's intuition and/or some experimental results about typical bond lengths and angles, enters the whole enterprise. In a proper DN-explanation *this knowledge should stand at the end of a deduction and not at its beginning*. It is an ironic consequence of this that the empirical validation of *ab initio* models has not been seen as a validation of the underlying theory but rather as a validation of the introduced approximations. This is how Redhead has put it in his seminal work, *Models in Physics*:

> The underlying theory, the N-particle Schrödinger equation, is believed because of the success of wave mechanics in explaining the properties of simple systems, the hydrogen atom, the hydrogen molecule and so on. For more complicated systems we use approximations—comparison with experiment is then regarded as a test of the approximation, not of the theory. We are doing here a remarkable sort of empirical mathematics, testing approximations not by examining the situation mathematically but by doing experiments! Of course the fact that the Schrödinger equation gives a good account of

> simple systems does not as a matter of pure logic entitle us to infer its accuracy in treating complicated systems, so there is an extra inductive step here leading us from simple to complicated systems. (1980, 156).[20]

This also puts in doubt the nomic character of the *ab initio* procedure. If, as it were, the final model that is used in a *ab initio* calculation, i.e., the Hartree–Fock operator applied to the LCAO model is not deducible from the primordial model, i.e., the Schrödinger-equation with the original Hamiltonian, then it seems that the "law" that is *de facto* operating in the *ab initio* derivation is contained in the Hartree–Fock model. But this simplified model is based on assumptions that in part are contradictory to the assumptions of the original model. These assumptions include:

(i) considering an isolated system (i.e., our molecule),
(ii) separating electronic and nuclei movement, and
(iii) splitting up our system into independent single-electron problems.

All contradict a central theorem of quantum physics—that the states of all *existing* particles are in fact correlated. The model used for a quantum chemical treatment of molecules therefore has to rely on a drastically *impoverished theory* (Redhead 1980, 146). As a consequence this impoverished model is not valid for all molecules under all circumstances but is valid only in combination with some "ceteris paribus" conditions, i.e., it is valid only "from case to case." The exact scope of this ceteris-paribus condition, in turn, again, is not deduced from the original model but is the result of an empirical investigation. It is one of the main thesis of Cartwright's work (1983, especially essay 2, 3, and 8) that these "ceteris-paribus generalizations" are not to be considered as "laws" in the classical sense. So *ab initio* methods neither fit the DN-model nor the causal model of explanation. In the case of the DN-explanation there is even double failure. Neither the deductive nor the nomological part of a DN-explanation is provided. At this stage even those chemists who were reluctant to admit that the molecular mechanics approach bears any explanatory relevance are faced with a *tu quoque*. Their favorable model, the quantum chemical approach is as useless in bringing about explanations in the senses required by either of the two accounts considered here. This last result also is at odds with the way the *ab initio* approach is judged in practice.

CONCLUDING REMARKS

Thus, from the perspective of either of two major and widely received accounts of scientific explanation, chemists do not provide proper explanations for one of their most central concepts, the concept of molecular structure. These classical accounts of scientific explanations were part of a general thread in the philosophy of science that pinpointed the search for scientific explanations as one of the major tasks of science in general [see for instance (Nagel 1961)]. Hence, taking these accounts seriously, it seems that we have come back to Kant, who maintained that a great part, if not the whole of chemistry does not deserve to be classified as scientific (Kant 1984). We have finally reached a point where it becomes clear that philosophy of science, even though having emerged from the anti-metaphysical positivistic school of the early 19th century, sometimes shares the metaphysical "*Weltfremdheit*" it criticized in its predecessors.

I think that the considerations above show that such a position is self-defeating. Chemistry *is* a science. Chemists *do* explain molecular structures. At least that is what chemists think. And who should know better? If we take this as a starting point it seems that we have to revise the philosophical account of scientific explanation. As I have argued elsewhere a good starting point is the pragmatic model of explanation of van Fraassen (1977).[21] According to this account, explanations are answers to why-questions. What counts as relevant to a why-question depends on its pragmatic embedding. It depends on the context in which the question was asked and on the interests of the questioner. This appears to be promising for the situation in chemistry. The choice of which molecular modeling method to use depends on the size of the molecule and on the different aspects of a molecule we are interested in. Each model represents responds to relevance relations that specify how to answer a why-question. If we were asked for the structure of a large molecule, we use a molecular mechanics model to give an appropriate answer. Structural features of the small methane molecule are better answered by referring the *ab initio* approach. Nye has nicely illustrated this pragmatic feature in the chemist's model-choice:

> Hoffmann and Laszlo have suggested that chemistry is more like music than like mathematics in its parting with deductive rigor. To "represent" or to "explain" a molecule, like camphor, for example, the chemist may call on different representations . . . Any of these figures, or a ball and stick model, a space filling model, or an electronic-distribution model, may be best for the occasion. Which of these representations . . . is right? Which is the molecule? . . . All of them are models, representations suitable for some purposes, not for others . . .
>
> Like the woodcut suite of Hokusai entitle "Thirty-six Views of Mount Fuji" (or like Claude Monet's multiple paintings of the cathedral at Rouen) there is no one rigorous answer or explanation to the "nature" of a molecule (1994, 279).

How the relevance relation represented by our model is structured, if it is deductive, probabilistic or causal, does not matter. Having said this, it also seems to me that van Fraassen's model is too liberal on this point. Recall that chemists do not regard neural network models explanatory. In van Fraassen's original account there is no way of ruling out such examples. In addition the evaluation criteria provided by van Fraassen, further criteria have to be introduced. These would allow for the comparative evaluation of different relevance relations for explanations of the same type of explanandum. Roughly speaking, what counts is first how many molecules could be successfully treated by the corresponding model (the *descriptive scope* of the model in respect to the specific type of explanandum) and second in how many questions for different types of explananda the model could be successfully referred to (this has been labeled the "type-unificatory" aspect).

Taking this into account the neural network model clearly appears inferior to the other two models discussed above. This is because the relevance relation represented by this artificial intelligent model is highly restricted. It can only be used to find answers to questions for only specific structural features (e.g., the folding pattern) of only a specific kind of molecules (proteins). *Molecular mechanics* and *ab initio* fare better in that respect. But applying these criteria to a comparison of these latter two models leads to a stalemate. Molecular mechanics models allow for a much wider descriptive scope than *ab initio* models if explanations for the structure of a molecule are required. *Ab initio* models, on the other hand, allow for a much higher degree of type unification; they enable the chemist to ask questions not only about structural properties but also about nearly all other kinds of physical properties of a molecule in general. I take this result to nicely reflect the pragmatic character of chemical explanations. Which explanation to prefer depends on the weighting scheme applied to these criteria. For a typical chemist, a "*voyeur of molecules*" as Hoffmann (1995) has it, the descriptive scope may be more important than for a theoretical chemist. The latter, on the other hand may rather emphasize the fact that his model can correctly explain a great manifold of physical properties, though only for a very limited number of smaller molecules. There is no a priori way to designate one or the other preference as right. As I said in the beginning chemists and physicist do not only start at different points they start in different races. But which race it is and what the rules are is not so much a matter of philosophy of science—but a matter of scientific practice itself.

NOTES

1. In the original text of the Nietzsche quote the word "conception" in "*we have perfected the conception . . . , but have not got a knowledge of what is above and behind the conception*" is "*Bild*." Literally "*Bild*" is translated with "*picture*" or "*drawing*."
2. In the early days of *ab initio* calculations even such a local optimization was impossible. To calculate the structure of such a simple molecule like methane, Janoschek et al. in their pioneer work had to use regression analysis in order to determine the closest minimum on the BO-surface. Only four stationary points on the BO-hypersurface lying around to the experimentally observed C–H distances were calculated and the minimum of the regression curve was taken as the result (Janoschek et al. 1967). This illustrates again that the success of an *ab initio* calculation largely depends on empirically determined starting structures.
3. A thumb rule is that only molecules containing less than 100 atoms can be treated by *ab initio* models; see for instance (Boyd 1990).
4. For classical parameterization approaches see for instance Bowen and Allinger (1991). For the use of a global optimization method for force field parameterization see Hunger et al. (1998).
5. See for instance Burkert and Allinger:

   > "One might ask at this point why we should be interested in calculating the structures and energies of molecules, since these can be determined by experimental methods? The answer to the question is at least twofold. *First, there is the matter of understanding*. A calculational method that gives good results contains within it a potential for understanding that does not come from a collection of experimental results" (1982, 9; emphasis added).

6. A "Hinton-diagram" is a diagram of the network's weight matrix.
7. The classical example is a sentence like "All men sitting on bank x in Hyde Park are bold." Hempel remains rather reluctant to specify what counts as a real scientific law and later hints at Goodman's criterion of the projectibility of the predicates as necessary for a proper scientific law (1956).

8. Later, influenced by Dowe's critique, Salmon revised this account and now maintains that the criterion for a causal process is its capability to transmit a conservative quantity (Salmon 1994). Applied to our example, the cave prisoners, in an anticipation of modern physics 2000 years later, would have had to measure the total energy of the shadow enemy at t_1 and t_2 to find out if it was a real causal process or not.
9. It should be mentioned however that another account of explanation, Churchland's pattern-activation model, explicitly refers to such back-propagation networks as promising cognitive models of explanatory processes (1989, 205). I discuss this remarkable tension between applying scientists and cognitive scientists on the explanatory value of Neural Networks in my *Why Churchlands Model of Explanation Does Not Explain* (unpublished manuscript).
10. See for instance Comba:

    "Isolated force field parameters are generally of little value. A force field parameter set is only reliable as a whole and together with a specified set of potential energy functions" (1995, 12).

    To this point also compare Bowen and Allinger (1991), Burkert and Allinger (1982), and Kunz (1991).
11. The same applies, in different form though, to a molecular dynamics procedure. The random elements here are the starting point and the initial velocities of the atoms. More or less arbitrarily chosen is the size of the time steps Δt.
12. The word "actual" is important, because one could object that the fact that the minima are stable in fact *is* deduced from statistical mechanics and that the conformational search is just a vehicle to find these minima. The search does not affect the deductive character of the derivation. (I am grateful to Ansgar Beckermann for this objection) Yet, this is the wrong perspective. My claim is that, in order to find out what the term "explanation" means one has to have a close look at what scientist *actually do* when they explain. In the case of Molecular Mechanics they do not derive, they *search*.
13. Of course we may have an analytical solution that is able to reproduce the system's state at any *point* in time. To calculate a whole trajectory however we are restricted to the above mentioned approach of piecemeal agglomeration.
14. I am deliberately sticking to the Molecular Mechanics model here. As we all know, from a quantum physical perspective, the process of a molecule absorbing and exchanging energy is in fact irreducibly discrete. The fact that quantum physics poses some severe problems for the process account of causality, however, is readily admitted by Salmon.
15. This line of critique is similar to a critique put forward by Kitcher (1989). As Kitcher pointed out in the case of marking a projectile with chalk there is a great number of interactions between the mark and the air molecules taking place. The above described case, however, is interesting in so far as it also applies to Salmon's objection to Kitcher's critique which, roughly speaking, maintained that the mark transfer criterion would be still applicable when the process was divided in microscopic parts (Salmon 1994). In the case of a chemical reaction, however, we can get as microscopic as we want. Still Kitcher's point remains a valid one.
16. I already mentioned the severe problem of having an appropriate measure for the similarity between two worlds. Here, I am, with Lewis, assuming that diversity in facts weighs less than diversity in laws. Hence, the closest possible world to W_1 in which A is true is a world in which as few facts as necessary have to be changed to make A true. This is simply the world where the geometrical arrangement of the atoms is preserved but one atom is changed by the mass of one neutron.
17. See for instance Rappé and Casewit 1997:

    A correct description of 'the' molecular structure ... for a molecule with several conformations must comprise a proper weighting of all the conformations .... Statistical mechanics provides the Boltzmann equation, which is used to obtain the probability or population of each conformation (1997, 22).

18. In fact, as already mentioned earlier, the molecular dynamics approach is not even a necessary part of such an explanans, but just one of many other algorithms that might be used to overcome the delicate problem of finding the energetically most favorable conformation (e.g., Monte Carlo, Grid Search or Genetic Algorithms).

19. A quite similar point has been made by Woodward (1989).
20. A typical example is an above cited work in which the modeling of the structure of the methane molecule is used not to validate the quantum chemical approach but to test how many basis set functions are sufficient to get a correct result (Janoschek et al. 1967). To this point also compare Nye:

> Quantum wave mechanics gave chemistry a new understanding, but it was an understanding absolutely depending on purely chemical facts already known. What enabled the theoretician to get the right answer the first time, in a set of calculations, was the experimental facts of chemistry, which, Coulson wrote, 'imply certain properties of the solution of the wave equation, so that chemistry could be said to be solving the mathematicians' problem and not the other way around' (1992, 277).

21. See Hunger "*The chemistry of pragmatics—van Fraassen's model of explanation reconsidered*," unpublished manuscript.

REFERENCES

Allinger, N.L. 1977. Conformational Analysis. 130. MM2. A Hydrocarbon Force Field Utilizing V_1 and V_2 Torsional Terms. *Journal of American Chemical Society* 99: 8127.

Bowen, J.P. and Allinger, N. 1991. Molecular Mechanics: The Art and Science of Parameterization. In: Lipkowitz, K.B. and Boyd, D.B. (eds.), *Reviews in Computational Chemistry*, Vol. 2. New York: VCH Publishers.

Boyd, B. 1990. Aspects of Molecular Modeling. In: Lipkowitz, K.B. and Boyd, D.B. (eds.), *Reviews in Computational Chemistry*. New York: John Wiley & Sons.

Burkert, U. and Allinger, N. 1982. *Molecular Mechanics*. Washington: American Chemical Society.

Cartwright, N. 1983. *How the Laws of Physics Lie*. New York: Oxford University Press.

Chandonia, J.-M. and Karplus, M. 1995. Neural Networks for Secondary Structure and Structural Class Prediction. *Protein Science* 4: 275.

Christen, H.R. and Vögtle, F. 1988. *Organische Chemie*, Band 1. Frankfurt am Main: Salle & Sauerlander.

Churchland, P.M. 1989. *A Neurocomputational Perspective: The Nature of Mind and the Structure of Science.* Massachusetts: MIT Press.

Coffa, Alberto J. 1974. *Hempel's Ambiguity.* Synthese 28, pp. 141–163.

Dirac, P.M.A. 1929. Quantum mechanics of many electron systems. *Proceedings of the Royal Society* (London) 123: 714.

Dowe, P. 1992. Wesley Salmon's Process Theory of Causality and the Conserved Quantity Theory. *Philosophy of Science* 59: 195.

Duhem, P. 1991. *The Aim and Structure of Physical Theories*. Weiner, P.P. (trans.). Princeton, New Jersey: Princeton University Press.

Gasteiger, J. and Zupan, J. 1993. *Neural Networks in Chemistry*. Weinheim: John Wiley & Sons.

Goodman, N. 1956. *Fact, Fiction and Forecast*. Indianapolis: Bobbs-Merrill.

Hehre, W.J., Radom, J., Schleyer, P.v.R., and Pople, J.A. 1986. *Ab initio Molecular Orbital Theory*. New York: John Wiley & Sons.

Hempel, C.G. 1977. *Aspekte wissenschaftlicher Erklärung.* Berlin: Gruyter.

Hempel, C.G. and Oppenheim, P. 1948. Studies in the Logic of Explanation. *Philosophy of Science* 15: 135.

Hoffman, Ronald. 1995. *The Same and Not the Same.* New York: Columbia University Press.

Hunger, J., Beyreuther, S., Huttner, G., Allinger, K., Radelof, U., and Zsolnai, L. 1998. How to Derive Force Field Parameters by Genetic Algorithms?—Modeling TripodMo$(CO)_3$ Compounds as an Example. *Journal of Computational Chemistry* 20: 455, (1999).

Janoschek, R., Diercksen, G., and Preuss, H. 1967. Wellenmechanische Absolutrechnungen an Molekülen und Atomsystemen mit der SCF-MO-LC(LCGO) Methode/VI. Das Methan (CH_4). *International Journal of Quantum Chemistry* 1: 373.

Jorgensen, W.L. and Tirado-Rives, J. 1988. The OPLS Potential Functions for Proteins. Energy Minimization for Crystals of Cyclic Peptides and Crambin. *Journal of American Chemical Society* 110: 1657.

Kant, I. 1984. *Metaphysische Anfangsgründe der Naturwissenschaft*. Berlin: Harald Fischer Verlag.

King, R.D. 1996. Prediction of Secondary Structure. In: Steinberg, M.J. (Hrsgb.), *Protein Structure Prediction—A Practical Approach*. Oxford: Oxford University Press.

Kitcher, P. 1989. Explanatory Unification and the Causal Structure of the World. In: Kitcher, P. and Salmon, W.C. (Hrsgb.), *Scientific Explanation (Minnesota Studies in the Philosophy of Science 13)*. Minneapolis: University of Minnesota Press.

Kunz, R.W. 1991. *Molecular Modelling für Anwender.* Stuttgart: Verlag.

Lewis, D. 1973. Causation. *Journal of Philosophy* 70: 556.

Müller-Herold, U. and Primas, H. 1984. *Elementare Quantenchemie.* Stuttgart: Tuebner Verlag.

Nagel, E. 1961. *The Structure of Science*, Routledge & Kegan Paul, London.

Quian, N. and Sejnowski, T. 1988. Predicting the Secondary Structure of Globular Proteins Using Neural Network Models. *Journal of Molecular Biology* 202: 865.

Rappé, A.K. and Casewit, C.J. 1997. *Molecular Mechanics Across Chemistry.* Sausalito: University Science Books.

Redhead, M. 1980. Models in Physics. *British Journal for the Philosophy of Science* 31: 145.

Salmon, W.C. 1977. A Third Dogma of Empiricism. In: Butts, R.E. and Hintikka, J. (eds.), *Basic Problems in Methodology and Linguistics*. Dordrecht: Kluwer Academic Publishers.

Salmon, W.C. 1981. Causality: Production and Propagation. In: Asquith, P.D. and Giere, R. (eds.), *PSA 1980*, Vol. 2. East Lansing: Michigan State University.

Salmon, W.C. 1994. Causality Without Counterfactuals. *Philosophy of Science* 61: 297.

Salmon, W.C. 1998. *Causality and Explanation.* New York: Oxford University Press.

Sklar, L. 1993. *Physics and Chance: Philosophical Issues in the Foundations of Statistical Mechanics*. New York: Cambridge University Press.

van Fraassen, B.C. 1977. The Pragmatics of Explanation. *American Philosophical Quarterly* 14: 143.

Whitehead, A.N. 1929. *Process and Reality.* New York: The MacMillan Company.

Woolley, R.G. 1985. The Molecular Structure Conundrum. *Journal of Chemical Education* 62(12): 1082.

Woolley, R.G. and Sutcliffe, B.T. 1977. Molecular Structure and the Born–Oppenheimer Approximation. *Chemical Physics Letters* 45(2): 393.

CHAPTER 8

PROFESSIONAL ETHICS IN SCIENCE

JEFFREY KOVAC
Department of Chemistry, University of Tennessee, Knoxville, TN 37996-1600; E-mail: jkovac@utk.edu

INTRODUCTION

In 1982, William Broad and Nicholas Wade published a book entitled *Betrayers of the Truth: Fraud and Deceit in the Halls of Science* which brought the issue of ethics in science to broad public attention (Broad and Wade 1982). Since then much has been written on scientific ethics, including a two-volume report by the National Academy of Sciences (Panel on Scientific Responsibility and the Conduct of Research 1992, 1993). While most of the public scrutiny has been directed at incidents of scientific misconduct in the biomedical sciences, some authors have begun to study ethics in the physical sciences (Dyson 1993; Coppola and Smith 1996; Kovac 1996, 1998, 1999, 2000, 2001; Coppola 2000). In this chapter, I outline a theory of scientific ethics based on two assumptions. First, scientific ethics is a subset of professional ethics, the special rules of conduct that should be followed by people engaged in one of those pursuits traditionally called professions (Callahan 1988). Consequently, understanding scientific ethics requires understanding science as a profession (Davis 1998). Second, ethics is an integral part of science. Questions of ethics arise as a natural part of the day-to-day internal practice of science and in the relationship between science and society. At a deeper level, I will argue that epistemology and ethics are so tightly interconnected that scientific rationality is dependent on the moral character of scientists. Since I am a chemist, whose scientific research lies at the interface of physics and chemistry, my focus will be on the physical sciences, especially chemistry.

Clearly, it is impossible to fully develop a theory of professional ethics in a brief chapter. My goal is to sketch the broad outlines of such a theory raising what I think are the most important issues and providing a preliminary discussion of each.

THREE MEANINGS OF ETHICS

To begin, we must distinguish among three uses of the word "ethics." The first and most common meaning is ordinary or common morality, those more or less universal rules of behavior that we expect every rational person to obey. The norms

D. Baird et al. (eds.), Philosophy of Chemistry, 157–169.

of ordinary morality are so broadly shared that they form a stable social consensus (Bok 1995; Beauchamp and Childress 2001). Ordinary morality can be summarized in terms of fundamental moral principles that express the general values that underlie more specific rules governing behavior. These moral principles include respect for autonomy, non-maleficence (avoid causing harm), beneficence (provide benefits), and justice (fair distribution of goods and services), among others. Moral rules are more specific guides to conduct, which can be expressed negatively ("Don't lie") or positively ("Tell the truth") (Gert 1988). The rules (and permissible exceptions) of ordinary morality are learned from parents, ministers, schoolteachers, and others who influence us as we grow into adulthood. We can assume that most scientists have a well-developed sense of ordinary morality.

The second meaning of ethics is ethical theory, a branch of philosophy. Ethical theory is the attempt to put ordinary morality into a systematic framework (Beauchamp 1991; Rachels 1999). An ethical theory is similar to a scientific theory, though there are differences. A scientific theory is supposed to provide some sort of explanation of a class of natural phenomena. An ethical theory is a systematic presentation of the basic components of ethics derived from an integrated set of principles that is supposed to provide ultimate reasons for moral decisions. The "experimental facts" of an ethical theory are what William Gass calls clear cases, behavior that is clearly right or wrong (Gass 1980). Some actions, such as torture or the abuse of children, are so clearly wrong that we would regard a person, who condoned them as morally deficient and would reject any moral principle that approved them. An ethical theory must account for these facts, but is also prescriptive in a way that a scientific theory can never be.

The third meaning, the one we are concerned with in this chapter, is professional ethics. Professional ethics is the special rules of conduct adhered to by those engaged in pursuits ordinarily called professions, such as law, medicine, engineering, and science. Professional ethics is specific. Legal ethics applies only to lawyers (and no one else); scientific ethics applies only to scientists. Professional ethics governs the interactions among professionals and between professionals and society. In many cases, it requires a higher standard of conduct than is expected of ordinary people, but it must be consistent with ordinary morality and with appropriate moral theories.

In recent years there has been an enormous interest in professional ethics, stimulated primarily by the complicated ethical questions raised by modern medicine (Beauchamp and Childress 2001). After Watergate, legal ethics became more prominent in legal education and recent changes in the ABET accreditation standards for engineering programs have resulted in active scholarship in engineering ethics (Davis 1998; Harris, Pritchard, and Rabins 1995). Scientists have lagged behind the other professions in understanding the importance of ethics in science, but there are signs that this attitude of neglect is changing (Resnik 1998; Kitcher 2001).

While there are a number of ways of understanding science, I have found that viewing science as a profession has been the most useful way to understand the ethics of science. While scientists do not have the autonomy of the practitioners of the classic learned professions, such as physicians and lawyers, in the past century the individual scientific disciplines, including chemistry, have come to regard themselves as professionals and have organized societies to protect their professional interests as well as further scholarship (Knight and Kragh 1998). For example, the American Chemical Society certainly regards chemists as professionals, both promulgating professional standards and protecting the interests of its members. Even if one cannot regard science, generically, as an independent profession, individual disciplines, especially chemistry, are certainly better examples. While scientists might be regarded as "captive" as opposed to autonomous professionals, many other professions, including medicine, law and engineering, are practiced by people who are employed by larger organizations. In the following section, I will outline the concept of a profession and show that science fits the model reasonably well.

THE CONCEPT OF A PROFESSION

A profession is more than a group of people engaged in a common occupation; it derives from two bargains or contracts: one internal and one external. The internal bargain governs the interactions among members of the profession while the external bargain defines the relationship of the profession to society. Professions develop through a historical process of self-definition. For example, although chemistry is an ancient science that began with the crafts of metal working, ceramics, dying, and tanning, it really began to define itself as an independent science in the 18th century and matured to a fully recognized independent field by the middle of the 19th century when chemists began to organize scientific societies (Brock 1992; Knight 1992). The American Chemical Society was founded in 1876 by a group of New York chemists who had attended the first American national meeting of chemists at Northumberland, Pennsylvania, held in 1874 to commemorate Joseph Priestley's isolation and characterization of oxygen in 1774 (Reese 1976). The ACS was originally founded as a scientific, rather than a professional, society. Its purpose was mainly to encourage research by holding scientific meetings and publishing journals, although some thought was also given to the training of students and improving the public image of chemistry. Professionalism became an explicit concern of the ACS in the 1930s largely due to the economic pressures of the Great Depression. Among other actions, the ACS developed standards for an "approved" degree in chemistry analogous to those in medicine and law (Committee on Professional Training 1999). Partly because of the large number of chemists employed by private industry, professionalism is a continuing concern of the ACS. In fact, there is considerable tension in the society between the interests of industrial and academic scientists.

Similar organizations of chemists were founded in Europe during the 19th century. In England, The Chemical Society was founded in 1841 and the Faraday Society, which was to be a bridge between science and technology, particularly electrochemistry, arose in 1902. The Deutche Chemische Gesellshaft was founded on the British model in Berlin in 1867. By 1900, chemists internationally were well organized. Similar patterns can be traced for the other sciences (Knight and Kragh 1998).

The various sciences do differ from the classic learned professions, medicine and law, because they are less client-focused. Scientists do not "hang out a shingle" and practice independently, but as they have matured, each of the disciplines, and sometimes subdisciplines, has developed an internal code of practice analogous to those of the classic learned professions and a professional identity.

INTERNAL CODE OF PRACTICE

As part of the process of self-definition, members of a profession must agree on an internal code of practice and negotiate the relationship between the profession and society. The internal bargain consists of several parts: standards of education and training, a formal or informal certification or licensing procedure, and a code of practice, which might include a formal code of ethics. Some professions, such as law, medicine, and engineering have well-defined standards of education enforced by accreditation boards. In science, the standards are more informal, although the American Chemical Society has a Committee on Professional Training, which develops standards for a professional baccalaureate degree (Committee on Professional Training 1999). Chemistry is the only scientific discipline with an explicit professional undergraduate degree. Although other disciplines do not have formal written standards, there is remarkable informal agreement on what comprises an acceptable education in any scientific field. This agreement can be seen by examining a selection of college catalogues and the contents of the textbooks used to teach the various courses.

The standards of training in science have evolved over the years, but currently an earned doctorate from a reputable university is the usual requirement for professional status in most fields. Scientists without a doctorate, however, can be recognized after publication of credible research in refereed journals. Formal certification is not common in the sciences, so recognition comes from accomplishments rather than a professional license. There are a few exceptions, such as clinical chemistry, mostly in fields closely related to medicine.

There have been a number of attempts to formulate the internal code of practice of science. Perhaps the most famous is that of Robert K. Merton (1973). Merton identified four principles of scientific practice:

(i) *Universalism:* Truth claims must be evaluated using pre-established impersonal criteria.
(ii) *Communism:* (Other authors have preferred the term communality.) This is the obligation of public disclosure of scientific findings. Science is public knowledge (Ziman 1968).
(iii) *Disinterestedness:* The advancement of science is more important than the personal interests of the individual scientist.
(iv) *Organized skepticism:* All scientific truth is provisional and must be judged based only on the evidence at hand. Scientific conclusions are always open to question. (This is similar to Popper's famous principle of falsifiability [1965].)

Merton's list has been modified and expanded by later writers to include ideas, such as objectivity, honesty, tolerance, doubt of certitude, selflessness, individualism, rationality, and emotional neutrality (Barber 1952; Cournand and Meyer 1976; Zuckerman 1977).

Merton's principles are analogs of the broad moral principles, such as justice, mentioned earlier. They are the basis of more specific moral rules that govern the day-to-day practice of science. Some of these more practical rules include:

(i) Experimental and theoretical procedures must be reported accurately so that independent investigators can replicate the work if they so choose.
(ii) The data reported must be complete and correct and the limits of error noted. Scientists are not supposed to suppress data that do not agree with their expectations.
(iii) The interpretation of the data must be done objectively. Prior expectations should not interfere with data analysis and non-scientific factors, such as politics or the expectations of the funding agency, should not influence the analysis.
(iv) Credit should be given where credit is due. Scientists are expected to cite previous work where appropriate and to give credit to those who have aided in the research. Conversely, it is assumed that all the authors of a scientific paper have contributed to the research.

A longer but similar list of ethical principles for science has recently been given by Resnik (1998).

In addition to these norms, there are more specific research practices that can vary depending on the discipline. Some of the criteria that distinguish "good physics" are quite different in kind from those that distinguish "good biology." Learning the techniques and standards of research in a particular discipline is a major part of the graduate education of a scientist. It is what Thomas Kuhn has called normal science (1962).

It is important to recognize that the internal code of science has evolved over time. While the broad principles of the code go back to the early days of the Royal Society of London, specific details and norms of scientific practice have changed significantly since Boyle and Newton (Holton 1994). Therefore, recent charges of scientific fraud directed at historical personages should be regarded with some skepticism.

EPISTEMOLOGY AND ETHICS

On one level, the internal code of practice of a science can be regarded merely as a social convention. On a deeper level, however, it has both ethical and epistemological significance. In several articles, John Hardwig (1985; 1991; 1994) has argued that in science (and other fields) epistemology and ethics are intimately intertwined. As a chemist, I claim to know things about chemistry and other sciences, which I have not thoroughly studied myself. In some cases, I do not have either the background or the ability to follow the detailed arguments that established the knowledge I claim as my own. As Hardwig points out, I believe many things are true merely because I trust that the scientists who report them did the appropriate experiments or theoretical calculations and interpreted them correctly; I accept their testimony as truth. Therefore, my knowledge depends on the moral character of other scientists. While I think all scientists tacitly understand how dependent our knowledge is on the integrity of others, this issue is rarely discussed.

Hardwig's analysis is based on the idea of epistemic dependence. Scientists are dependent on each other for knowledge, which leads to an unequal power relationship; one person becomes the "expert," the other a "layperson." Each has ethical responsibilities. Experts must be careful in what they say and laypersons must be careful in evaluating and using the information they receive. Further, the community of experts has ethical responsibilities to ensure that its members behave responsibly. Hardwig presents a preliminary set of maxims, which clarify the ethics of expertise. These maxims are organized into four categories: maxims for experts, maxims for those who rely on experts, maxims for the community of experts, and maxims for a society or group that relies on experts. Many of these maxims are consistent with the informal internal code of practice of science, while others address the appropriate relationship between science and society. For example, among the maxims for experts are the following:

(i) Do not overestimate the scope or certainty of your knowledge, or the inferences that can be validly drawn from it.
(ii) Tell the truth as you see it in your professional judgment, even if you have to tell your employers, clients, or those in power things that they do not want to hear.
(iii) Recognize the human propensity to rationalize.
(iv) Know your own ethical limits.

All of these are part of the internal code of practice of science. As noted above, it is essential for scientists to report data completely and to interpret them objectively. One of the subtle ethical dangers in scientific practice is self-deception, misinterpreting, or overinterpreting data to meet prior expectations. The history of science is filled with examples of self-deception; N-rays and polywater are two of the most famous. Maxims 1, 3, and 4 above speak directly to this danger. Maxim 2 is more directly related to Merton's principle of organized skepticism. It is essential that scientists tell each other the truth, as they see it, even if it means challenging a well-established idea.

The maxims for the community of experts are also important. These include:

(i) Never use rewards and punishments to stifle dissent within the community of experts.
(ii) Take steps to ensure that your members are worthy of the social trust placed in them.
(iii) Recognize the obligation to be a "whistleblower."

From my perspective, probably the most important maxim for the scientific community is maxim 2 because it epitomizes the essence of the bond of trust that should exist within the scientific community and between science and society, which will be discussed more fully in the next section.

Beyond the social aspects of trust in science, there is the question of trusting oneself. As chemist and philosopher Michael Polanyi has so nicely explained, the acquisition of knowledge is a skillful act of personal commitment: the art of knowing,

> To affirm anything implies, then, to this extent an appraisal of our own art of knowing, and the establishment of truth becomes decisively dependent on a set of personal criteria of our own which cannot be formally defined (Polanyi 1964, 70–71).

Although Polanyi does not explicitly say so, among those personal criteria must be ethical standards. Polanyi reinforces Hardwig's maxim for experts: know your ethical limits. As noted above, failure to properly apply personal ethical standards to the act of knowing results can lead to self-delusion. Any theory of ethics in science that does not address the question of the ethics of personal knowledge in science would be, in my view, incomplete.

EXTERNAL BARGAIN: SCIENCE AND SOCIETY

The external bargain addresses the relationship of the profession to society. In general, the profession lays claim to a body of specialized knowledge and skill not easily attainable by the majority of people. In return for a monopoly on the practice of those skills, the profession agrees to use them to serve society and to render professional judgment when asked. For some professions, such as law, medicine, and engineering, the bargain with society is highly structured; parts are even written into law. For science, the agreement is more informal. A brief historical sketch will help to clarify the relationship of science and society.

Perhaps the first agreement between science and government came with the establishment of the Royal Society of London by Charles II. The Royal Society was given the right to publish without censorship and pursue the new specialty of natural philosophy. In return, the Royal Society was to avoid the study of politics, morality, and religion. In the words of Robert Hooke, the "Business and Design" of the society was "To improve the knowledge of natural things, and all useful Arts, Manufactures, Mechanics, Practices, Engynes, and Inventions by Experiments (not meddling with Divinity, Metaphysics, Moralls, Politicks, Grammar, Rhetoric or Logick)" (Proctor 1991, 33). The gentlemen who founded the Royal Society established the early standards for scientific practice. The central question was what should be considered scientific truth? Robert Boyle was a central figure in this development (Shapin 1994).

There was an interesting tension in the early Royal Society between what we would now call pure and applied science and it can be exemplified by Boyle and Hooke. While Boyle was the paradigm of the Christian gentleman scientist, Hooke, the curator of experiments, was considered the "greatest mechanick this day in the world" (Shapin 1989). Boyle's disinterested gentility contrasted sharply with Hooke's protection of patent rights. While both men were more complex than this polarized comparison suggests, the division between pure and applied science is a key issue in science policy, and the tension between the scientific ideal of open communication and personal economic gain is an important contemporary issue in professional ethics in science (Baird 1997). While the practical aspects of science have always been important, the so-called German model of pure research was the dominant theme in the development of science in the U.S. Science in the universities and research institutes was pursued for its own sake. Practical applications were certain to follow as the secrets of nature were revealed. American science could point to outstanding examples of the practical utility of pure science, for example, the work of Irving Langmuir at General Electric and John Bardeen at Bell Laboratories.

The Second World War changed the nature of research in America forever. The Manhattan Project and the development of radar showed how science, with generous government support, could make significant accomplishments in a short time. The new bargain between science and society was outlined in two postwar reports: Vannevar Bush's *Science: The Endless Frontier* (1990) and John R. Steelman's *Science and Public Policy* (1947). These reports led to our current system of research funding centered on the National Science Foundation.

The essence of this bargain can be summarized in a few words:

> Government promises to fund the basic science that peer reviewers find most worthy of support, and scientists promise in return that the research will be performed well and honestly and will provide a steady stream of discoveries that can be translated into new products, medicines or weapons (Guston and Kenniston 1994, 2).

While there has been considerable recent discussion of the suitability of this form of the contract (Brown 1992; Guston 1999), essentially all practicing scientists would agree with this statement of the bargain between science and society, at least as an ideal.

A provocative and useful way of thinking about science and society has recently been developed by Donald Stokes in his book, *Pasteur's Quadrant* (1997). Traditionally, we have thought of pure and applied science as being the two ends of a linear continuum. Stokes points out that this way of classifying science is neither accurate nor particularly useful in thinking about science policy. There is a lot of science that cannot be neatly classified either as pure or applied. He cites the work of Pasteur as one example. While much of Pasteur's research was directed at fundamental questions in what has come to be called microbiology, it was inspired by very practical problems raised by medicine or by French industry. To account for this "use-inspired basic research," Stokes has developed a two-dimensional classification scheme. One axis classifies the research based on whether or not it is directed at fundamental scientific questions. The second axis asks whether or not the research is inspired by a well-defined use or application. The two axes define four quadrants. The first, which Stokes calls Bohr's quadrant, encompasses research that is directed at fundamental understanding and has no immediate use: classic pure research. Bohr's work on quantum mechanics and atomic structure is an example. Pasteur's quadrant houses research that aims at fundamental understanding in the context of a well-defined application, what Stokes calls use-inspired basic research. The third quadrant, named for Edison, is applied research where there is no concern for deepening our fundamental understanding of nature. This is what has usually been called applied research. The fourth quadrant, which has no name, involves research that is neither of particular use, nor does it aim at fundamental understanding. An example might be the *Peterson's Guide to Birds*, which is a systematic collection of information, essential to the bird watcher, but designed neither to provide evidence for the deeper theories of biology nor to meet an important societal need.

Stokes's scheme is particularly interesting, because much of contemporary science is firmly in Pasteur's quadrant. As stated by Guston and Keniston, much of the federal funding for science is based on the expectation that research will provide useful products and processes. Increasingly, research proposals are judged on the potential applicability of the results and university scientists are encouraged to apply for patents and start small companies to commercialize their discoveries. Many of the ethical tensions in contemporary science are a result of the emphasis on doing research in Pasteur's quadrant.

A MORAL IDEAL FOR SCIENCE

On one level, the internal code of practice and the code of ethics of a profession are the result of a social contract entered into by the members of the profession and tacitly or formally ratified by society. On another level, it seems that professional ethics has a deeper source. In examining the authority of a professional code, Michael Davis suggested that professions are based on a moral ideal, an ideal of service that goes beyond the requirements of ordinary morality and law, and the demands of the market (Davis 1987). The moral ideal for attorneys is the pursuit of equal justice under law; for physicians, it is curing the sick, protecting patients from disease and comforting the dying. The existence of a moral ideal fits with the old idea of a profession as a calling. If members of a profession share a moral ideal, then the internal code of practice and code of ethics, which develops out of that ideal, have an authority that goes beyond mere social convention or fear of sanctions; they represent the core values of the profession.

Can we identify a moral ideal for science? This is a complicated question because science is not a monolith. The disciplinary and professional culture of chemistry is quite different from that of physics or biology. The moral ideal of an academic scientist working in Bohr's quadrant might differ from the industrial scientist working in Pasteur's or Edison's quadrant. While I do not yet have a complete answer, I have identified two important parts of a moral ideal for science: the habit of truth and the ideal of the gift economy.

As I have written elsewhere, I think an essential part of the moral ideal of science can be found in Jacob Bronowski's book, *Science and Human Values*, "the habit of truth" (1956). Science is the dispassionate search for deep knowledge of the natural world, what Einstein called, "the secrets of the Old One" (French 1979, 275). The best scientific research is driven by an insatiable curiosity about the way the world works. And because scientific knowledge is severely constrained by experiment, scientists are bound by what Richard Feynman called "a principle of scientific thought that corresponds to a kind of utter honesty—a kind of leaning over backwards" (1985, p. 341).

Truth matters in science, at the very least for practical reasons. Scientific principles underlie engineering and other practical pursuits. A false physics will result in unsafe bridges and buildings. Lysenko's incorrect theory of inheritance resulted in a disastrous Soviet agricultural policy that led to widespread hunger because of crop failures. At a deeper level, science is a tightly coupled intellectual system. For example, the principles of thermodynamics are important in the understanding of phenomena as disparate as the efficiency of steam engines and the equilibrium of chemical reactions. Scientific theories are tightly constrained by both experimental facts and their relationship to other theories. Finally, as Susan Haack has so persuasively argued, the pursuit of truth is the hallmark of any intellectually honest pursuit (Haack 1997).

The second part of the moral ideal concerns the relationships between scientists: the principle of the gift economy. Because scientific research is so difficult and because science is public knowledge, the scientific community is bound by an ideal of completely open communication exemplified by the gift economy (Hyde 1979).

The concept of a gift economy is best introduced by contrasting it with the commodity economy, which governs our day-to-day economic interactions. Transactions in the commodity economy are closed, mutually beneficial exchanges: fee for goods, fee for service. Other than perhaps a well-defined warranty, nothing more is expected in a commodity transaction. No further relationship is expected or, in many cases, desired. None of us expects that the purchase of an item of clothing will lead to a lasting friendship with the sales clerk. Each of us has many such transactions each day.

On the other hand, gift exchanges are intended to open or maintain human relationships. We give gifts to those who are closest to us and those gifts are often very personal. When we give a gift there is no expectation that anything will be given in return, though the recipient might reciprocate to show that he or she also wants to maintain the relationship. There is no *quid pro quo*. Gifts create a community of obligation, initiating, and strengthening social ties.

In its ideal, the scientific community is a gift economy. Individual researchers contribute their gifts, the results of their research, for others to use. In return, they receive the gifts that others have contributed and use them in their own work. There is a continuous cycle of giving. Just as in the gift economy, the most respected scientists are those who contribute the most. Scientists who take from the community and return nothing are not respected, particularly if they turn the gifts of the scientific community into saleable commodities for personal profit. For science to prosper, it is essential that new findings become part of the open literature where they can be tested, used, and extended.

The gift economy is an ideal of pure science, science in Bohr's quadrant, but in Pasteur's and Edison's quadrant there is a tension between the ideal of open communication and the application of scientific knowledge to the production of useful commodities (Baird 1997). This tension goes back to the time of Boyle and Hooke and is particularly important in chemistry, perhaps the most useful of sciences. Successful commercialization of scientific discoveries often requires that information be withheld from the open literature, at least until a patent information is filed, or even kept as a "proprietary knowledge." Scientists who see commercial applications for their discoveries can be tempted to keep them secret rather than contribute them to the community. There are at least two dangers. First, a discovery that is not made public will not be scrutinized, so there is the possibility that the original investigator will try to commercialize something that is incorrect, wasting both time and money. Second, the discovery is not available for others to use and extend, retarding the overall progress of science.

An interesting discussion of the tension between the gift and the commodity economies can be found in the recent book, *Who Owns Academic Work*, by Corynne McSherry (2001). McSherry traces the history of the concept of authorship and the relationship between the commonwealth of ideas and an individual creation. This tension also exists in science where every discovery is based on a wealth of prior research. In contemporary society, intellectual property right is an area where ethics, public policy, and law become strongly entangled (Nelkin 1984).

The concept of science as a gift economy is an important part of the moral ideal of science, but it is incomplete, particularly for science done in Pasteur's quadrant. One of the challenges for a theory of ethics in science is to find additional principles to govern applied science. Such principles will necessarily involve the complicated relationship between science and society.

CONCLUDING REMARKS

The study of professionalism and ethics in science is still in its infancy. In this chapter, I have tried to outline a philosophy of the profession of science and show how this view provides deep insight into ethical questions. While the various scientific disciplines have much in common, each is unique. My own field of chemistry has its own history and traditions, which are different from those of physics or biology. The general study of professionalism and ethics in science will be enriched by examining each of the disciplines individually to see how their internal and external bargains have developed historically, and how they function in contemporary practice. Since ethics is an integral part of science, such studies should reveal much about the history and philosophy of science.

ACKNOWLEDGMENTS

I am grateful to the Camille and Henry Dreyfus Foundation for their support of my work in scientific ethics. The ideas in this article have been developed over many years in discussions with Roger Jones, Brian P. Coppola, Michael Davis, Donald Gotterbarn, and Susan Davis Kovac.

REFERENCES

Baird, D. 1997. Scientific Instrument Making, Epistemology and the Conflict between Gift and Commodity Economies. *Technè: Electronic Journal of the Society for Philosophy and Technology* (http://scholar.lib.vt.edu/ejournals/STP/stp.html) 2(3–4): 25–46.

Barber, B. 1952. *Science and the Social Order*. New York: Free Press.

Beauchamp, T.L. 1991. *Philosophical Ethics*, 2nd ed. New York: McGraw-Hill.

Beauchamp, T.L. and Childress, J.F. 2001. *Principles of Biomedical Ethics*, 5th ed. Oxford: Oxford University Press.

Bok, S. 1995. *Common Values.* Columbia: University of Missouri Press.

Broad, W. and Wade, N. 1982. *Betrayers of the Truth: Fraud and Deceit in the Halls of Science*. New York: Simon & Schuster.

Brock, W.H. 1992. *The Norton History of Chemistry*. New York: Norton.

Bronowski, J. 1956. *Science and Human Values* (revised edition). New York: Harper Torchbooks.

Brown, Jr., G.E. 1992. The Objectivity Crisis. *American Journal of Physics* 60: 779–781.

Bush, V. 1990. *Science: The Endless Frontier*. Washington, D.C.: National Science Foundation (reprint of 1945 edition).

Callahan, J.C., ed. 1988. *Ethical Issues in Professional Life*. New York: Oxford University Press.

Committee on Professional Training. 1999. *Undergraduate Professional Education in Chemistry: Guidelines and Evaluation Procedures.* Washington D.C.: American Chemical Society.

Coppola, B.P. 2000. Targeting Entry Points for Ethics in Chemistry Teaching and Learning. *Journal of Chemical Education* 77: 1506–1511.

Coppola, B.P. and Smith, D.H. 1996. A Case for Ethics. *Journal of Chemical Education* 73: 33–34.

Cournand, A. and Meyer, M. 1976. The Scientist's Code. *Minerva* 14: 79–96.

Davis, M. 1987. The Moral Authority of a Professional Code. In: Pennock, J.R. and Chapman, J.W. (eds.), *Authority Revisited, Nomos XXIX*. New York and London: New York University Press.

Davis, M. 1998. *Thinking Like an Engineer*. New York: Oxford University Press.

Dyson, F. 1993. Science in Trouble. *The American Scholar* 62: 513–522.

Feynman, R.P. 1985. *Surely You're Joking Mr. Feynman*. New York: W. W. Norton & Co.

French, A.P. 1979. *Einstein: A Centenary Volume*. Cambridge: Harvard University Press.

Gass, W. 1980. The Case of the Obliging Stranger. In: Gass, W. (ed.), *Fiction and the Figures of Life*. Boston: Godine, 225–241.

Gert, B. 1988. *Morality: A New Justification of the Moral Rules*. New York: Oxford University Press.

Guston, D.H. 1999. *Between Politics and Science: Assuring the Integrity and Productivity of Research*. Cambridge: Cambridge University Press.

Guston, D.H. and Keniston, K. 1994. Introduction: The Social Contract for Science. In: Guston, D.H. and Keniston, K. (eds.), *The Fragile Contact: University Science and the Federal Government*. Cambridge: MIT Press.

Haack, S. 1997. Science, Scientism, and Anti-Science in the Age of Preposterism. *The Skeptical Inquirer* (http://www.csicop.org/si/9711/preposterism.html) 21(6): 37–42.

Hardwig, J. 1985. Epistemic Dependence. *Journal of Philosophy* 82: 335–349.

Hardwig, J. 1991. The Role of Trust in Knowledge. *Journal of Philosophy*. 88: 693–708.

Hardwig, J. 1994. Toward an Ethics of Expertise. In: Wueste, D.E. (ed.), *Professional Ethics and Social Responsibility*. Lanham, MD: Rowman & Littlefield.

Harris, C.E., Jr., Pritchard, M.S., and Rabins, M.J. 1995. *Engineering Ethics: Concepts and Cases*. Belmont, CA: Wadsworth.

Holton, G. 1994. On Doing One's Damnedest: The Evolution of Trust in Scientific Findings. In: Guston, D.H. and Keniston, K. (eds.), *The Fragile Contract*. Cambridge: MIT Press.

Hyde, L. 1979. *The Gift: Imagination and The Erotic Life of Property*. New York: Vintage.

Kitcher, P. 2001. *Science, Truth and Democracy*. Oxford: Oxford University Press.

Knight, D. 1992. *Ideas in Chemistry*. New Brunswick, NJ: Rutgers.

Knight, D. and Kragh, H. (eds.). 1998. *The Making of the Chemist*. Cambridge: Cambridge University Press.

Kovac, J. 1996. Scientific Ethics in Chemical Education. *Journal of Chemical Education* 74: 926.

Kovac, J. 1998. The Ethical Chemist, *CUR Quarterly* 13: 109–113.

Kovac, J. 1999. Professional Ethics in the College and University Science Curriculum. *Science* and *Education* 8(3): 309–319.

Kovac, J. 2000. Science, Law, and the Ethics of Expertise. *Tennessee Law Review* 67: 397–408.

Kovac, J. 2001. Gifts and Commodities in Chemistry, *Hyle* 7: 141–153.
Kuhn, T.S. 1962. *The Structure of Scientific Revolutions*. Chicago: University of Chicago Press.
Panel on Scientific Responsibility and the Conduct of Research. 1992. *Responsible Science*, Vol. I. Washington, D.C.: National Academy Press.
Panel on Scientific Responsibility and the Conduct of Research. 1993. *Responsible Science*, Vol. II. Washington, D.C.: National Academy Press.
McSherry, C. 2001. *Who Owns Academic Work: Battling for Control of Intellectual Property*. Cambridge: Harvard.
Merton, R.K. 1973. The Normative Structure of Science. In: Merton, R.K. *The Sociology of Science*. Chicago: University of Chicago Press. 267–278.
Nelkin, D. 1984. *Science as Intellectual Property*. New York: Macmillan.
Polanyi, M. 1964. *Personal Knowledge*. New York: Harper Torchbooks.
Popper, K. 1965. *The Logic of Scientific Discovery*. New York: Harper and Row.
Proctor, R.N. 1991. *Value-Free Science*. Cambridge: Harvard University Press.
Rachels, J. 1999. *The Elements of Moral Philosophy*, 3rd ed. Boston: McGraw-Hill.
Reese, K.M. (ed.). 1976. *A Century of Chemistry*. Washington, D.C.: American Chemical Society.
Resnik, D.B. 1999. *The Ethics of Science: An Introduction*. London: Routledge.
Shapin, S. 1989. Who Was Robert Hooke? In: Hunter, M. and Schaffer, S. (eds.), *Robert Hooke: New Studies*. Woodbridge, Suffolk: Boydell Press, 253–285.
Shapin, S. 1994. *A Social History of Truth*. Chicago: University of Chicago Press.
Steelman, J.R. 1947. *Science and Public Policy*. Washington, D.C.: The President's Scientific Research Board, U.S. Government Printing Office.
Stokes, D.E. 1997. *Pasteur's Quadrant: Basic Science and Technological Innovation*. Washington, DC: Brookings Institution Press.
Ziman, J.M. 1968. *Public Knowledge*. Cambridge: Cambridge University Press.
Zuckerman, H. 1977. Deviant Behavior and Social Control. In: Sagarin, E. (ed.), *Deviance and Social Control*. Beverly Hills: SAGE Publications.

SECTION IV

CHEMISTRY AND PHYSICS

CHAPTER 9

IS THERE DOWNWARD CAUSATION IN CHEMISTRY?

ROBIN FINDLAY HENDRY
Department of Philosophy, University of Durham, 50 Old Elvet, Durham DH1 3HN UK; E-mail: r.f.hendry@durham.ac.uk

INTRODUCTION

Unless the future of chemistry holds some great ontological revision, it seems safe to assume that whenever there is chemical change there is physical change. For instance, on the assumption that an object's membership of a natural kind is determined by its possession of certain physical properties, there cannot be change in chemical kind membership without change in these kind-constituting physical properties. There is a consensus in recent philosophy of mind that this kind of relationship—known as supervenience—is, however, compatible with a range of views on the ontological relationship between two domains.[1] It is, for instance, a commitment both of physicalism—the view that physical laws and facts determine all laws and facts—and of some forms of emergentism, the view that there are *autonomous* facts associated with the physical systems of higher orders of complexity studied by some of the special sciences.

Now chemistry is of multiple interest here. To some of us, it is the central science of matter whose methods, products, and relationship to physics are of interest in themselves, but the relation between chemistry and physics is of wider resonance. As characterized by Oppenheim and Putnam (1958, 407), classical reductionism is *explicitly* hierarchical and cumulative, and so must be any *non*-reductive physicalism, insofar as the dependence relation on which it turns is also transitive. Appeal to chemical theories figures large in biochemistry, so establishing the dependence of the chemical on the physical would make it that much easier to establish the physical dependence of the biological (and after that the mental). But that is not all, for in assessing the evidence for the universality of physical law, claims about chemistry must be central (see Hendry 1999). Non-physicalists typically suspect that the universality of physical law is a philosophers' fiction: a story of conquest that has been extrapolated from a few minor imperialist skirmishes. Since chemistry is right on the border of the empire of physics, this is where the impartial observer might reasonably expect some real applications of physical theory, involving detailed and rigorous treatments of chemical problems, rather than programmatic sketches, promissory notes or claims of reducibility "in principle."

D. Baird et al. (eds.), Philosophy of Chemistry, 173–189.

In this paper, I will be investigating the extent to which there are such treatments, and whether the applications that *are* available can figure as evidence for the completeness of physics. In the next two sections, therefore, I will examine the role of completeness as an element in the various physicalist positions, and also in arguments for these positions. I will then go on to assess some of the evidence for completeness itself. Very briefly, my overall argument with respect to the latter is as follows: (i) No completeness thesis can be non-trivial if it fails to rule out downward causation. (ii) Given some plausible argument strategies that are available to the physicalist, the available evidence does not support completeness theses that rule out downward causation. The crucial stage of my argument, stage (ii), I illustrate with examples from quantum chemistry.

PHYSICALISM AND THE COMPLETENESS OF PHYSICS

Schematically, physicalism can be thought of as the claim that the physical facts determine all the facts. In explicating a claim of this sort, we need to do two things: first decide what kinds of fact are to count as physical, then provide some dependence relation that explicates the thought that one set of facts "determines" another.[2]

What counts as physical? In the philosophy of mind, a broad construal is often at work: roughly, one according to which the "physical" includes anything that takes up space (hence the special puzzles about mental causation). The broad construal may be harmless in that context, but is obviously too broad for our present purposes, since it fails to exclude the chemical. If chemical properties are a subset of physical properties, dependence follows automatically for any physicalist position based on a reflexive dependence relation: we know right away that the chemical depends on the physical, but only because it depends on the chemical. This would be a terminological answer to the dependence question, and would leave open what dependence relations hold between different subsets of "physical" properties. A narrower, more informative, conception of the physical might proceed in terms of the *discipline* of physics, but this would not suit physicalist intuitions that physical properties are basic. Physics itself studies a heterogeneous array of entities and properties, and it is hard to see why theories constructed within, say, fluid dynamics or astrophysics should be expected to be more fundamental than chemical theories.[3] Nor would a physicalist position of this sort survive serious thought as to how historical contingencies determine which domains of phenomena came to be studied in physics departments, rather than (say) departments of engineering or chemistry.

Perhaps a well-motivated conception of the physical will proceed in terms of the laws and the categories associated with a few "fundamental"—for which read, general, and abstract—theories in physics, especially quantum mechanics, relativity and their descendants. This seems fair, since presumably it is the explanatory performance of these great theories that underwrites the evidential standing of physicalism. This is indeed how physicalists have tended to identify the physical: in a sense that allows it to contrast with the chemical, and be correspondingly informative (see for instance Quine 1981; Papineau 1990; Field 1992).

So to the second dimension of physicalism: physicalist positions have been articulated in terms of a variety of dependence relations, including supervenience (there can be no change without physical change), realization (higher-level properties are causal roles played by physical properties), and token identity (everything concrete that instantiates a non-physical property also instantiates a physical property) to name but a few.[4] It is customary to distinguish reductive from non-reductive physicalist positions, depending on whether or not the dependence relation involved implies that higher-level properties, facts, or entities are "nothing more than" lower-level properties, facts, or entities. Historically, the question of whether the sciences offer a unified picture of the world was investigated first in terms of logical relations between their theories. In what sense do the special sciences add to the predictive and the expressive power enjoyed, in principle, by physics? Thus it was that the question of whether one group of entities or properties is reducible to another became a question of what kind of logical relations hold among the *theories* that describe them.[5] If we are interested in the ontological question of how the subject matter of chemistry (say) is dependent on the subject matter of physics, the logical relations between chemical and physical theories, though still important, can at best provide evidence for settling the question, rather than constitute an answer in themselves. The logical and ontological questions come apart if it is possible that the subject matter of chemistry is "nothing more than" the subject matter of chemistry, while for practical reasons (to do with complexity and mathematical intractability), chemical theories will never be deducible from physical theories.

How is the vague-sounding phrase "nothing more than" to be understood? Kim (1997, 279–286) has articulated a functionalist physicalism: for an object to have a higher-level (for instance mental or macroscopic) property is for it to have a physical property that plays some causal role. Thus, for a body to be transparent is for it to have some physical property that causes it to transmit light. The physical property that meets the causal condition is the *realizer* of the causal role in that body. Although a functional property can be realized differently in different kinds of macroscopic body, it makes sense, Kim argues, to identify the transparency of glass with the microstructural property that realizes it: being transparent is, for glass, "nothing more than" its microstructural realizer in glass. But as Kim admits, this last implication only follows if it is accepted that "the microstructure of a system determines its causal/nomic properties" (283), for the functional role is specified causally, and so the realizer's realizing the functional property that it does (i.e., the realizer–role relation itself) depends on how things in fact go in a particular kind of system. For a microstructure to determine the possession of a functional property, it must completely determine the causal/nomic properties of that system. Thus, Kim's argument from realizationism to reductionism turns on, but offers no independent support for, a thesis that has come to be of central importance to physicalism: the causal completeness of the physical. Note also that in order for a physicalist position to be *reductive*, it must be committed to the causal completeness of the physical (I will henceforth refer to "strict" and "liberal" physicalist positions). Papineau (1990) formulates this thesis as follows:

> all physical events are determined (or have their chances determined) entirely by prior physical events according to physical laws (Papineau 1990, 67).

Papineau's version of the completeness of physics concerns the *science* of physics, while Kim's version concerns the causal relations which are its subject matter. The relation comes via Papineau's contention that the science of physics aims at completeness, in the sense that the causal processes it describes—or in the ideal limit of physical inquiry *would* describe—are closed with respect to the non-physical: there are no causal factors which are not described by physics, or for whose description physics defers to another science. In this respect, physics is, he argues, quite different from, for instance, meteorology, chemistry, biology, and psychology, all of which admit external causal factors.

The completeness thesis seems to be central to current physicalism, constituting a litmus test commitment for a thoroughgoing acceptance of the position. Thus McLaughlin (1992), Horgan (1993), and Kim (1997) object to versions of physicalism based on supervenience as dependence relation, because they fail to rule out downward causation, as any physicalism worth the name ought. One way of putting the problem is as follows: suppose that some group, A, of special science properties, supervenes on some group of physical properties. If there are law-like connections among the A-properties (special science laws), these will be reflected in relations among the physical properties, although the latter relations may well be messy and disjunctive. Given the supervenience, any A-changes will be accompanied by physical changes: should we regard the A-changes as determining the physical changes, or vice versa? If the former, then there is a sense in which the physical properties are being "pushed around" by the A-properties (hence there is "downward causation" from the A-properties to the physical properties on which they supervene). Only the causal completeness of the physical rules out this possibility. One last reason for the importance of completeness to physicalism is that, insofar as it appears (acknowledged or unacknowledged[6]) as an element in arguments for physicalism, it itself appears to be an empirical thesis. Not only does the closure of the physical explain the privileged role that physics has in ontology, but also it is the kind of principle for whose support we can appeal directly to discoveries and explanations within science itself.

BROAD'S EMERGENTISM

Brian McLaughlin (1992) has recently told the historical story of what he calls "British emergentism," a strand of non-reductive materialist thinking that he traces from John Stuart Mill through to C.D. Broad, on whom I will concentrate here. Roughly, the British emergentists held that things in the world are composed only of material stuff, but that new, emergent kinds of behavior can be displayed by complex aggregates of material stuff. So far so vague: what does "new" mean here? Broad and the other emergentists canvassed various formulations, but perhaps the most perspicuous is as follows: suppose that emergent behavior is law-like, as presumably it must be, if a special science is devoted to its study. A law exhibited by some complex system is emergent if the behavior it describes is not determined to occur by the laws governing the parts of the system in isolation. If aggregates of material stuff, and their law-like patterns of behavior, are stratified into "orders" of complexity, an emergent law is a primitive matter of fact concerning the behavior of aggregates at its level of application. Broad's presentation of these possibilities sometimes appears to proceed in logical or epistemic terms: an emergent law could not be deduced or predicted from the underlying laws governing the constituents of the system it covered (for discussion, see Stephan 1992). An epistemic formulation of emergent behavior of this sort would be unsatisfactory because (i) it would relativize emergence to the state of knowledge from which the prediction is made and (ii) although non-predictability might be evidence for emergence, it does not constitute emergence itself, since there are unpredictable but non-emergent kinds of behavior.

By way of explicating emergentism, Broad (1925, 44ff) articulated a position he called "Pure Mechanism," a variety of what would now be called microdeterminism. This "ideal of a mechanical view of the material realm" (44) would suppose that there is one fundamental physical stuff of which everything is made, and one fundamental intrinsic quality (e.g. mass or charge). One fundamental law (e.g. the law of gravitation or Coulomb's law) would determine interactions between pairs of particles, given their possession of relevant intrinsic properties, such as having a particular mass. Further suppose that there is some law of combination that determines a resultant force, given the action of a number of component forces. Pure Mechanism is committed to the claim that motions within any aggregate system of particles must arise from the interactions of its parts, as determined by the laws of interaction and combination.

Broad notes that mechanism *must* fail, owing to the existence of irreducibly macroscopic qualities like colors and temperatures:

> The plain fact is that the external world, as perceived by us, seems not to have the homogeneity demanded by Pure Mechanism. If it *really* has the various irreducibly different sensible qualities which it *seems* to have, Pure Mechanism cannot be true of the whole of the external world and cannot be the whole truth about any part of it (1925, 50–51).

The failure of Pure Mechanism on account of the existence of irreducibly macroscopic qualities need not bear directly on the completeness of physics. It may be possible, after all, that the mechanistic laws governing the interactions of microscopic particles are causally complete, and the macroscopic qualities consequently causally inert with respect to the behavior of systems of microscopic particles. Broad seems to have been aware of this possibility, for given the assumption that sensible objects really do have the "irreducibly different sensible qualities" they appear to have,[7]

> [t]he best that we can do for Pure Mechanism . . . is to divide up the external world first on a macroscopic and then on a microscopic scale; to suppose that the macroscopic qualities which pervade any region are causally determined by the microscopic events and objects which exist within it; and to hope that the latter, in their interactions with *each other* at any rate, fulfil the conditions of pure mechanism (1925, 51).

Broad goes on to note that

> there is no *a priori* reason why microscopic events and objects should answer the demands of Pure Mechanism even in their interactions with each other; that, so far as science can

tell us at present, they do not; and that, in any case, the laws connecting them with the occurrence of macroscopic qualities *cannot* be mechanical in the sense defined (1925, 51).

There follows a distinction between "intra-physical" and "trans-physical" laws (1925, 52); the intra-physical laws relate qualities that may be possessed by microscopic physical bodies, while the trans-physical laws relate physical qualities to (for instance) irreducibly macroscopic qualities. Thus, for instance, a law according to which all bodies with a particular physical composition have a particular color would be trans-physical in this sense. The intra-physical laws, Broad thought, could possibly be (though he ventured the opinion that they are in fact not) as conceived of by Pure Mechanism, while the trans-physical laws necessarily could not.

Broad drew a twofold distinction between ontological theories based on how they sought to account for the "differences of behavior between different things" (1925, 58). On the one hand is the kind of theory that appeals to the presence of a particular component or substance as explaining a particular kind of behavior (e.g. vital or teleological). On the other hand is the kind of theory that appeals only to difference in *structure*, but this second kind of theory may also come in two forms:

(i) On the first form of the theory the characteristic behavior of the whole *could* not, even in theory, be deduced from the most complete knowledge of the behavior of its components, taken separately or in other combinations, and of their proportions and arrangements in this whole...
(ii) On the second form of the theory the characteristic behavior of the whole is not only completely *determined by* the nature and arrangements of its components; in addition to this it is held that the behavior of the whole could, in theory at least, be deduced from a sufficient knowledge of how the components behave in isolation or in other wholes of a simpler kind (1925, 59).

The first form of this second kind of theory Broad called the "Theory of Emergence," while the second form he called "Mechanistic." The quote makes clear that the deducibility of the behavior of the whole from the behavior of its components is a *sign* of the nomological determination on the mechanistic theory. Similarly, the *non*-deducibility of the behavior of the whole posited by the emergentist is not peculiar to any particular state of knowledge, because emergent behavior cannot "be deduced from the *most complete knowledge* of the behavior of its components" (1925, 71).

It has already been noted that emergence comes in different modal grades, from the necessary emergence of the trans-physical laws (given the kinds of qualities they relate) to the merely contingent emergence (given the kinds of qualities they relate) of emergent intra-physical laws, if there are any. For example, Broad presents breathing as a process that can be described physically (and hence is intra-physical), but one that may well meet his criterion for emergence:

The process of breathing is a particular kind of movement which goes on in living bodies. And it can be described without any essential reference to secondary qualities. Yet in its details it may be such that it could not be deduced from any amount of knowledge about non-living wholes and the movements that take place in them. If so it is an "ultimate characteristic" of the vital order... But this law is not trans-physical, in the sense defined (1925, 80–81).

Trans-physical laws are "necessarily of the emergent type" (80), because they connect physical properties with (for instance) secondary qualities. In contrast, breathing is a movement,

> hence it cannot be positively proved that breathing is an "ultimate characteristic" or that its causation is emergent and not mechanistic (1925, 81).

A mechanistic account of breathing would show it to be a kind of behavior that is determined to occur by deeper physical laws. The absence of such an account may result *either* from (i) the incomplete nature of our knowledge (either of the fundamental laws themselves or of their consequences), *or* (ii) the fact that breathing is not a kind of behavior that is determined to occur by deeper physical laws. So there is also an epistemic difference between trans-physical laws and emergent intra-physical laws: the former can be known to be emergent *a priori*, by reflecting on the kinds of qualities related. The latter, however, cannot, and moreover "it cannot be positively proved that any intra-physical law is emergent" (1925, 80), because:

> Within the physical realm it always remains logically possible that the appearance of emergent laws is due to our imperfect knowledge of microscopic structure or to our mathematical incompetence (1925, 81).

The emergentist and the mechanist differ over the extent to which nomological unity underlies the manifest differences in the behavior of things:

> On the emergent theory we have to reconcile ourselves to much less unity in the external world and a much less intimate connexion between the various sciences (1925, 77).

The difference is a matter of degree. At one end of the spectrum, Pure Mechanism allowed only a single fundamental (e.g. gravitational) law of interaction. Looser forms of mechanism would result where independent (e.g. electromagnetic) kinds of interaction are acknowledged to complicate the picture. Hence the mechanist's claim must be made relative to a hypothesized set of fundamental interactions, rather than to a state of knowledge. At the other end of the spectrum, the emergentist posits in addition a range of irreducible laws covering the behavior of specific kinds of aggregate system, which could be either of the trans-physical or the intra-physical type. Leaving aside the trans-physical emergent laws, the emergentist allows that the *physical* behavior of aggregate physical systems may fail to be determined completely by the general dynamical laws governing their parts in isolation. Put another way, while the mechanist posits a single fundamental law, or just a few such laws, the emergentist allows that there could be a great many.[8] Turning to chemistry in particular, the mechanist claims that just a few laws of microscopic dynamics suffice to determine the behavior of every atom or molecule (Broad 1925, 70). Hence, given those few laws and the physical constitution of a chemical species, it will be possible in principle (though perhaps not in practice) to deduce a complete account of its behavior. The emergentist will allow that for some atoms and molecules, there will be fundamental laws describing only their behavior, which are not instances of more general laws.

Broad's different versions of mechanism are more elegantly set out within the Hamiltonian formulation of mechanics, which rolls the law of combination for forces up into an expression for the energy of a composite system. A full list of the "fundamental" forces operating in a system is sufficient to determine a "resultant" Hamiltonian function for that system, which represents its energy. Broad's mechanist opponent is committed to the claim that only Hamiltonians that are "resultant" with respect to this list of forces govern real physical motions. Note that the emergentist and the mechanist do not disagree on whether every system has a Hamiltonian, but only on whether every system's Hamiltonian is determined by the specification of a few interactions that occur very generally. Every motion can be conceived of as arising from some Hamiltonian or other, but if the behavior of some systems is governed by "non-resultant" Hamiltonians,[9] then there is a precise sense in which the behavior of those composite systems is not determined by the more general laws governing their constituents. Furthermore, to the extent that the behavior of any subsystem is affected by the supersystems in which it participates, the emergent behavior of complex systems must be viewed as determining, but not being fully determined *by*, the behavior of their constituent parts. And that is downward causation.

McLaughlin (1992, 89) concedes that emergentism is a perfectly coherent position, and downward causation a perfectly coherent possibility. It is just that emergentism is factually mistaken, for there is no evidence for downward causation from chemical or biological systems, no evidence for "configurational" forces or Hamiltonians. If there had been any such thing as configurational forces governing chemical bonding, one would expect that they would have been cited when chemical bonding came to be explained by quantum mechanics in the years after 1925 (see also Papineau 2000, 197–202). McLaughlin does not himself investigate the empirical details, but rather asks a historical question: why was Broad's *The Mind and its Place in Nature* (1925) the last great emergentist tract? His answer is that the advent of quantum mechanics ushered in a new era of scientific explanation, in which all the supposedly *sui generis* laws of chemical bonding were accommodated within quantum mechanics. Kim (1997, 290) makes a slightly different point. Some of Broad's own candidate cases of emergence, such as the transparency of water, were particularly unfortunate: once we realize that transparency just is a functional property—that is, the property of certain physical properties to enter into certain causal relations with light—then the detailed quantum mechanical account of those causal interactions assures us that there is nothing mysteriously "non-physical" about the transparency of water.[10]

However, to the extent that these are empirical arguments for strict versions of physicalism, they must *ipso facto* constitute arguments against downward causation. Now Broad's position is perfectly compatible with the existence of successful accounts within physical theories of chemical bonding. Where Broad's emergentist differs from the strict physicalist (or pure mechanist) is in the nature of those physical accounts: are the forces or Hamiltonians they cite resultant or configurational? If resultant, then the argument for strict physicalism is strong. If they are configurational, then *prima facie* we have examples of downward causation and evidence *against* strict physicalism. Since the central empirical examples cited by strict physicalists involve chemical bonding, the prospects for their position would not be good, set against the various weaker physicalist positions that admit the possibility of downward causation. It is on this genuinely empirical question that the strength of the argument for strict physicalism must turn, and to which I will now turn.

DOWNWARD CAUSATION FOR MOLECULES

Nancy Cartwright has long argued for three claims about the application of general physical theories. Firstly, the applicability of an abstract physical theory like quantum mechanics to a concrete system requires a model of the type of system in question. Secondly, not every kind of physical system is represented by such a model. Thirdly, the models that represent some complex kinds of system, such as lasers or benzene molecules, may be relatively autonomous, in that they are not related in any systematic way to more generally applicable or fundamental models.[11] Elsewhere, I have developed an argument for this third thesis with respect to quantum mechanical molecular models (see Hendry 1998). In this section, I want to develop a slightly different argument, relating these claims directly to physicalism, emergence, and downward causation. The result of our survey of Broad's emergentist position was that even if there is, in principle (or in the mind of God), an accurate quantum mechanical treatment for every material system, this is not sufficient to establish the completeness thesis required by strict physicalism. The emergentist also expects a Hamiltonian for every system, but holds that some of these will be "configurational." If the emergentist is right, then even if quantum mechanics has been successfully applied to atomic hydrogen and other simple systems, the motions of electrons in more complex systems are adequately described only by configurational Hamiltonians. In fact, the situation is slightly worse within molecular quantum mechanics: there does seem to be a way to generate a genuinely resultant Hamiltonian for every molecule, but for good reasons these are ignored in explanation in favor of what appear to be configurational Hamiltonians. My central point is that the resultant Hamiltonians are ignored not only for the pragmatic reasons of their mathematical intractability, but also for the epistemic reason that they are quite unsuitable for describing molecules.

Some stage setting is required first, involving the adequate formulation of a completeness claim for a theory like quantum mechanics. Suppose we start with the following:

(1) For every physical system, there is a descriptively adequate Hamiltonian.

Let us suppose that a Hamiltonian is descriptively adequate for a system if it possesses a range of eigenfunctions whose eigenvalues correspond to the energy states of that system. Now (1) will not do: the existence, for each system, of just *any* old Hamiltonian, even if descriptively accurate, is not sufficient for a completeness thesis that is of any use to strict physicalism, for two reasons. Firstly, if no constraints are put on the acceptability of Hamiltonians beyond the usual mathematical ones, the requirement is presumably vacuous. Mere mathematics would assure us that (1) is true: given any spectrum of energy states, just work backwards to construct a Hermitian operator with a set of eigenfunctions with the required spectrum as eigenvalues. Equate this operator with the sum of the kinetic energy and the potential energy operators (the kinetic energy operator is determined for each system, the potential operator is just the *ad hoc* "Hamiltonian" minus the kinetic energy operator). Presumably this is not what is meant when the physicalist requires that only physical events determine (the chances of) physical events *in accordance with* physical laws. The lesson of Broad's emergentism is that the mere existence of a force function for every system, satisfying no further constraints on its construction, fails to rule out downward causation, for the force function for some complex systems might be "configurational." The same holds for Hamiltonians and quantum mechanics. In short, an argument for strict physicalism requires an argument for the completeness of physics that also constitutes a refutation of emergentism.[12] So strict physicalism, of the kind that is incompatible with downward causation, requires something like the following:

(2) For every physical system, there is a descriptively adequate "resultant" Hamiltonian.

What is required now is a perspicuous rendering of the distinction between "resultant" and "configurational" Hamiltonians, which turns out to be easy, at least for molecular quantum mechanics. The resultant Hamiltonian for any particular system is that which arises from the following method:[13]

(i) Specify a list of fundamental physical interactions (gravitational, electromagnetic, strong- and weak-nuclear).
(ii) Enumerate the microparticles present in the relevant system and list their charges, masses, and values of any other relevant quantities.
(iii) Using only the approved "fundamental" forces in (i), list the interactions occurring between the microparticles enumerated in (ii).
(iv) Using the results of steps (i)–(iii), write down the kinetic and potential energy operators and add them.

Happily for the strict physicalist, something very much like this method is taught to every student of quantum chemistry, with the proviso that particle enumeration remains at the level of electrons and nuclei, and only electrostatic terms are usually included in the potential energy operator. It really does yield a Hamiltonian for every molecule, with the slight difficulty that it yields the same Hamiltonian for isomers, which are typically very different from the chemical point of view. So the physicalist's universalist intuitions would seem to be vindicated, at least within the realm of molecular quantum mechanics. Unhappily for the strict physicalist, however, these Hamiltonians play very little part in quantum chemical explanations (except, predictably, for very simple systems). Instead, quantum chemical explanations use model Hamiltonians that are not obtainable from the above algorithm. Of course, the strict physicalist will argue that everything that is explained by reference to the model Hamiltonians could (in principle) be explained using the "resultant" Hamiltonians, because the former approximate to the latter. I will consider that claim shortly: first let us examine the cases.

Historically, the development of quantum theory was associated closely with spectroscopy, essentially because classical mechanics failed repeatedly to provide adequate explanations of the spectroscopic behavior of molecules. But if steps (i)–(iv) are ignored, how does a quantum mechanical account of the spectra of a simple molecule really begin? Here is how one textbook of spectroscopy describes carbon dioxide:

The CO_2 molecule is linear and contains three atoms; therefore it has four fundamental vibrations ... The symmetrical stretching vibration is inactive in the infrared since it produces no change in the dipole moment of the molecule. The bending vibrations . . . are equivalent, and are the resolved components of bending motion oriented at any angle to the internuclear axis; they have the same frequency and are said to be doubly degenerate (Silverstein, Bassler, and Morrill 1981, 96).

The next step is to apply quantum mechanics. There are models in quantum mechanics for simple rotating bodies, and for simple oscillators, usually to be found in the chapter of the textbook on quantum mechanics *after* the chapter in which the Schrödinger equation was introduced. With some adjustments, we can view parts of the molecule as quantum mechanical harmonic oscillators and rigid rotators, allowing us to quantize the rotational and vibrational motions that background chemical theory already tells us that the carbon dioxide molecule must exhibit. This provides the energy levels: differences between these energy levels correspond to spectral lines (in the infrared region in the case of CO_2's vibrational modes). For a more accurate account, consider the *coupled* vibrational and rotational modes. To a first approximation, vibrational and rotational modes are taken to be *additive*, but finer structure can be explained in terms of anharmonicity and other effects of the distortion of the molecule away from its equilibrium geometry (see for instance Steinfeld 1985, Chapter 8).

Let us review the structure of the explanation just given: we use quantum mechanics to explain the motions of parts of the molecule *within the context* of a given structure for the molecule as a whole. The emergentist will see this as a case of downward causation: we did not recover the CO_2 structure from the "resultant" Hamiltonian, given the charges and masses of the various electrons and nuclei; rather we viewed the motions of those electrons and nuclei as constrained by the molecule of which they are part. Of course, the physicalist will say that steps (i)–(iv) are ignored only for practical reasons of their intractability. In principle, the same explanation could be given using the exact treatment. One way to support that contention would be to have a *theoretical* justification for the molecular structure attributions, which allows us to link the "model" Hamiltonian with the "exact" one: what we in fact explain with the former, we could have explained with the latter. Elsewhere (1998, Section 2), I have called this the "proxy defence" of model Hamiltonians, for they stand in for the exact ones.

As it is usually presented in textbooks of theoretical chemistry, the Born–Oppenheimer approximation plays just such a role. It is usually given a heuristic justification, as in this explanation of how it applies to the hydrogen molecule-ion H_2^+, which consists of two hydrogen nuclei and one electron:

The Born–Oppenheimer approximation supposes that the nuclei, being so massive, move much more slowly than the electrons, and may be regarded as stationary. This being the case, we can choose the nuclei to have a definite separation (i.e., we can choose a definite *bond length*) and solve the Schrödinger equation for the electrons alone; then we can choose a different bond length and repeat the calculation. In this way, we can calculate how the energy of the molecule varies with bond length (and in more complex molecules with angles too), and identify the equilibrium geometry of the molecule with the lowest point on this curve. This is far easier than trying to solve the complete Schrödinger equation by treating all three particles on one footing (Atkins 1986, 375).

But it is difficult to see how this gets us back to the resultant Hamiltonian of steps (i)–(iv). The Born–Oppenheimer wavefunction looks more like the solution to an altogether different equation: the nuclei are treated classically, and we view the electrons as constrained by the resultant field.

If the non-quantum mechanical treatment of the nuclei is the issue, we need a mathematical proof that this does not make much difference to the energy, which can, perhaps, be had from what is sometimes called the adiabatic approximation, to which the Born–Oppenheimer approximation can be viewed as a further approximation. Let us suppose that we can de-couple nuclear and electronic motions. Born–Oppenheimer calculations at all nuclear configurations would yield a series of electronic fields, which can be used as a constraint on a subsequent quantum mechanical treatment of the nuclear motions. But could the product of this calculation reasonably be said to describe a molecule? In the above calculation, the quality of the approximation (i.e., its closeness to the "exact" treatment) is tracked via the value it gives for the overall energy of the molecule. But in optimizing the energy, we remove the symmetry properties of the molecule on which explanations of its chemical and spectroscopic behavior depend. The "resultant" molecular Hamiltonian of steps (i)–(iv) enjoys nuclear permutation and rotational symmetries, but Born–Oppenheimer structures—and real molecules—do not (see Woolley 1976, 34 and Hendry 1998, Section 3 for discussion). Not only that: the lower symmetry of, say, hydrogen chloride is central to the explanation of its acidic behavior and its boiling point (see also Woolley and Sutcliffe 1977). So perhaps the "exact" Hamiltonians that issue from (i) to (iv) cannot make the fine-grained distinctions among molecules that are required for chemical explanation, which is why they do not appear in them. Thus, Woolley argues that the "full, spin-free molecular Hamiltonian . . . describes the interactions of an assembly of electrons and nuclei; it does not describe a particular molecular species" (1991, 26).

The problem arises from the *nature* of the "resultant" Hamiltonians, not their mathematical tractability. Taking as his example the molecular formula C_3H_4, Woolley argues that

> this presents a collection of 3 carbon nuclei, 4 protons and 22 electrons. For quantum chemistry we easily imagine the nuclei to be placed in the arrangements corresponding to the three distinct stable molecules of this formula, and then apply quantum mechanics (the Schrödinger equation) to the electrons to obtain the total electronic energy in the three cases. Other nuclear arrangements lead to different electronic energies and we represent the whole set of such energies as a potential energy surface (including of course the classical Coulomb energy of the nuclei). Suppose we apply quantum mechanics to *all* the particles in one go, what do we get? It is easier to say what we have never found so far—no suggestion of three distinct isomers for the molecules of allene, cyclopropene and methyl acetylene (1998, 11).

To get the quantum mechanical description of the "particular molecular species" allene, cyclopropene, and methyl acetylene, we need to put the nuclear configuration in by hand. This, according to Woolley, is precisely the role played by the Born–Oppenheimer procedure, which in his view is not really an approximation. The complaint is not that there are *no* explanations of empirically determined molecular shapes, or even that the explanations are *ad hoc*, or of poor quality. Rather it is that the explanation is conditioned on determinate nuclear positions: *if* electronic motions are constrained by a stable nuclear backbone, *then* the energy dependence is such that such-and-such is the lowest energy configuration.

Of course, none of this rules out a different "proxy-defence," turning on a different mathematical relationship between model Hamiltonians and the exact "resultant" Hamiltonian. Nor does it rule out a different way of generating a physicalistically respectable "resultant" Hamiltonian for every molecule, one which could be explanatorily useful. This is true, but the burden of proof is firmly back on the strict physicalist's side.

THE METHODOLOGICAL ARGUMENT

This disunified scene, may, of course, be temporary. Physics may be working towards a more complete and unified picture of molecular reality. What evidence is there for this? On one argument, the requirement of universal coverage is built into the very practice of physics. Thus, Quine has it that if the physicist suspected there was any event that did not consist in a redistribution of the elementary states allowed for by his physical theory, he would seek a way of supplementing his theory. Full coverage in this sense is the very business of physics, and only of physics (1981, 98).

This gets us supervenience directly out of the scientific method: the physicist would not settle for anything less. How does a physicist go about showing that some event *does* consist in a redistribution of the elementary states allowed for by his physical theory? By giving a suitable application of the theory in question that *does* imply that the event occurred. So the claim is that physicists have a duty to provide applications of their theories to every kind of physical situation, which seems to be about as false as philosophical claims about scientific practice ever are. What serious effort has *any* physicist ever put into checking whether the observed motions of, say, the Forth Road Bridge consist in redistributions of elementary states allowed for by current physical theory? This is not just the Lakatosian point that there are "recalcitrant instances" for every theory, that scientific research tends to proceed in an "ocean of anomalies" (Lakatos 1970, 138). Nor is it the pessimistically inductive point that we can expect *any* particular theory, in the fullness of time, to be overthrown. If true, both of these claims apply equally to sciences other than physics within *their* domains, and could apply to physics even if physical theories were subject to a duty of universal coverage that is, even if the domain of physics were all-encompassing. My point is just that there are large classes of events for which there is no tendency for physicists even to *begin* to construct detailed applications, which are therefore not part of the "business of physics," but which *are* the business of other sciences. In any case, it is far from clear that the application of quantum mechanics to molecules from the 1930s onwards was perceived to be the job of physicists, or that the results of its application by chemists would be of the kind required by any non-trivial physicalism (see Hendry 2001).

Of course, Quine's claim is a conditional one: *if* the physicist suspects that a particular event constitutes a counterexample, *then* he would supplement his theory. Perhaps physicists have no such suspicions. But this equanimity must be well grounded. To adapt a well-known example of Clifford's, suppose the Captain of a sunken ship points out that he did not suspect that his craft was not sea-worthy (1879). This does not absolve him of responsibility for the disaster if he made no effort to check its sea-worthiness. One need not posit general epistemic duties to see that the Captain has an (epistemic) duty to ensure that his equanimity is justified, for it is a specific duty that he bears in virtue of his position as Captain. Similarly, if physicists are practitioners of a science for which closure is claimed, then their assuredness that the special sciences offer no counterexamples to this closure (i.e., no configurational Hamiltonians) must be grounded in evidence of some kind. So their equanimity would be unreasonable without the kind of evidence we failed to find in the last section.

Field (1992, 283) and Smith (1992, 40) make weaker methodological claims than Quine's, giving physicists much less onerous *explanatory*, rather than predictive duties. Field argues that physics is in the business of ensuring that its theories "mesh" with those of higher level sciences, and that 'successful meshing" between physical and higher-level theories requires explanation: the explanations can be provided only by reductionism. A "mesh" consists of a sketched and approximate microreduction. Thus the explanatory duties of physics push it towards reduction of other sciences. Smith has it that any physicalist worth his salt will insist that, where a low-level theory interfaces with a higher-level theory, we should be able to use the lower-level theory to explain why assumptions of the higher-level theory actually obtain (1992, 39–40).

This looks far more plausible: the whole enterprise of quantum chemistry, after all, is to use quantum mechanics to recover facts about chemical bonding that are well known to the chemist. If there is meshing between chemistry and quantum mechanics, do the real meshes and interfaces we saw in the last section really support strict physicalism? Surely they cannot, if they are consistent with downward causation. In any case, it is not as if, in the explanation of the spectrum of carbon dioxide, there are two independent theoretical accounts that were compared and found to be consistent. Perhaps that *would* constitute an explanation of the less fundamental models. But neither chemistry nor the quantum mechanics of "resultant" Hamiltonians have the resources for independent accounts of the spectrum of carbon dioxide. Rather than an explanation *of* chemical structure *by* physical theory there was a joint venture: the explanation of various facts by the use of quantum mechanics applied to a given molecular structure. There was no mesh or interface *between* the quantum mechanics and chemistry, at least none that required explanation. What we had was an instance of quantum chemistry, the quantum theory of atoms and molecules.[14]

Perhaps this is too easy. Perhaps the importance of the duty of universal applicability is as a motivation, rather than as an achievement: attempts to unify disparate domains have motivated some of the most ambitious and successful episodes in the history of physics. Newtonian mechanics, we are often told, was the synthesis of terrestrial and astronomical physics. More poignantly for the present discussion, in the early 1920s—the last years of the old quantum theory—attempts to fit atomic models to spectroscopic data required a diverse battery of inexplicable and mutually incompatible quantum conditions. Pauli and Born, among others, saw in this chaos the need for a radical departure. Hindsight tells us that it was quantum mechanics that they foresaw, a theory whose appeal, initially at least, lay in its unifying power. A *methodological* objection to the acceptance of disunity in science follows: so much the worse for subsequent progress had *Heisenberg* been content with the disunified scene that prompted his efforts. But consider the details: firstly, quantum mechanics merely ushered in a new set of disunities, as we have seen; secondly, the historical claim that important advances in physics arise only from the unifying impulse is surely false. In normal science, the aims are different. But there is some justification in the complaint, for whether or not quantum mechanics *really* unified physics, it was an important advance. With hindsight, would we really have counseled Pauli, Heisenberg, and Born to be content with the old quantum theory? This is *only* a methodological complaint, however. If expectations of unity are sometimes fruitful, this does not imply the truth of the underlying reductionist metaphysics. The fruitfulness of an aim does not imply its achievement.[15] Nor should it blind us to disunities in science.

CONCLUSION

I will end with a familiar contrast between two pictures of the relation between physics and the special sciences. On the one hand, we have the physicalist story: the special sciences study regularities of limited scope, and even if the relationships are rarely clear, the entities and properties they study are dependent on physical entities and properties. But the underlying causal processes are all physical, and fall under a few physical laws. On the other hand is the pluralist version, in which physical law does not fully determine the behavior of the kinds of systems studied by the special sciences. On this view, although the very abstractness of the physical theories seems to indicate that they could, in principle, be regarded as applying to special science systems, their applicability is either trivial (and correspondingly uninformative), or if non-trivial, the nature of scientific inquiry is such that there is no particular reason to expect the relevant applications to be accurate in their predictions. Although the pluralist picture is a modern one, Broad's characterization of emergentism is surprisingly close to it.

Now chemistry might seem to be the kind of discipline for which physicalist dependence claims are at their most plausible, but this appearance is deceptive: reasons for thinking that the chemical depends on the physical are *at best* only *as good as* the reasons for thinking that the biological, the mental, and so on are so dependent, just so long as "physical" is taken as a contrastive term, marking off the physical from (say) the biological or the mental. In that context, it is natural if the physical is construed broadly, to mean the "physico-chemical" (rendering "physicalist" dependence claims about the chemical trivially true). Dependence claims targeted on these other disciplines may enjoy extra support from such principled claims, as are embodied (for instance) in the functionalist conception of the mental, about concepts that characterize discourse within these disciplines, concepts that also help to mark their domains off from the (broadly) physical in principled ways. In the absence of arguments appealing to such characteristic concepts, the dependence of the chemical on the physical must rely solely on the plausibility of claims about the generality of physics. The burden of my argument has been that strict physicalism fails, because it misrepresents the details of physical explanation.

NOTES

1. See Horgan (1993, Section 1) for an influential statement of this view.
2. I will treat dependence and determination as converses: I also apologize in advance for invoking dependence and causal relations among facts, events, properties, and entities indifferently, as seems to fit the context.
3. For discussion of these issues, see Crane and Mellor (1991 Sections 1 and 2), Papineau (1990, 1991), and Crane (1991). David Knight (1995, Chapters 5 and 12) charts the changing views of chemistry's place in the hierarchy of the sciences.
4. It has been argued, though, that supervenience is not itself a dependence relation, but only a modalized covariance of properties. On this view, it is at best the *sign* of ontological dependence, related to it as correlation is to causation, see Horgan 1993.
5. A related but distinct notion of reduction is as a dateable achievement of intertheoretic explanation. For this usage, see for instance Scerri 1994.
6. In the ubiquitous arguments from overdetermination, it is often an unacknowledged premise: the existence of genuine mental causation could not imply *over*determination unless the physical structure of the world is already sufficient to determine the course of physical events, to the extent that they are determined.
7. Broad also argued that even if macroscopic bodies only *appeared* to have irreducibly macroscopic qualities, Pure Mechanism would still have a problem in accounting for this appearance (1925, 49).
8. The emergentist's special science laws are fundamental in that the behavior they describe is not determined to occur by more general laws. The very general laws of mechanics retain their special status, for they retain their applicability to the systems whose emergent behavior they fail to determine completely.
9. From here, I will use McLaughlin's term "configurational" as the contrast to "resultant" (1992, 52).
10. Note, however, that Kim's property identities are in themselves insufficient to rule out downward causation. What is required in addition is the causal completeness of physics with respect to systems within which the physical properties that realize these second-order properties are instantiated.
11. See Cartwright (1983) for arguments for the first and third claims and Cartwright (1999, Chapter 1) for the second claim.
12. A constraint that would seem to be accepted by McLaughlin 1992 and Papineau 2000.
13. See Woolley (1976) for a discussion. The distinction is particularly simple in quantum mechanics, for the Hamiltonian formulation already provides a law of combination for the different kinds of interaction. So there is no need for a separate parallelogram of forces.
14. Quantum chemistry indeed is a distinct theory, if complex theories are partly individuated by their models (as for instance Cartwright 1983 and Giere 1988 argue), and quantum chemistry deploys its own distinctive set of Hamiltonians, which is what I have been arguing. Elsewhere, I have sought a methodological understanding of how this distinctiveness arose (see Hendry 2001).
15. I have argued elsewhere for a similar separation of methodological claims and philosophical conclusion in the case of scientific realism, see Hendry, 1995.

REFERENCES

Atkins, P.W. 1986. *Physical Chemistry*, 3rd ed. Oxford: Oxford University Press.
Broad, C.D. 1925. *The Mind and its Place in Nature*. London: Kegan Paul, Trench and Trubner.
Cartwright, N. 1983. *How the Laws of Physics Lie*. Oxford: Clarendon Press.
Cartwright, N. 1999. *The Dappled World*. Cambridge: Cambridge University Press.

Clifford, W.K. 1879. The Ethics of Belief. In: *Lectures and Essays*, Vol. II. London: MacMillan.
Crane, T. 1991. Why Indeed? *Analysis* 51: 32–37.
Crane, T. and Mellor, D.H. 1990. There is No Question of Physicalism. *Mind* 99: 185–206.
Field, H. 1992. Physicalism. In: Earman, J. (ed.), *Inference, Explanation and Other Frustrations: Essays in the Philosophy of Science*. Berkeley: University of California Press, 271–291.
Giere, R.N. 1988. *Explaining Science: A Cognitive Approach*. Chicago: University of Chicago Press.
Hendry, R.F. 1995. Realism and Progress: Why Scientists Should be Realists. In: Fellows, R. (ed.), *Philosophy and Technology*. Cambridge: Cambridge University Press, 53–72.
Hendry, R.F. 1998. Models and Approximations in Quantum Chemistry. In: Shanks, N. (ed.), *Idealization in Contemporary Physics*. Amsterdam/Atlanta: Rodopi, 123–142.
Hendry, R.F. 1999. Molecular Models and the Question of Physicalism. *Hyle* 5: 117–134.
Hendry, R.F. 2001. Mathematics, Representation and Molecular Structure. In: Klein, U. (ed.), *Tools and Modes of Representation in the Laboratory Sciences*. Dordrecht: Kluwer, 221–136.
Horgan, T. 1993. From Supervenience to Superdupervenience: Meeting the Demands of a Material World. *Mind* 102: 555–586.
Kim, J. 1997. Supervenience, Emergence and Realization in the Philosophy of Mind. In: Carrier, M. and Machamer, P.K. (eds.), *Mindscapes: Philosophy, Science, and the Mind*. Konstanz: Universitatsverlag Konstanz, 271–293.
Knight, D.M. 1995. *Ideas in Chemistry: A History of the Science*, 2nd ed. London: Athlone.
Lakatos, I. 1970. Falsification and the Methodology of Scientific Research Programmes. In: Lakatos, I. and Musgrave, A. (eds.), *Criticism and the Growth of Knowledge*. Cambridge: Cambridge University Press, 91–196.
McLaughlin, B. 1992. The Rise and Fall of British Emergentism. In: Beckermann, A., Flohr, H. and Kim, J. (eds.), *Emergence or Reduction? Essays on the Prospects for Non-Reductive Physicalism*. Berlin: Walter de Gruyter, 49–93.
Oppenheim, P. and Putnam, H. 1958. Unity of Science as a Working Hypothesis. In: Feigl, H. Scriven, M. and Maxwell, G. (eds.), *Minnesota Studies in the Philosophy of Science*, Vol. II. Minneapolis: University of Minnesota Press, 3–36. [Page references are to the reprint in Boyd, R., Gasper, P. and Trout, J. (eds.), 1991. *Philosophy of Science*. Cambridge, MA: MIT Press, 405–427.]
Papineau, D. 1990. Why Supervenience? *Analysis* 50: 66–71.
Papineau, D. 1991. The Reason Why. *Analysis* 51: 37–40.
Papineau, D. 2000. The Rise of Physicalism. In: Stone, M. and Wolff, J. (eds.), *The Proper Ambition of Science*. London: Routledge, 174–208.
Quine, W.V. 1981. Goodman's Ways of Worldmaking. *Theories and Things*. Cambridge, MA: Harvard University Press, 96–99.
Scerri, E. 1994. Has Chemistry been at least Approximately Reduced to Quantum Mechanics? *PSA 1994*, Vol. 1. East Lansing, MI: Philosophy of Science Association, 160–170.
Silverstein, R.M., Bassler, G.C. and Morrill, T.C. *Spectrometric Identification of Organic Compounds*, 4th ed. New York: Wiley.
Smith, P. 1992. Modest Reductions and the Unity of Science. In: Charles, D. and Lennon, K. (eds.), *Reduction, Explanation, and Realism*. Oxford: Clarendon, 19–43.
Steinfeld, J. 1985. *Molecules and Radiation*, 2nd ed. Cambridge, MA: M.I.T. Press.
Stephan, A. 1992. Emergence: a Systematic View on its Historical Facets. In: Beckermann, A., Flohr, H. and Kim, J. (eds.), *Emergence or Reduction? Essays on the Prospects for Non-Reductive Physicalism*. Berlin: Walter de Gruyter, 25–48.
Woolley, R. 1976. Quantum Theory and Molecular Structure. *Advances in Physics* 25: 27–52.
Woolley, R. 1991. Quantum Chemistry Beyond the Born–Oppenheimer Approximation. *Journal of Molecular Structure* (THEOCHEM) 230: 17–46.
Woolley, R. 1998. Is there a Quantum Definition of a Molecule? *Journal of Mathematical Chemistry* 23: 3–12.
Woolley, R. and Sutcliffe, B. 1977. Molecular Structure and the Born–Oppenheimer Approximation. *Chemical Physics Letters* 45: 393–398.

[illegible] 1992. [illegible]

[illegible] 1978. [illegible]

[illegible]

[illegible]

[illegible]

[illegible]

Hendry, R.F. 2001. [illegible] and Molecular Structure. In: Klein, U. (ed.), [illegible]

Hoffmann, R. [illegible]

[illegible]

[illegible]

[illegible]

McLaughlin, B. 1992. The [illegible] In: Beckermann, A., [illegible] (eds.), Emergence or Reduction? [illegible] Walter de Gruyter, [illegible]

Oppenheim, P. and Putnam, H. 1958. Unity of Science as a Working Hypothesis. In: [illegible] and Maxwell, G. (eds.), [illegible]

[illegible]

[illegible]

[illegible]

[illegible]

[illegible]

[illegible]

[illegible]

[illegible]

[illegible]

[illegible]

CHAPTER 10

PHYSICS IN THE CRUCIBLE OF CHEMISTRY

Ontological Boundaries and Epistemological Blueprints

G.K. VEMULAPALLI

Department of Chemistry, The University of Arizona, Tucson, AZ 85721, USA, gkv@u.arizona.edu

INTRODUCTION

Physics and chemistry are two closely related areas of science. Developments in physics have had profound influence on chemistry. Even a cursory examination of the history of the two sciences shows that many important developments in chemistry came in the wake of physics. This should come as no surprise considering that physics is concerned with the laws and theories that govern matter and energy while chemistry is concerned with the properties of materials and transformations of matter and energy. Yet the developments in chemistry are not mere extensions of ideas forged in physics. Further, the physico-chemical theories used to explain these new developments are not logical extensions of the theories in physics. As shown by several studies in recent years (Scerri 2000 and references cited in the article), chemistry does not reduce to physics, except in a restricted ontological sense. Chemistry, paradoxically, is both strongly dependent on and yet independent of physics. This raises an important question about the fundamental laws of physics. What is the nature of the laws of physics, which are at the root of many important developments in chemistry (the study of matter), and yet which cannot subsume the chemical concepts that arise in the application of these very laws to the behavior of matter?

My primary aim is to show how physics and chemistry interact in giving us our current understanding of chemical bond. I want to establish the "factual" aspects of the relation between the two sciences without any reference, initially, to what has been speculated about that relation by other investigators. Once this is done, I will show that the laws of physics relevant to chemistry are statements of limitations. They define a boundary within which chemical laws and theories are valid and beyond which they become meaningless. Within the boundaries, the laws of physics do not dictate what is actual; nor do they provide a clear path connecting the theories and the laws of physics with those applicable to chemistry. The latter have to be established, among the many possibilities within the boundaries set by physics, by chemical research.

I will return to these points after examining a few examples of interdependence of chemistry and physics, examples that bring the question raised above into focus.

D. Baird et al. (eds.), Philosophy of Chemistry, 191–204.

HEGEMONY OF PHYSICS

Major developments in physics often change the direction of chemical research. In particular they affect (i) the types of experiments, (ii) the vocabulary used in describing chemically significant phenomena, (iii) the explanations given for the phenomena, and even (iv) the topics of research considered to be important. Let us consider typical examples.

(i) Some of the most important data in chemistry are the relative molar masses (RMM), since no quantitative conclusions about chemical composition or reactivity can be drawn without them. RMM are now determined by mass spectroscopy, from colligative properties, and by the application of the ideal gas law. Each of these techniques depends on concepts and theories of physics. Indeed, these techniques would not have found their way into the chemistry laboratory except for the interpretation the physical theory provides. What incentive would there be for chemists (or biochemists) to determine, for instance, the osmotic pressure of a solution if there were no thermodynamically established connections between it and the molar mass of the solute?

RMM of small and medium size chemical may be determined by purely chemical methods that do not depend on physical theory. The highly reliable gravimetric technique provides an example of such methods. That a diligent practitioner of this technique, T.W. Richards, was awarded the 1914 Nobel Prize in chemistry illustrates the crucial role it has played in the development of chemistry. At present, however, physical methods are the ones that are most commonly used in the chemistry laboratory.

In recent years, chemical research has become increasingly dependent on instrumentation. One cannot, in many cases, tell the difference between physics and chemistry research laboratories from casual observation. Among the plethora of instruments that crowd the modern chemical laboratories are various types of spectrometers for different regions of the electromagnetic spectrum, mass spectrometers, high-vacuum apparatus for molecular beam studies, x-ray and electron diffractometers, machines for surface studies, and detectors for varieties of signals. Unlike computers in business applications, these instruments are not used as mere tools in the chemical laboratory except in the routine determination of the quantity of materials by standard methods. On the contrary, theories of physics play a central role in the design of the experiments with these instruments, and in interpretation of the data gathered from the experiments.

(ii) Chemists no longer use words like "affinity" and "valence" to describe the chemical bond. Such words are replaced by "orbital overlap," "sigma," and "pi" bonds, words that cannot be defined outside the quantum theoretical context. Even a term like "resonance," whose use in chemistry predates quantum theory, has taken on a very different meaning since the inception of this theory. Resonance, as first envisioned by organic chemists, is the possibility of simultaneous structures for a molecule. Resonance in quantum mechanics, on the other hand, is a consequence of indistinguishability of elementary particles (electrons in bonding theory). Indeed, if we were to list the words that are connected with essential concepts in analytical, inorganic, or physical chemistry, we will find that many of them owe their current meaning to theories of physics and, devoid of that theoretical context, they loose their meaning.

(iii) Modern explanation of chemical bonding depends on the wave nature of electron and often starts with atomic orbitals (AO). (We are concerned only with elementary description of bonding. Readers unfamiliar with the material will find Coulson's excellent book informative and rewarding.) In the molecular orbital method, AO from different atoms are added (with suitable coefficients) and then products of resulting molecular orbitals (MO), called configurations, are generated. Bonding and spectra are explained with reference to these configurations (or to their linear sums, in case of degeneracy). In the valence bond method products of AO from different atoms, the products supposed to represent different resonance structures, are added to give the wave functions. Both MO and valence bond theories call for a very different type of explanation for chemical bond than the one given by theories based on particle model for electron, developed before the advent of wave mechanics by Lewis, Langmuir, and Kossel. The current preference of chemists for wave theoretical explanation over the particle theoretical explanation is justified by pointing out that the former gives a quantitative account of bonding. However, the wave theory is used also for qualitative explanation of bonding, even when it does not give a better insight than the particle theory. The nature of explanation of bonding has certainly changed following the development of quantum theory. Part of the motivation to do so, it would appear from the above reasoning, was to make chemical explanation consistent with physical theory.

(iv) In recent years, the interest in chemical kinetics has moved to the area of state-to-state dynamics where dissociation of a single molecule or reaction of pairs of molecules prepared in specific quantum states are investigated. This research proceeds in a markedly different direction from that of the classical investigations (which continue to interest a majority of chemists), in which dynamics of molecules distributed among many energy states are studied. The impetus for the state-to-state kinetics comes from the development of molecular beam techniques and scattering theories in physics. While these investigations may give a theory-reductionist explanation for a few gas phase reactions between relatively simple molecules, they are not likely to play a significant role in the far wider areas of large molecule kinetics or reactions in solution where many significant questions are yet to be answered. This is because the density of energy states for large molecules, and for molecules in solution, is so high that these systems are essentially classical in nature, devoid of quantized energies. Thus, it is clear that the interest in the state-to-state kinetics arose as an important area of investigation chiefly because of the developments in physics.

The above examples amply illustrate the strong influence physics exerts on chemistry. Bunsen was perceptive in remarking: "A chemist who is not at the same time a physicist is nothing at all." (Schlag 1998). Of the major areas of chemistry (molecular structure, properties of substances, reactivity and synthesis), only synthesis is least affected by physics. When chemists synthesize new molecules, particularly in the area of organic chemistry, they frequently follow their own set of rules rather than to depend on the laws and theories of physics.

The influence of chemistry on physics is less direct. There have been important investigations in chemistry that led to developments in physics. The discovery of the third law of thermodynamics was a result of low temperature chemical equilibrium studies by Nernst. Chemical studies undoubtedly played a significant role in the early stages of the development of electromagnetism. In fact, it is electrochemical investigations by Faraday that led G.J. Stoney to coin the word "electron" and to estimate its charge (1874) before it was detected by J.J. Thomson in the gas phase (1897). To this list, we should add the modern atomic theory, which took root in chemistry before it found its way into physics.

Investigations in chemistry, however, have generally raised questions in physics without markedly altering the direction of research in that science.

DISJOINTED DISCIPLINES

The asymmetric relation between physics and chemistry has led some scientists and philosophers to the think that chemistry is reducible to physics. It is only a question of time before some equations in physics will explain all chemical phenomena. What is left for chemists is to simply follow the rules of physics as they slowly move into the yet unknown parts of chemistry. This is the impression one gets from the often quoted statement by Dirac (1929): "The underlying physical laws necessary for mathematical theory of large part of physics and the *whole of chemistry* (emphasis added) are thus completely known, and the difficulty is only that the exact applications of these laws leads to equations much too complicated to be soluble."

The first part of the statement may not be false, even though it might be offensive to some chemists. However, it is not a helpful statement for investigating the relation between physics and chemistry. It is something like telling a tourist trying to find his destination in a foreign land that the *necessary* law—that he has only two degrees of freedom—is beyond dispute. We know the law will not be violated, but it is of little consolation in finding the target. The second part of Dirac's statement is misleading. It implies that chemical phenomena will be fully explained (or explained away) if the equations deduced from the necessary laws can be solved. That this idea is seriously flawed becomes clear when we look more closely at the relation between physics and chemistry. I will consider one area of contact between chemistry and physics—quantum chemistry—to show the complex relation between the two disciplines. A previous article (Vemulapalli and Byerly 1999) examined classical and statistical thermodynamics.

The concept of chemical bond is central to chemistry. Its importance becomes obvious when we consider that the number of molecules known to chemists is $>10^6$ while the types of bonds needed to explain the variation in their properties (approximately but quantitatively) are very few, perhaps a dozen or two—even when we consider rare types. Given the number and types of atoms in a molecule, a chemist can usually predict (from an idea of bonds between atoms) the molecular geometry and classify the molecule either as a polar or non-polar molecule. This may not be as sophisticated as predicting the orbit of a planet from Newton's laws. However, when one considers the number of cases subsumed by the theory of chemical bond, it becomes an impressive achievement, rivaled by very few in the annals of scientific research.

Yet "chemical bond" can never be raised to the level of a precise quantitative concept for the simple reason that the total molecular energy is not a sum of the nearest neighbor interactions. Consider the bonding picture that chemistry books give for the water molecule. Oxygen and the two hydrogen atoms form a symmetric triangle with the O–H distances of 95.8 pm and the H–H distance of 152 pm. The two chemical bonds are indicated by lines joining the central oxygen to each hydrogen atom. As every first-year student in chemistry is led to believe, there is no bond between the two hydrogen atoms. Yet it is impossible to defend the assumption that there is a strong interaction between two atoms 95.8 pm apart and none between two atoms 152 pm apart.

The picture, unfortunately, is not improved by adding a bond between the two hydrogen atoms. The total energy of a molecule, which may have many more than three atoms, cannot be expressed as a sum of the energies of pairs of atoms or even clusters of atoms within a molecule. The potential energy between molecules (or free atoms) can be expressed, at least in principle, as a sum of the energies of interacting clusters (Meyer's cluster expansion) because the molecules are localized. Within a molecule, however, the electrons are delocalized. Hence, even in principle the energy cannot be partitioned.

Interestingly, chemists—unlike physicists and biologists who analyze at length the important approximations in their theories—have not tried to justify the assumption of bonds between only a limited number of atom pairs in a molecule, an assumption so central to chemistry. In view of this and considering its immense usefulness to chemical research, the idea of a distinct, clearly defined chemical bond is perhaps best considered a benevolent prejudice.

Quantum theory had and continues to have a crucial role in explaining the nature of chemical bond. Indeed, the first quantitative theory of the chemical bond, developed by Heitler and London, immediately followed the advent of the Schrödinger equation. Since then quantum theory has given valuable quantitative estimates of bond energies and knowledge of molecular geometries. It also led to the refinement of important concepts such as free valence, resonance energy and bond order (Coulson 1962). However, the application of quantum theory to chemistry is quite different in its methodology from the application of, say, Newton's laws to planetary orbits. In the latter case, one calculates, using some approximations, the numerical values of one of the variables that appear in the theory. For instance, the radial distance of the planet appears as a variable in the mathematical formulae for potential energy and force. Thus, we can say that Newton's theory projects, through calculations, the information on quantities that are part of the theoretical framework. By contrast, when quantum theory is applied to chemistry, the quantities for which we seek information are not those that appear in the Schrödinger equation. Positions and momenta, even in the probabilistic sense, have no significance in chemical bonding theory. The wave function in the Schrödinger equation for many-electron systems is a complete unknown.

Neither bond energy nor bond length are quantized properties. There are no representative variables for them in the Schrödinger, Heisenberg, or Dirac formulations of quantum theory.

The main point is this: applications of quantum theory to chemical bonding involve calculation of quantities that have no representation or status in the original theory. Hence, many extra-quantum assumptions must be made to adapt the basic theory to the form that is useful in chemistry. Broadly speaking, there are three types of assumptions in applying quantum theory to the chemical bond. The first type may be justified on theoretical grounds, even though they are not part of the original theory. The second set of the assumptions is theory-neutral. Theory allows several equivalent styles for interpreting bonding but chemists choose one particular style. The third set of assumptions actually contradicts the original theory, at least in spirit. Let us examine each case separately.

(i) The wave functions used in quantum chemistry are not the analytical solutions of the correct Schrödinger equation, the basic law of quantum theory. They are obtained by judiciously adding a set of mathematically convenient functions. In the language of quantum theorists, they are superpositions of familiar basis sets of functions that may or may not be eigenfunctions of any Hamiltonian, the quantum operator for energy that is at the heart of the Schrödinger equation. This procedure is justified because any desired function in a region of space may be approximated by a sum of other functions (basis set) in the same region. (If the basis set is a complete set, the desired function becomes identical to the accurate solution.) In practice, however, calculation with a complete basis set is an impossibility. Hence, quantum chemical calculations are done with a highly truncated set often with functions that do not have all the required mathematical properties and almost always with an approximate energy operator. Beyond this there are several assumptions made in getting the weighting factors in the summation (superposition).

As the calculations are done with the approximate functions, the variation principle is used as a yardstick for judging the merit of quantum chemical calculations. According to this principle, the energy calculated can never be lower than that of the lowest energy state of a given symmetry. Because of this principle, chemists were able to "mathematically experiment" with a variety of arbitrarily chosen functions and with different approximate Hamiltonians, and find those techniques that give the lowest possible energy. The function that gives the lowest energy is then assumed to be the best approximation to the true wave function and is further used to compute other properties of molecule. The experimental chemist's interest is with the properties and not the total energy of the molecules. There is no guarantee that a wave function that gives the best energy also gives best estimates of other properties, but this procedure is followed for lack of alternatives. I mention this to indicate that quantum chemical calculations involve not only many approximations but also further assumptions in connecting the theoretical results with experimental observations.

It is not my intention to denigrate the quantum chemical calculations, or the qualitative and quantitative interpretations based on those calculations, since they depend on some necessary assumptions. The importance of these calculations to chemistry cannot be overestimated. Quantum calculations now rival precise experimental measurements and quantum chemistry provides an invaluable resource for data that cannot be easily obtained in the laboratory. Quantum chemistry also provides the theoretical framework to interpret the otherwise diverse data.

Nevertheless, it should be emphasized that the epistemological and the theoretical connections between quantum laws and quantum chemical explanations of bonding are far removed from deductive reasoning; they are not mere extensions of quantum rules to chemistry. They do not conform to a reductive model for relation between physics and chemistry.

The methodology behind the development of quantum chemistry is quite dissimilar to the development of physics starting with Newton's laws. The oxymoron "inductive theory" best describes the methodology one follows in adapting quantum mechanics to chemistry. For a few molecules in each class, quantum chemists find by trial and error the best set of approximations for the Hamiltonian, for the truncated basis set, for the evaluation of matrix elements, and for various other theoretical entities. After that they use the thus developed "theory" (their word for a set of approximations) for other molecules in that class. (For instance, the popular Hückel theory involves ignoring almost all but the minimum number of integrals and approximating the remaining from empirical considerations.) This reasoning is not dissimilar to the one that leads to the conclusion that all swans are white-based on repeated observations, except that induction is on the set of approximations that give reliable results for a class of molecules rather than on the complexion of a species of birds.

Max Born wrote: "But I believe that there is no philosophical highroad in science, with epistemological signposts. No, we are in a jungle and find our way by trial and error, building our road *behind* us as we proceed. We do not *find* signposts at crossroads, but our own scouts *erect* them, to help the rest" (Born 1956). This statement, with the words emphasized by Born, comes very close to describing the actual methodology used in developing the quantum theory of chemical bonds.

(ii) Mixing of AO on the same atom is known as "hybridization" in chemistry. The hybridization concept plays a very important role in organic and inorganic chemistry, since it allows the chemists to visualize directed bonds and thus the geometry of a molecule. Electron charge distribution in an atom is spherically symmetric. However, we can induce directed charge distribution suitable for formation of spatially oriented bonds by judiciously adding AO. However, there is some confusion about the theoretical status of the concept of hybridization. (Coulson is a good source for readers interested in knowing the mechanics of hybridization.) Consider the set of hydrogen orbitals with $n = 2$ quantum number. Three of p and one of s symmetry constitute the set of four functions. Since these functions are degenerate, any linear combination of these functions is an equally good representation of the $n = 2$ level in the hydrogen atom. Thus, in case of degeneracy, hybrid orbitals are equivalent to the usual hydrogen orbitals, which are the products of spherical harmonics and associated Laguerre polynomials. No causal agent is needed to hybridize the orbitals. Whether we consider a set of orbitals, hybrids or not is a matter of choice.

Consider now a closed shell configuration such as that of a neon atom. The total electronic wave function is a Slater determinant with 10 rows and 10 columns, if spin is explicitly shown. Now suppose we write the Slater determinant such that each row corresponds to one of the spin orbitals. Since adding rows (with proper weighting factors) does not change the determinant, the hybrid and non-hybrid orbitals are equivalent representations of the total electronic wave function. Similar arguments can be used to show that there are several equivalent representations for orbitals in a bonded atom. The use of hybrid orbital representation, while not inconsistent with the basic quantum theory, does not follow from it. It is of course convenient to start with hybrid orbitals for describing bonds between atoms. It also fits the traditional views of directed bonds developed by van't Hoff and Le Bel. The chemical bond is not a causal agent for hybridization as textbooks often labor to explain. Nor is hybridization a consequence of quantum laws.

(iii) Chemists find concepts like "percentage ionic character of the bond" and "electronegativity of atoms" useful in ordering molecules according to their polarity. These concepts apply to models that are not consistent with quantum logic, even though they are often described in quantum chemistry books along with wave functions, matrix elements, and other tools for quantum calculations. Consider what is required to define percentage ionic character purely in quantum mechanical terms. All the information about a system, according to quantum theory, is coded in its wave function. Hence, in order to identify the ionic character of a bond we must find the weighting factor for the ionic component of the wave function. When this is done, however, we find that it is not orthogonal to the covalent component in which electrons are shared equally between the two atoms. Thus, ionic and covalent characters are not independent of each other, one implies the other. The main point is that chemists find it very useful to assume distinct ionic and covalent characters for a bond to describe properties of a molecule and for comparing them with properties of other molecules. In quantum theory, however, the concepts of covalent and ionic bonds cannot be separated into mutually exclusive concepts. There is no consistent way to patch up this difference. Given this situation, chemists have not abandoned the notion of distinct ionic character of a bond. Instead, they have adapted the theory—it is more appropriate to say they have twisted the theory—for their purpose by re-defining rules of interpretation, keeping in mind that quantum theory, in this context, can at best give guidelines for qualitative interpretation. This is precisely the methodology followed by the L. Pauling, R.S. Mulliken, C.A. Coulson and other pioneers who adapted quantum theory to chemistry.

RELATION BETWEEN CHEMISTRY AND PHYSICS

When examining the relation between two disciplines of science, it is useful to contrast what is provisionally postulated from what has been accepted as fundamental. Science, after all, is a dynamic enterprise that gathers new information while discarding or modifying once accepted hypotheses, and changing its interpretation of observed data. Scientists make many highly speculative hypotheses during the course of theory development, some of these survive and others fall by the wayside. Feyerabend's (1970) colorful statement that "anything goes" is not far from reality during the initial development of scientific hypotheses. However, after the hypotheses and the theories are accepted by the scientific community—always tentatively—one begins to distinguish spurious connections from systematic correlations, logically dictated relations from accidental parallels.

During the last half of the past century scientific knowledge expanded at a rate unprecedented in the history of human thought. As a consequence, it is now virtually impossible for anyone to have a bird's-eye view of even a small part of science, let alone the whole of it. The constant change makes it a struggle even to keep up with one's own specialty. Because of this, scientists necessarily specialize narrowly and isolate themselves from all but a handful of other scientists working in closely related areas. If a scholar in the humanities is likened a person with an impressionistic view of an open landscape, a scientist is like a superbly skilled miner confined to exploring a small region.

Many ideas on chemical bond are rather firmly established now, after more than a century of investigations following the discovery of electron; they are accepted by scientists in widely different areas of specialization. We can now confidently distinguish between transient hypotheses and those with a level of stability. When this is done, we will not fail to notice the pattern of interaction between chemistry and physics that I have detailed earlier. Among the chemical bond concepts that scientists (not just chemists) hold indispensable both for explaining what has been experimentally established and for guiding what is yet to be discovered, some can be derived (with appropriate assumptions) from quantum theory, some are conceptually incompatible with quantum theory, and the rest independent of quantum theory. One finds similar relationship in the overlapping between chemistry and physics in thermodynamics and statistical thermodynamics (Vemulapalli and Byerly 1999).

The above analysis shows that the relation between chemistry and physics cannot simply be one of reduction, except in a restricted ontological sense. Nevertheless, it is not difficult to find statements in the literature to the effect that chemistry either has been reduced or potentially reducible to physics in articles that show skepticism about generality of reduction. For instance, Klee (1997) claims: "If, for example, chemistry is reducible to physics, as indeed it seems to be, then if we were to make an ontological inventory of all the kinds of objects and properties there are in the world, we would not in addition to all the physical kinds of objects and properties (i.e., all the kinds of physical subatomic particles and their subatomic properties) need to list any chemical objects and properties." He concludes later, "There have been a few successful reductions in the history of science . . . . The reduction of chemistry to physics has already been mentioned." Klee does not give any support for this claim, except that explanation of chemical phenomena by physics is supposed to follow naturally, given the ontological dependence of chemistry on physics in the sense that the objects and the properties that appear in chemistry can be defined in terms of quantities in physics.

Derry (1999) writes: "Chemistry studies reactions between atoms and molecules, forming and breaking chemical bonds. The bonds are formed by electrons, subject to the equations of quantum physics. So chemistry is reducible to physics." Chemical bonds are not, contrary to this claim, "subject to" in the sense of their being "reducible to" quantum physics. Thus, we cannot conclude that chemistry reduces deductively to physics.

It is true that the same elementary particles appear as ultimate constituents of matter in both chemistry and physics, and that the states of chemical systems are determined by (supervene on) the behavior of these particles. I do not believe that there is a single chemist who does not accept the idea that the "particles" of physics and chemistry are the same and that chemistry need not, and should not, introduce material entities that are not composed of entities postulated by physicists. However, this realization alone is insufficient to explain chemical phenomena, to develop chemical theories with predictive powers, or even to provide models for visualization.

Given, the strong ontological connection between chemistry and physics and only tenuous epistemological relations between the two disciplines, it may be futile to look for a simple or elegant model that would characterize the relation between the two. Indeed, the only relevant question may be the one I posed earlier. What is the nature of the laws and theories of physics that have such strong influence on chemistry and yet cannot subsume the concepts and models that arise in chemical systems in accordance with these very laws of matter?

The use of the word "law" in scientific literature is both ambiguous and vague. Thermodynamics and classical mechanics are said to be grounded in a set of fundamental laws. The nature of these laws is very different from that of the laws of observed experimental regularities, for example, Dulong and Petitt's law for heat capacities of monatomic crystals and Raoult's law for the vapor pressure of solutions. Quantum mechanics and statistical mechanics are identified as theories that are introduced through postulates sharing the universal character of the laws of thermodynamics and the classical mechanics. Generally speaking, what are called laws in science include statements of experimentally observed regularities in a class of systems, theoretical principles that govern such regularities, and idealized models that explain the regularities. In this inclusive sense, there are many laws both in chemistry and physics. However, only a few of them are fundamental, in that they subsume other laws or provide rationalizations for their validity. In the following discussion, I assume that the laws associated with thermodynamics, statistical mechanics, quantum mechanics, and classical mechanics are the fundamental laws of physics. These laws have, historically, shaped chemical theory to be what it is at present. Physical laws such as those of relativity have not played a significant role in chemical theory.

Now we are in a position to examine the answer to the questions raised earlier. The fundamental laws mentioned above posit restrictions on what can happen in nature. They are, on the whole, silent about which possibilities actually occur in nature, even when they provide the theoretical tools for analysis of natural processes and predictions of outcomes of measurement. They are not epistemological blue prints but ontological limits. Only if we accept this view we can understand how they have been effective in initiating new directions in chemistry and while allowing independent developments in chemistry along its traditional lines. A chemist devising new theories along with new experiments or simply providing explanation for observed chemical behavior cannot violate the boundaries set by the fundamental laws. Within these boundaries, however, there are many possibilities not forbidden by the basic laws, but only a particular mechanism fits the specific chemical phenomena under investigation. For a chemist, the mechanism by which a bond forms is as important as the general properties of the bond between them. This mechanism must be discovered by chemists as they try different hypotheses within the broad framework of the physical laws. The application of physical laws into new areas is not generally a simple, direct extension of the existing physical theory.

Quantum theory, as pointed out earlier, has been successful in providing valuable insight into chemical bonds. It has also provided a basis for interpretation of molecular properties. Among the ideas that form the logical foundations of quantum theory are the wave–particle duality, the Born interpretation of P, and the uncertainty and commutation relations. However, these are of only peripheral interest to a chemist interested in the application of quantum theory to chemical bonds. Indeed, many practicing quantum chemists rarely concern themselves with the foundations of quantum theory or how chemical information is related to them. There is no possibility of finding exact solutions to the Schrödinger equation for molecules using the correct Hamiltonian. Thus, a direct extension of wave theory into chemistry to obtain eigenfunctions is impractical. Further, lack of symmetry in molecules makes it impossible to derive the eigenvalues from operator algebra.

The extension of quantum laws into chemistry, as described earlier, depends on the variational principle. The essential part of quantum calculations, as far as chemistry is concerned, is the lower limit that variation principle puts on energy. Instead of trying to derive the principles of bonding from quantum theory—which would have been futile—chemists have explored many likely mechanisms with the confidence that their calculations will not yield energy below the limit set by the variation principle. Imagine what it would be to develop chemical theory without the variation principle. Since, in that case, the calculated energy could be above or below the true value, which is an unknown; there could be no way to judge the merit of theoretical models and no indication on how to refine them. It is the lower limit on energy set by the variation principle—the ontological boundary—and the chemists' efforts to reach that boundary, and not the deductive apparatus of quantum mechanics, which have contributed to the development of chemical bonding theory. That quantum chemical calculations may not reach that boundary with true Hamiltonians is of little consequence to chemists. The major gift of physics to chemistry is the boundary delineated by the former and not the paths toward that boundary.

POSTULATES OF IMPOTENCE

Some philosophers have wondered about the distinction between science and, say, such areas as astrology and voo doo. Popper came to the conclusion that only scientific theories are falsifiable and therein lies their authenticity. Feyerabend championed the anarchist idea that there is no privileged position for one "discipline" over the other. My study of relation between physics and chemistry leads to an alternate view. There is a distinction and that stems from the fact that science, based on theory and experiment, discovers the ontological boundaries in nature. Scientists working within these boundaries gain insight about nature, sometimes using orthodox methods and sometimes deviating from them. Voodoo and astrology, on the other hand, have whims for boundaries. Science recognizes and admits a perimeter beyond which we cannot reason: voodoo and astrology, do not.

The features of the relationship between physics and chemistry, described above, are not restricted to quantum theory. Let us consider the second law of classical mechanics. Its main application to chemistry is in the study of transport (in gaseous and liquid phases) through which chemists derive information on the size and shape of molecules. The chemical application of the second law depends on the assumption that

$$f \geq ma$$

In the ideal case where a molecule is studied in isolation (e.g., in the kinetic theory of ideal gases) the equality sign holds. In all other cases, the inequality sign is what governs the application of the law. In studying sedimentation and diffusion of macromolecules in solution, for example, one assumes that force of acceleration is greater than the product of mass and acceleration. In adapting the above equation to actual problems opposing force terms are added on the right hand side. The parameters in these terms are almost always obtained from extra-theoretical considerations. We see from this that the chemical applications of the second law depend on it being a limiting law—a law that sets the boundary for the permitted dynamic behavior.

This last point suggests an interesting analogy between chemistry and engineering in their relation to physics. Although the question whether chemistry has been or could be reduced to physics has often been raised, one rarely hears a debate on the question whether engineering reduces to physics. This lack of curiosity seems due to the tendency to separate science and engineering into two different cubbyholes. The engineer's dependence on physics is rather similar to that of the chemist. One does not, in general, use the laws of physics (e.g., Newton's second law or the second law of thermodynamics) in engineering to predict the behavior of devices by trying to deduce that behavior. Instead, the laws are considered limiting laws, benchmarks that allows the optimization of a device. Even though the aims of chemistry and engineering are quite different, physics plays a similar role in both areas.

Though there is no simple model for the complex relation between physics and chemistry or between physics and engineering, as a rough description it might be said that physics is like the constitution of a nation, a constitution that prohibits certain activities but does not give any prescription as to the allowed activities that should be pursued. Physics, in a similar vein, places limits. Within those limits, there is a rich variety of possibilities, which cannot be derived simply from the fundamental laws.

Whittaker (1951) identifies laws of physics that have "permanent and absolute character" as *Postulates of Impotence.* They are negative statements in the sense they posit that something is impossible. It is well known that the laws of thermodynamics are qualitative assertions of what cannot happen. Whittaker successfully argues that the fundamental laws of electromagnetism, relativity, and quantum mechanics are also based on negative statements. He then claims, "The postulates of this kind already known have proved so fertile in yielding positive results—indeed very large part of modern physics can be *deduced* from them—that it is not unreasonable to look forward to a time when entire science can be *deduced* by syllogistic reasoning from postulates of impotence (emphasis added)." I concur with the view that it is the negative character of the laws of physics that give them the permanent and absolute status. I disagree, however, that any science (including physics applied to materials) can be deduced from the postulates of impotence, if by deduction one means development of quantitative theory from the laws, without further assumptions and without input from experiment. Indeed, careful examination of branches of physics, for example, solid state physics, shows that many theoretical strategies are tried within the framework of the laws, and the relation between them and the postulates of impotence are very much like what I have documented for the chemical bond.

This loosely defined relation between the disciplines of chemistry and physics—and the difficulties of deducing the former from the latter—makes some scientists, both physicists and physical chemists, conceive chemistry as an approximate and imperfect branch of physics. Until a rigorous theory is developed, they think, we must make do with what we have. It is only when better computers and better mathematical techniques evolve that we can really solve the problems of chemistry. This is the message one gets from the second part of Dirac's statement quoted above. I believe that this is a totally wrong view. Better computers and better mathematical techniques might allow accurate numerical match between experiments and computations, but in that very process the mechanisms that interpret the data and give insight into the nature of chemical processes will be lost. Consider the case of non-ideal gases. We now have the computer power to fit the pressure–volume–temperature data to any level of accuracy desired. But in that process we lose all insight into molecular behavior. The resulting "theory" is far inferior to the approximate van der Waals equation, which relates the behavior of gases to the size of the molecules and the forces between them. It is worth noting that this equation, while consistent with the laws of physics, is not derived from them. Gain in formalism does not necessarily lead to improvement in understanding; it may even prevent understanding and thus further investigations.

CONCLUSION

Well-established concepts in one area of science often find natural applications in other areas of science. In this regard, physics plays a pivotal role. Investigations in physics focus, for the most part, on systems with small number of actual or reduced variables and lead to firmly established concepts and fundamental laws. It is thus natural for chemistry to look into physics for guidance as it explores the unknown territories. The main thesis of the article is that physical laws are about limitations. While they set the boundaries within which chemical mechanisms must fall, they cannot dictate the actual mechanisms. Thus, chemistry has to be viewed as both dependent and independent of physics. The relation between physics and other areas of scientific inquiry is likely to follow the pattern described here: within the possibilities allowed by physics, scientists have to look for explanations that depend on factors outside physics.

ACKNOWLEDGMENT

I am deeply indebted to Professor Henry C. Byerly. This article owes much to his advice and encouragement.

REFERENCES

Born, M. 1956. *Experiment and Theory in Physics*. New York: Dover Publications.

Coulson, C.A. 1962. *Valence*. New York: Oxford University Press.

Derry, G.N. 1999. *What Science is and How it Works.* Princeton: Princeton University Press.

Dirac, P.A.M. 1929. Quantum Mechanics of Many-Electron Systems. *Proceedings of the Royal Society (London)*. A123: 714–733.

Feyerabend, P. 1970. Against Method: Outline of an Anarchistic Theory of Knowledge. In: Radner, M. and Winokur, S. (eds.), *Minnesota Studies in the Philosophy of Science*, Vol. IV. Minneapolis: University of Minnesota Press.

Klee, R. 1997. *Introduction to the Philosophy of Science*. New York: Oxford University Press.

Popper, K. 1959. *The Logic of Scientific Discovery*. New York: Basic Books.

Scerri, E.R. 2000. Realism, Reduction and the 'Intermediate Position'. In: Bushan, N. and Rosenfeld, S. (eds.), *The Philosophy of Chemistry*. New York: Oxford University Press.

Schlag, E.K. 1998. *Zeke Spectroscopy*. New York: Cambridge University Press.

Vemulapalli, G.K. and Byerly, H. 1999. Remnants of Reductionism. *Foundations of Chemistry* 1: 17–41.

Whittaker, E. 1951. *Eddington's Principle in the Philosophy of Science*. New York: Cambridge University Press.

SECTION V

CHEMICAL THEORY AND FOUNDATIONAL QUESTIONS

CHAPTER 11

SOME PHILOSOPHICAL IMPLICATIONS OF CHEMICAL SYMMETRY

JOSEPH E. EARLEY

Department of Chemistry, Georgetown University, 502 W Broad St, Suite 501, Falls Church VA 22046, USA; E-mail: earleyj@georgetown.edu

INTRODUCTION

This paper deals with a fundamental insight that was long in development and slow in diffusion—but which has deep philosophical implications. The insight is that the symmetry properties of any object are not mutually independent—they generally come in bundles. The collection of symmetries of some object (those that preserve aspects of the object's structure that are relevant to the problem at hand) is considered to constitute a *group*. The branch of mathematics that deals with such bundles of symmetries is designated *group theory*.

Philosophers have often discussed symmetry (and group theory) in systems treated by quantum mechanics, but have shown less interest in systems mainly considered by chemists. This paper deals with two types of chemical symmetry, and suggests that *closure* (the feature that distinguishes groups from other sets) may deserve increased philosophical attention.

Symmetry may be defined as "immunity to a possible change" (Rosen 1995). If a planar BF_3 molecule (on the left in Figure 1) were to undergo a rotation of 120° around a line that passes through the B atom and is perpendicular to the plane of the molecule, the molecule would assume a configuration indistinguishable from the original configuration. The BF_3 molecule is immune to the possible change of rotation about that "threefold" axis. In contrast, if the $BClF_2$ molecule (on the right in Figure 1) were to undergo a similar rotation, the resulting configuration of atoms would be quite different from the original one. Every statement about molecular symmetry must include specification of an operation and of some geometric entity—rotation and a line (symmetry axis) in the example.

Since all four atoms lie in the same plane in the lowest energy form of the BF_3 molecule, the molecule is also symmetric with respect to three other rotations (one about each of the B–F bonds) and it is also symmetric with respect to reflection through the plane of the molecule. These six symmetries (as a group) specify much of the structure of the molecule.[1]

D. Baird et al. (eds.), Philosophy of Chemistry, 207–220.

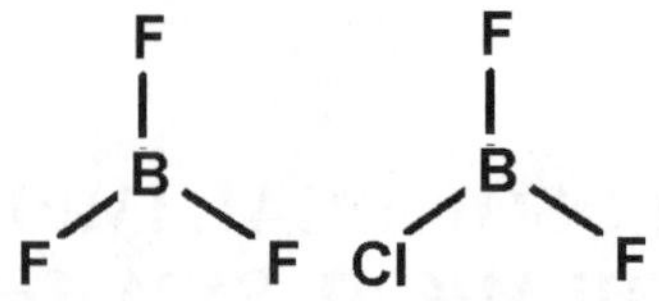

Figure 1: Geometric structure of two planar molecules.

Astrophysicist Arthur Eddington made a remarkable claim:

> What sort of thing is it that I know? The answer is *structure*. To be quite precise, it is structure of the kind defined and investigated by the mathematical theory of groups (1939, p. 147).

That is to say, whatever Eddington—and presumably the rest of us as well—might possibly know is said to be somehow covered by group theory. Group theoretical reasoning was first applied to physics only in the years just before this claim was made—at that time, many physicists did not share Eddington's enthusiasm for that approach (see below). Each subsequent decade has seen an expansion of the applicability of group theoretic reasoning. The recent revival of philosophical structuralism (in several versions) (French and Ladyman 2003) may be seen as a further expansion of appreciation for the power of group theoretic approaches. This chapter introduces some basic concepts of group theory in the section on Some Concepts of Group Theory; the section on Development of Group Theoretical Concepts gives an account of the origin and spread of group theoretic ideas; the section on Chemical Substances: What "Is" Might Mean deals with chemical substances of the usual sort; and the section on Groups of Processes and Chemical Reaction Networks concerns consequences of symmetries in collections of chemical reactions.

SOME CONCEPTS OF GROUP THEORY

Formally, any *collection* of elements (of whatever sort) constitutes a *set*. (Let X_n designate the set of *n elements*, x_i.) A *groupoid* involves a set S (*the carrier*) and a *binary operation*, $\cdot$ (some procedure that can be applied to two elements of the carrier set to yield a single result). If the operation is associative, i.e., $a \cdot (b \cdot c) = (a \cdot b) \cdot c$, the groupoid merits the designation *semi-group*. Semi-groups may or may not contain an *identity element*, a unique element that, when coupled with a second element (via the group-defining operation $\cdot$), regenerates the same (second) element, i.e., $e \cdot a = a = a \cdot e$. The identity element is often designated as 1 (Baumslag and Chandler 1985).

Elements of sets can be of any sort whatsoever. Some sets have *elements* that are *mappings* (transformations) of other sets. A particularly interesting kind of mapping, when applied to members of a set X, generates results that are *themselves members of the same set* X. (This is a mapping of X onto itself, sometimes called an *endomap*.) If, for a certain set X, several different mappings (endomaps) of this sort (p, q, r, . . .)

exist, *those mappings may themselves be elements of another set, M_x*—the set of *mappings of set X onto itself.* A semi-group can be built on set M_x, using *composition of mappings* as the semi-group-defining operation, $\cdot$. That is, $p \cdot q$ corresponds to the composite mapping obtained by first applying mapping p to member of X (generating another member of X) and then applying mapping q—so that $x_{(p \,\cdot\, q)} = (x_p)_q$. Since each of the initial mappings, when applied to any member of X, generates another member of X, all composites of such mappings also generate members of X. The general existence of such sets of endomappings is asserted by *Cayley's theorem*: for any semi-group S, a one-to one mapping of S onto M_x exists, for some suitable X. This type of semi-group-endomap operates under an important kind of *closure*.

A *group* is a semi-group that has an identity element, and for which each element has *an inverse*—an element of the group that yields the identity element when coupled with the given element, i.e., $a \cdot b = 1 = b \cdot a$. (The inverse of element a is a^{-1}.) In summary, a *group* consists of a set of elements and a binary operation such that: (i) application of the operation is associative, (ii) an identity element exists, and (iii) every element has an inverse. Groups may contain *subgroups*, smaller collections of elements that themselves fulfill the criteria of being a group.

A set, S_x, of all *one-to one mappings of X onto itself* is called a *symmetric group on X*. Application of the defining operation (composition of mappings) to any pair of elements of the group (mappings of X onto X) *generates a member of the group, i.e.*, $a \cdot b = c$. Both a and b are elements of the group and c is also an element of the group (all of these are mappings). That is to say, symmetric groups have the property of *closure*. Closures of related sorts abound in nature. (A main point of this paper is that achievement of closure merits increased philosophical attention.) The members of symmetric groups are the possible *permutations* (rearrangements) of the elements of the set on which the group is based. Cayley's theorem shows that every group has the same structure as (is *isomorphic* to) a group of permutations. That is, every group is isomorphic to a subgroup of the symmetric group that is based on some suitable set.

It seems that some further illustration of what is meant by *closure* would be in order. If n is the *number of members* of a certain set X, then there are $n!$ (n factorial: $n \times n-1 \times n-2 \times \ldots 1$) *elements* of the corresponding symmetric group S_n. (These $n!$ variations are the *permutations* of X.) If there are *three* members of X (1, 2, 3, say), there are *six* permutations of X:

$$i = 1\,2\,3, \sigma_1 = 2\,3\,1, \sigma_2 = 3\,1\,2, \tau_1 = 1\,3\,2, \tau_2 = 3\,2\,1, \text{and } \tau_3 = 2\,1\,3$$

There are then six (only six) equivalent members of the group S_3—the six *mappings* that yield each of the six permutations from the starting configuration, 1 2 3.

For example, consider BF_3-like molecules. For a planar equilateral triangle with vertices labeled 1, 2, and 3 (Figure 2), rotation by 120° about the threefold axis (the line perpendicular to the plane of the triangle and intersecting the center of the triangle) corresponds to σ_1. Rotation about the same axis by 240° corresponds to σ_2. The mappings τ_1, τ_2, and τ_3 correspond to rotations by 180° about each of the three twofold axes that connect the vertices and the center of the triangle. Successive application of any number of these mappings (the defining operation of the group)

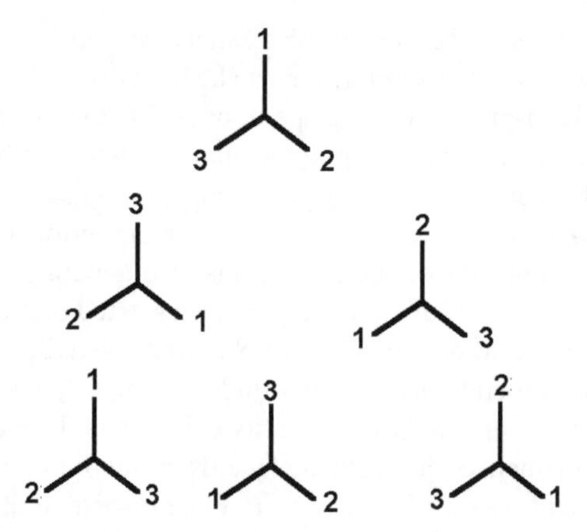

Figure 2: Possible permutations of a planar structure with threefold symmetry.

to any permutation of X generates another permutation of the set X. The identity element i (leave it alone) exists. Each element Z has an *inverse* Y such that $Z \cdot Y = 1$. (On this basis S_3 is a group, not a semi-group.)

There turn out to be only six different ways for a molecule (or another kind of object) to exemplify the group S_3—there are just six "realizations" of the group. In every case, there are a number of ways in which a particular group can be realized. For most groups that number is quite large. Such realizations usually cluster into types (as the two σ and three τ representations in S_3). The S_3 group and its properties are relevant to all planar trigonal molecules, and also to *any question whatsoever* for which three *fully equivalent* entities interact. The abstract S_3 group is relevant to a vast variety of problems—this group summarizes what must be the case in any situation to which it applies.

DEVELOPMENT OF GROUP THEORETICAL CONCEPTS (AFTER WUSSING 1984)

Traditional artists in many cultures were quite familiar with spatial symmetries, and explored the wide (but limited) range of possible two-dimensional geometric arrangements of objects. Formal geometric reasoning was highly developed by mathematicians in the ancient world, particularly in Greek-speaking areas. Dramatic mathematical developments in the late 18th century led to several types of abstract geometry. Although group theory is now regarded as the study of symmetry, the first steps that we now see as related to that approach did not come from geometry. The initial insights were gained in investigations of J.-L. Lagrange (1736–1813) and others as to why certain techniques involving extracting roots (radical methods) that were successful for solving algebraic equations of the third and fourth degrees did not work for equations of higher degree. In 1799, P. Ruffini (1765–1822) published a demonstration that it was *not* possible to solve equations of the fifth degree by radical methods. He used an approach that involved systematic changes (permutations) and explicitly used a closure property. A.-L. Cauchy (1789–1857) and N.H. Abel (1802–1829) extended and systematized Ruffini's insights. Early in the 19th century, it came to be recognized that these results were related to the work of L. Euler (1707–1783), C.F. Gauss (1777–1855), and L. Kronecker (1823–1891) on permutations as a part of number theory.

A major advance (arguably, the *discovery* of group theory) was made in 1832 by E. Galois (1811–1832), who attacked the problem of the solvability of equations using reasoning that would now be considered as of an explicitly group theoretic kind. Galois is considered to be the first to understand that the algebraic solution of an equation was connected with a group of permutations related to the equation. But the work of Galois was so concise as to be nearly unintelligible. In 1852, E. Betti (1823–1892), relying on the Liouville's 1846 edition of the work of Galois, applied group concepts from permutation theory to the solution of algebraic equations. Commentaries by C. Jordan (1838–1922) in 1865, 1869, and 1870 indicate that the wide significance of group theoretic methods was beginning to be appreciated by mathematicians. The work of J.-A. Serrat (1819–1885) indicates that by 1866 permutation group theory was an independent area of investigation. A.J. Cayley (1821–1895) connected permutation group theory with geometric studies. In 1872, F. Klein (1849–1825) initiated his *Erlangen Program* to apply the notion of a *transformation group* to a wide spectrum of problems in "geometries." In the early 1900s, mathematicians and physicists, especially at Göttingen, applied these methods to a variety of scientific problems. By 1920, group theory had reached its mature form.[2]

In 1926, E. Wigner (1902–1995) submitted a paper to the *Zeitscrift der Physik* showing that group theoretical approaches were applicable to certain quantum mechanical problems, specifically to understanding the spectra of polyelectronic atoms. He expanded that paper into a book that was published in German in 1931. In 1927, H. Weyl pointed out that, although formulation of quantum mechanics in terms of Hilbert space [due to J. von Neumann (1903–1957)] clarified how physical statements could be deduced from (self-adjoint) operators corresponding to physical observables, the deeper question of how one arrives at such operators could only be adequately treated using group theory.

In the preface to the 1959 English edition of his book, Wigner wrote:

> When the original German version was first published, in 1931, there was a great reluctance among physicists toward accepting group theoretical arguments and the group theoretical point of view. It pleases the author that this reluctance has virtually vanished in the meantime and that, in fact, the younger generation does not understand the causes and basis of this reluctance. (Wigner 1959, v)

(Eddington's sweeping statement, quoted at the beginning of this paper, was published in 1939.) In the 1940s and 1950s, physical chemists used group theory in dealing with solid-state structures and vibrational (IR and Raman) spectra. In 1958, Basolo and Pearson showed how an approach based on group theoretical reasoning (crystal field theory) rationalized the colors, magnetic properties, and reaction mechanisms of inorganic complexes. Cotton's 1963 primer *Chemical Applications of Group Theory* completed the naturalization of group theoretic methods in the chemists' domains. In 1965, Jaffe and Orchin published *Symmetry in Chemistry*, bringing application of group theory to spectroscopy of organic compounds to a wide chemical audience.

Suppes (1988) explicitly employed concepts of group theory in his philosophical discussion of models in science. He took the theories being considered as pre-existing, and also assumed that the elements of each of those theories constitute a *closed* set (corresponding to a mathematical group). He considered that two or more models (used in scientific theories) are *isomorphic* if they correspond to the same mathematical group. van Fraassen (1989) discussed several types of arguments involving symmetry—mainly to assist scientists and philosophers to resist "the siren song of empirical probabilities determined *a priori* on the basis of pure symmetry considerations." Like Suppes, he assumes, on occasion, that closure of elements is a widespread feature of entities and theories encountered in science—but he provides no account of the basis of those types of closure.

The question of how a collection of parts might possibly come to constitute a coherent whole—how closure might be obtained—is discussed obliquely in Rescher (1973) and van Fraassen (1991). Neither author gives an adequate basis for assumptions like those made implicitly in Suppes' treatment of scientific theories. Recent discussions of emergence (e.g., Klee 1984; Newman 1996; Humphreys 1997; Wilson 1999; Kronz and Tiehen 2002) do not explicitly make the important point that *closure* of networks of relationships (either in space or time) has important consequences. Silberstein and McGeever (1999) distinguish between *ontological emergence*—"features of systems or wholes that possess intrinsic causal capacities not reducible to any of the causal capacities of the parts nor to any of the (reducible) relations between the parts"—and *epistemological emergence*. The latter is "merely an artifact of a particular model or formalism generated by macroscopic analysis, functional description or explanation." That paper does not mention the significant difference between collections of parts that participate in a closure of relationships that correspond to a mathematical group, and aggregations of fragments that do not display such closure—this is the "group theoretical point of view" mentioned by Wigner.

There has been a recent increase in interest in philosophical structuralism (e.g., French and Ladyman 2003). Epistemological structural realism (ESR) holds (with Eddington) that our knowledge of nature is restricted to its structural aspects. Ontic structural realism (OSR) adopts the more radical position that nothing other than structures exist (suggesting the slogan "relationships without relata"). van Fraassen (1999 and forthcoming) has pointed out serious problems with each of these approaches and proposed an alternative, but related, conceptual scheme. All these discussions have principally been concerned with quantum mechanics. This paper considers some *chemical* applications of what Wigner called "the group theoretical point of view."

CHEMICAL SUBSTANCES: WHAT "IS" MIGHT MEAN[3]

Many chemical species are members of sets of *isomers*—two or more chemical substances that are made up of exactly the same atomic constituents but differ in properties. Although the number and identity of constituent atoms is the same for all members of a set of isomers, the various members of such sets often have quite diverse effects—for instance, on biological organisms—in some cases one isomer is therapeutic while another is quite toxic. Such divergence in behavior derives from differences in the symmetry of the units (often, molecules) that comprise those stuffs. Clearly, the formal treatment of symmetry—group theory—is highly relevant to chemistry. That point of view is especially useful in treating mutual inter-relations between the properties of composite units (e.g., molecules) and the characteristics of the components (e.g., atoms) of which those composite entities are made up.

The designation "mereology" [coined by S. Leśniewski (1886–1939)] is sometimes "used generally for any formal theory of part-whole and associated concepts." Martin (1992) claimed that chemical combination is not adequately dealt with by standard mereology. It may be that the group theoretical point of view has not yet been adopted by those that specialize in mereology. Lewis' book (1990) on the logic of set theory does not discuss group concepts. Simons (1987) considers integral wholes—the sort of coherence that molecules and other chemical entities exemplify—but only in a brief final chapter mainly based on an early article by Rescher and Oppenheim (1955), which derived from C. von Ehrenfels' 1890 discussion of the *Gestalt* concept.

A "mereological whole" is an individual that is comprised of parts that are themselves individuals (Simons 1987). Any two (or more) individuals can comprise a mereological whole—the star Sirius and your left shoe, for instance. Concerning such loose aggregates, D.M. Armstrong's "doctrine of the ontological free lunch" seems valid:

> ... whatever supervenes, or ... is entailed or necessitated, ... is not something ontologically additional to the supervenient, or necessitating, entities. What supervenes is no addition of being ... Mereological wholes are not ontologically additional to all their parts, nor are the parts ontologically additional to the whole that they compose. This has the consequence that mereological wholes are identical with all their parts taken together. Symmetrical supervenience yields identity (1997).

This principle may apply when detailed interactions between and among the components of an aggregate are negligibly small: it definitely does not apply to most chemical combinations. N. Goodman [1951 (1977)] discusses how various methods of specification of the individuals that are to be used in a mode of discourse will generate diverse "systems." He implies that—no matter what the system may be—once certain individuals have been identified that identification will also be valid when those individuals are included as parts in some whole. In chemical combination, this is not the case—chemical entities in combination are not precisely the same as those entities out of combination. For example, all hydrocarbons are composed (in some sense) of carbon and hydrogen atoms—but NMR spectroscopy clearly shows that, even in relatively small hydrocarbons, there are many different types of hydrogen atoms, and various kinds of carbon atoms. Each of these types and kinds exhibits different properties. Which specific position a given chemical entity (atom, ion, etc.) occupies in a molecule or other aggregate influences the detailed properties of that given entity, as well as contributing to the characteristics of the molecule as a whole.

In general, properties of chemical components are seriously changed by the fact of their being in combination (Earley 2005). Chemical combination is not well understood in terms of vectorial addition of properties of components. Regarding chemical combination, there is no "ontological free lunch"—*pace*, Armstrong.

Molecules (as well as molecular ions and related but more complex chemical entities) are composed of fairly small numbers of components that have inter-relationships of just a few types. Electrons, and nuclei of a particular kind, are indistinguishable from each other (in isolation). All components of chemical entities participate in continual thermal motion. Values of all structural parameters (bond lengths, angles) of each molecular entity oscillate incessantly. The geometric structure of each chemical molecular entity is far from a rigid arrangement of atoms in space. Stable chemical species are those that persist in a specific pattern of internal vibratory motion so that average distances and angles are reasonably well defined. (Varieties of internal vibration that do not conform to that pattern tend to fade away.) The identity of chemical individuals does not depend on single fixed static structure, but rather on the fact that continual return to prior arrangements of parts occurs (Earley 2003c). This indefinitely repeated recurrence is the effect (and signature) of the closure that establishes the collection of spatial symmetries of that chemical individual as a group, rather than some other kind of set. Every chemical entity, insofar as it persists in time, exemplifies some group theoretical structure, corresponding to the collection of spatial symmetry properties of the average arrangement of parts that persists through thermal motion. It is in virtue of the closure of relationships that grounds the dynamic stability of such spatial structures that molecules may act as units in interactions with the rest of the world (Earley 1998a; 1999).

Heelan (2003) reports that the young E. Husserl (1859–1938), as a junior faculty member in the philosophy faculty in the university at Göttingen, had been deeply impressed by the work on applications of group theory then being carried out by the mathematicians and physicists that were his colleagues in that faculty (including D. Hilbert and F. Klein). According to Heelan, Husserl came to hold that human identification of any object as a concrete individual is warranted only when the influence of a specific spatial region on the observer remains consistent under the set of transformations corresponding to the symmetry of three-dimensional space.[4] That is to say, the identification of any object as an individual is based on the recognition, by some interaction partner, that that object realizes some group. In favorable cases, the group of symmetries that accounts for the stability over time of that unit may be the same group involved in the recognition of individuality.

In a contemporary discussion of the meaning of the concept of "substance," Ruth Millikan holds:

> Substances ... are whatever one can learn from given only one or a few encounters, various skills or information that will apply to other encounters ... Further, this possibility must be grounded in some kind of natural necessity ... The function of a substance concept is to make possible this sort of learning and use of knowledge for a specific substance (2000).

On this basis, Millikan maintains that physical objects, historical individuals, stuffs, natural kinds, symphonies, familiar stories, and many other sorts of things, all merit the designation "substance." The way the many forms of verb corresponding to "to be" are used in the most ancient Indo-European sources (Greek epic poetry of Homer and Sanskrit *Vedas*) suggests that *persistence* (e.g., living) was the original characteristic meaning of "being" (Kahn 1993). This indicates that *stability over time* ought to be taken as a central ontological notion. Millikan's concept of substance includes the notion of persistence, characteristic of the ancient Indo-European sources, and adds the notion of interaction with some individual that is capable of learning. This latter aspect seems to make contact with Heelan's group theoretical interpretation of Husserl's doctrine.

The criteria of substantiality advanced by both Millikan and Husserl seem to require interaction with individuals that are capable of learning. This is consistent with the humanistic interests of those philosophers—but the main point they make does not seem to require a sentient observer, or even to be dependent on the detailed nature of the interaction partner of the putative substance. Elsewhere, I have argued that whenever a collection of individuals (of any sort) attains sufficient coherence that the collection, as a unit, has causal influence of any particular kind, then that aggregate may well be considered to have attained ontological status, at least with respect to that causal transaction (Earley 2003b). On this basis, the existence of chemical substances (and their interaction with the rest of the world, including their own components) can be seen to depend on *closure* of sets of relationships involving the components of those substances, which closure corresponds to transmutation of that set into a group that is characteristic of each substance, as such.

GROUPS OF PROCESSES AND CHEMICAL REACTION NETWORKS

The symmetries involved in the molecular chemical entities considered in the previous section were spatial, like those mentioned in the section on Introduction. But chemistry is mainly concerned with changes of some chemical substances into others—that is, chemistry deals with processes extended in time. *Temporal* symmetries are of central importance to reaction chemistry.

When a number of chemical species interact in some medium (an aqueous solution, a dilute gas, a volcanic magma) myriads of reactions may occur, each more or less rapidly. Many, usually most, of these reactions play themselves out rather quickly, as some necessary reactants are used up. However, it is possible that some set of reactions may happen to regenerate all necessary components, so that the set of reactions continues to function after other reactions have ceased. There is good reason to hold that such closure of reaction systems becomes more likely as the number of chemical species involved increases. For a sufficiently complex aggregate of reactive chemicals, the existence of one or more of such cyclical reaction networks becomes highly probable (Kauffman 1993; Earley 1998b).

Approximate stability (or regular oscillation) arising from many ongoing chemical reactions indicates that *closure* of a network of dynamic processes has occurred.[5] (Every biological organism owes its life to many such coherent networks of processes operating at diverse timescales—many internal to the organism and others external to it.) When a network of reactions in an open chemical system repeatedly gives rise to approximately the same sequence of states—as when a flame burns quietly in a Bunsen burner—it seems that the criteria of substantiality of Millikan and Husserl have been fulfilled, so that the coherence has attained ontological status[6]—as an alternative sort of chemical substance. On this basis, a particular Bunsen burner flame could qualify as a "substance" (in the Millikan and Husserl sense) over and above the hydrocarbon and oxygen molecules that are its feed and the water and the carbon dioxide molecules that are its products. (Learning to recognize a flame, as such, is useful for young animals.)

In each such case of coherence arising from a collection of chemical reactions, a set (usually quite large) of stoichiometric reaction equations like Eq. (1) connect the state of the system at an earlier time with the state of the system at a later time.

$$aA + bB + \cdots \rightarrow pP + qQ + \cdots \quad (1)$$

Here, A, B, ... P, Q, ... refer to molecules or ions, or (alternatively) to concentrations (activities[7]) of such entities, and a, b, ..., p, q, ... are "stoichiometric coefficients" that insure that the products of the reaction contain the same total number of each kind of atom as the reactants. A set of stoichiometric equations like Eq. (1), supplemented by the appropriate kinetic equations—"the complete rate laws"[8] would constitute a mapping from an earlier to a later state of the system.[9]

Frequently, chemical concentrations monotonically increase or decrease until a condition of chemical equilibrium has been reached. That is, the point in concentration space[10] describing the system moves toward a unique point that corresponds to the equilibrium condition. If the state that would correspond to chemical equilibrium should happen to be unstable, concentrations of chemical species may undergo oscillations around that state.[11] Systems that involve two significant (reference) reactants,[12] and only bimolecular and unimolecular steps, always have stable equilibrium states.[13] Systems with three reagents may or may not have stable equilibrium states. M. Eiswirth et al. (1991) argue that all complex mechanisms can be reduced to two-reactant or three-reactant processes.

King (1980) examined structures of chemical reaction systems involving three reference reactants. Each reactant can influence (increase or decrease) the rate of change of either or both the other two reagents, or could have no influence on one or both. It turns out that it can be shown that for three reagents there are 729 distinguishable patterns of such influence. This number can be divided into 16 families (having unequal numbers of members) of similar influence patterns. It can be shown that, in order to generate a chemical oscillation, each of the three reference reagents must influence all the others either directly or indirectly. Only five of the 16 families (including a total of 416 different patterns of mutual influence) satisfy that requirement. One family of influence networks that can support oscillations contains 16 possible patterns, two of which are shown in Figure 3.

King used switching circuit theory to explore the dynamics of motion around the point in concentration space that corresponds (or would correspond) to chemical equilibrium for each of the patterns in Figure 3. Each concentration is considered to have (only) two possible conditions, high (1) or low (0). On this basis, there are eight possible overall states of the three-reagent system. Those states can be represented

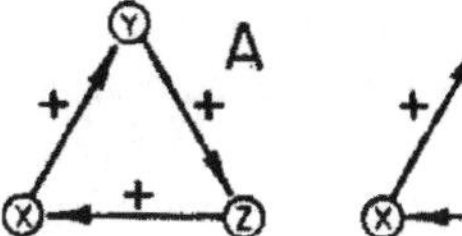

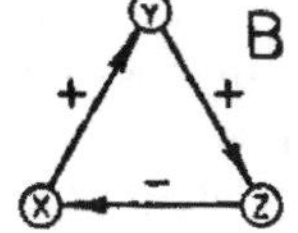

Figure 3: Influence diagrams for two reaction networks involving three reactants, X, Y, and Z. Arrows indicate the direction of influence: plus signs correspond to activation, minus signs correspond to inhibition. In diagram A, X activates Y, Y activates Z, and Z activates X. In diagram B, X activates Y, Y activates Z, but Z inhibits X. (After King 1980.)

as occupying the corners of a cube (Figure 4). Arrows lying along each edge of the cube show which state would succeed each of those states at the next increment of time. [An arrow from state 101 towards state 100 indicates that if a system starts in, or arrives at, a condition of high X and Z concentration and low Y concentration (110), then in the next time period it will change to a state that has high X concentration but low Y and Z concentrations (100).]

Figure 4A (corresponding to Figure 3A) indicates that, whatever the starting condition might be, the system will eventually come to reside in one of two non-equilibrium steady states (000 or 111)—that is, the reaction network corresponding to the A diagrams in both figures is *bistable*. Figure 4B, corresponding to Figure 3B, also has two special non-equilibrium steady states (101 and 010), but any system starting in one of those states will leave it, and no system will enter either of those states from any other—both states are *unstable*. A system starting at either one of those states will quickly move to the six-member cycle (000, 100, 110, 111, 011, 001), which corresponds to an oscillation around the unstable (virtual equilibrium) state corresponding to the center-point of the cube. No matter how the system starts up, it will eventually reach that closed path and will remain on that same sequence of states

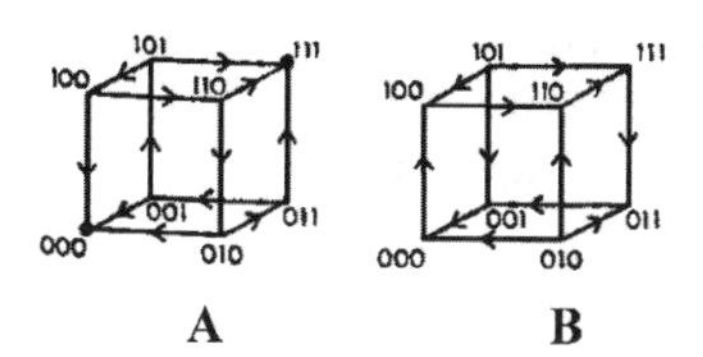

Figure 4: Switching diagrams corresponding to the two influence diagrams shown in Figure 3. Corners of each cube correspond to instantaneous states of the reaction system, with concentrations of Z, Y, and Z either high (1) or low (0). (State 000 has all three concentrations low, in state 010, X and Z are low and Y is high.) Arrows on the cube edges indicate the direction of change at the next increment of time that is predicted by the influence diagram. States with three incoming arrows are stable (sinks); states with three outgoing arrows are unstable (sources). In diagram B, a path with six connected arrows corresponds to a limit cycle oscillation around an unstable (virtual equilibrium) state represented by the center of the cube. That path is "an attractor." A corresponding six-member cyclical path in A is unstable (a repeller). (After King 1980.)

indefinitely (Earley 2003a). (That path describes a *limit cycle.*) Taking the concentrations of each of the three significant reference reactants (X, Y, Z) as the members of a basis set, and the chemical reactions summarized in the influence diagram as the appropriate mapping operations, the existence of the closed cycle of states indicates that the collection of (temporal) symmetries characteristic of this aggregate of chemical reactions constitutes a group, rather than a set of some other sort. Once a real system corresponding to this simplified model had reached the closed cycle of states, it seems that the criteria for substantiality (for properly constituted interaction partners) put forward by Millikan and by Husserl would be fulfilled: an ontological emergence would have occurred.

King's analysis of this three-reactant case clearly shows that the symmetry properties of chemical reaction networks provide the basis for understanding what must be the case in order for such systems to generate a closed sequence of states, and thereby provide the basis of the coherent interaction necessary to qualify as a substance in Millikan's (and Husserl's) sense.

This outlook would be consistent with what might be called a *moderate ontic structural realism* (MOSR). Absent the closure of the network of relationships that give rise to long-term persistence, the collection of reactions would not have sufficient integrity to warrant identification as an effective agent with respect to some interaction partners (sentient or not). In that sense, the existence of the coherence is totally dependent on the (group theoretic) structure, as Eddington and later structural realists maintain. But also, without the (lower-level) entities (A, B, etc.) that are involved in the component reactions, such closure would not occur. An appropriate material substrate (A, B, etc.) is also indispensable—but isomorphic coherences may result even from quite varied substrates. (Several different chemical species might function as A, say.)

Consideration of molecular chemical substances and coherent networks of chemical reactions in open systems does not raise the subtle problems that seem to be specific to quantum mechanics. But examining those chemical systems does suggest that fuller *philosophical* attention to the critical importance of the *closure* of relationships between components of compound individuals (i.e., fuller appreciation of the discovery of group theory that Galois made two centuries ago) seems likely to help resolve some current philosophic problems, specifically those connected with emergence.

NOTES

1. That group of symmetries does not determine the length of the B–F bond, a scale factor of the BF_3 molecule.
2. In the last half of the 20th century, further generalization has been achieved in the *theory of categories* (Lawvere and Schanuel 1997). This shows that group theory may be regarded as a special case of an even more inclusive approach.
3. Recall the widely quoted observation of William Jefferson Clinton: "It depends on what the meaning of the word "is" is" (Kaplan 2002).
4. Designated "the Galilean transformation."

5. As a crystal can fill all three-dimensional space, so such an oscillating chemical system can fill all time (Earley 1998, 1999a–e).
6. At least with respect to some interactions.
7. This assumes that reaction media are effectively constant, so that concentrations may be used rather than "activities" (effective concentrations).
8. Eq. (1) can be written more compactly as Eq. (2).

$$\sum_i \mu_i S_i = 0 \tag{2}$$

Here S_i refers to any of the i species involved in the reaction, either as reactants or products. The stoichiometric coefficient of the ith species is μ_i—taken with opposite signs for reactants and products. Using this more compact notation, the kinetic equations that describe how rapidly the reaction occurs can be written as:

$$\frac{1}{\mu_i}\frac{\mathrm{d}S_i}{\mathrm{d}t} = \sum_1^j k_j \prod_i S_i^{\upsilon_{ij}} \tag{3}$$

S_i is any specific component of the system and μ_i is the corresponding stoichiometric coefficient. The k_j are rate constants for the j mechanistic pathways that contribute to the overall reaction and the υ_{ij} specify how the rate of each reaction pathways depends on each concentration. The υ_{ij} need not be directly related to the μ_i.

9. Since it is not possible to reverse the direction of time so as to convert a later state into an earlier one, no inverse operation is possible in the chemical reaction system: such systems are described by semi-groups rather than by groups.
10. This is an n-dimensional space in which each dimension corresponds to a concentration of one of the n components of the reaction mixture.
11. In what follows, such a state will be designated by the neologism "virtual equilibrium state."
12. These are components of the reaction mixture that change concentration during the course of the reaction. Concentrations of other components are treated as approximately constant.
13. Addition of autocatalysis can destabilize that equilibrium state, but autocatalysis requires either a third reference reactant or a trimolecular reaction.

REFERENCES

Armstrong, D.M. 1997. *A World of States of Affairs*. Cambridge: Cambridge University Press, 12.

Baumslag, B. and Chandler, B. 1985. *Schuam's Outline of Group Theory*. New York: McGraw-Hill.

Basolo, F. and Pearson, R. 1958. *Mechanisms of Inorganic Reactions: A study of Metal Complexes in Solution*. New York: Wiley.

Cotton, F.A. 1963. *Chemical Applications of Group Theory*. New York: Wiley.

Earley, J.E., Sr. 1998a. Modes of Chemical Becoming. *Hyle* 4(2): 105–115.

Earley, J.E., Sr. 1998b. Naturalism, Theism, and the Origin of Life. *Process Studies* 27(3–4): 267–279.

Earley, J.E., Sr. 1999. Varieties of Chemical Closure. In: Chandler, J. and Van De Vijver, G. (eds.), *Closure: Emergent Organizations and their Dynamics*, *Annals of the New York Academy of Science*, New York: New York Academy of Science. Vol. 191, 122–131.

Earley, J.E., Sr. 2003a. Constraints on the Origin of Coherence in Far-From-Equilibrium Chemical Systems. In: Eastman, T., et al. (eds.), *Physics and Process Philosophy*. Albany: SUNY Press, 63–73.

Earley, J.E., Sr. 2003b. How Dynamic Aggregates may Achieve Effective Integration. *Advances in Complex Systems* 6(1): 115–126.

Earley, J.E., Sr. 2003c. On the Relevance of Repetition, Recurrence, and Reiteration. In: Sobczyńska, D., Zielonaka-Lis, E. and Kreidler, P. (eds.), *Chemistry in the Philosophical Melting Pot*. Frankfurt: Peter Lang, 171–185.

Earley, J.E., Sr. 2005. Why There is No Salt in the Sea. *Foundations of Chemistry* 7, 85–102.

Eddington, A. 1939. *The Philosophy of Physical Science*. New York: MacMillian.
French, S. and Ladyman, J. 2003. Remodeling Structural Realism: Quantum Physics and the Metaphysics of Structure. *Synthese* 136: 31–56.
Heelan, P. 2003. Paradoxes of Measurement. In: Earley, J. (ed.), *Chemical Explanation: Characteristics, Development, Autonomy*, *Annals of the NY Academy of Science*, Vol. 988.
Humphreys, P. 1997. How Properties Emerge. *Philosophy of Science* 64: 1–17.
Jaffe, H.H. and Orchin, M. 1965. *Symmetry in Chemistry*. New York: Krieger.
Kahn, C. 1973. The Verb 'Be' in Ancient Greek. Boston: Reidel.
Kaplan, J., ed., 2002. Bartlett's Familiar Quotations, 17th Edition. Little Brown, New York. (Cf. William Jefferson Clinton. Grand jury testimony, August 17, 1998.)
Kauffman, S.A. 1993. *The Origins of Order: Self-Organization and Selection in Evolution*. New York: Oxford University Press.
King, R.B. 1980. Chemical Applications of Group Theory and Topology. VIII. Topological Aspects of Oscillating Chemical Reactions. *Theoretica Chimica Acta (Berlin)* 56, 269–296.
Klee, R.L. 1984. Micro-determinism and Concepts of Emergence. *Philosophy of Science* 51(1): 44–63.
Kronz, F.M. and Tiehen, J.T. 2002. Emergence and Quantum Mechanics. *Philosophy of Science* 69(2): 324–347.
Lawvere, F.W. and Schanuel, S.H.1997. *Conceptual Mathematics: A First Introduction to Categories*. Cambridge: Cambridge University Press.
Lewis, D.K. 1990. *Parts of Classes*. Oxford: Blackwell Press.
Martin, R.M. 1992. *Logical Semiotics and Mereology*. Philadelphia: Benjamins.
Newman, D.V. 1996. Emergence and Strange Attractors. *Philosophy of Science* 63: 245–261.
Rescher, N. and Oppenheim, P. 1955. Logical Analysis of Gestalt Concepts. *British Journal of the Philosophy of Science* 6(22): 89–106.
Rescher, N. 1973. *The Coherence Theory Of Truth*. Oxford: Clarendon Press.
Rosen, J. 1995. *Symmetry*. New York: Springer-Verlag.
Silberstein, M. and McGeever, J. 1999. The Search for Ontological Emergence. *The Philosophical Quarterly* 49(195): 182–200.
Simons, P. 1987. *Parts: A Study in Ontology.* Oxford: Clarendon Press.
Suppes, P. 1988. Representation Theory and the Analysis of Structure. *Philosophia Naturalis* 25: 254–268.
van Fraassen, B. 1989. *Laws and Symmetry*. Oxford: Clarendon Press.
van Fraassen, B. 1991. *Quantum Mechanics: An Empiricist View*. Oxford: Clarendon Press.
van Fraassen, B. 1999/forthcoming. Structure: Its Shadow and Substance. In: Roger Jones and Philip Ehrlich (eds.), *Reverberations of the Shaky Game: Essays in honor of Arthur Fine*. Oxford: Oxford University Press, forthcoming.
Wigner, E.P. 1959. *Group Theory and its Application to the Quantum Mechanics of Atomic Spectra (expanded edition)*. New York: Academic Press.
Wilson, J. 1999. How Superduper does a Physicalist Supervenience need to be? *Philosophical Quarterly* 49(194): 33–52.
Wussing, H. 1984. *Genesis of the Abstract Group Concept* (translated by Shenitzer, A.). Cambridge, MA: MIT Press.

CHAPTER 12

THE PERIODIC SYSTEMS OF MOLECULES

Presuppositions, Problems, and Prospects

RAY HEFFERLIN

Southern Adventist University, Collegedale, Tennessee 37315, USA

Dass er [Koenig Salomo] mir die Geheimnisse der Natur diktiert?
Den Zusammenhang aller Dinge?
Das System aller moeglichen Erfindungen?
Möbius, in "Die Physiker" by Friedrich Durrenmatt,
Robert E. Helbling, editor, Oxford University Press, 1965

INTRODUCTION

We present a taste of the philosophical and historical aspects of the scientific enterprise for professional chemists and a glimpse of scientists' ontology and epistemology for philosophers and historians. The narrative is based on personal acquaintance with living participants in what is a relatively small and well-defined area of chemistry. The area is that of creating, testing, and utilizing organizations of molecular symbols embedded in two or more dimensions. These periodic systems play the same role for molecules as the chart of the elements does for individual atoms. The presuppositions, on several levels, that contributed to the creation of molecular periodic systems are explored. Problems faced by some systems are exposed, including the difficulty of visualizing them in multiple dimensions and the challenge of finding not molecular symbols, but linear combinations of symbols, in the compartments. New phenomena are found to emerge in going from atoms to diatomic to larger molecules. Prospects for further maturation and usefulness of the periodic systems are presented.

FOR WHOM IS THIS CHAPTER WRITTEN AND HOW IS IT ORGANIZED?

The purpose of this chapter is to investigate the presuppositions, problems, and prospects of periodic systems of molecules. It should present no problem for readers who have a broad knowledge of chemistry, have some acquaintance with these periodic systems, and desire to understand them from a philosophical or historical point of view. But the chapter was designed with readers in mind who have less background

D. Baird et al. (eds.), Philosophy of Chemistry, 221–243.

in chemistry, no acquaintance with the periodic systems, or little familiarity with mathematics, but who have an interest in the philosophy or history of science. It is the author's intent that the assumptions, obstacles, and future directions of research are explained while not requiring a detailed grasp of each research project. It is an advantage that periodic system research is a small field; the number of anecdotes is limited and most of them can be included.

Sections on "Why Would One Construct or Use a Molecular Periodic System?" and "What Presuppositions are Common to Work on All Periodic Systems?" take up the importance of periodic classifications and the presuppositions that appear to have been held by all of their creators. Section on "How Advanced is the Field of Molecular Periodic Systems?" is about how the research into the two classes of periodic systems has advanced and Table 1 shows where it stands at the present time.

Then, the narrative divides into two streams. One stream in sections on "What Presuppositions are Revealed by Physical Periodic Systems?"; "Prospects for the Best N-atomic Physical Molecular Periodic Systems"; and "What Problems Arise in Testing Against Data, and in Predicting New Data?" consists of the presuppositions, problems, and prospects for one class of molecular classifications, i.e., physical periodic systems. The second stream in sections on "Prospects for Chemical Periodic Systems of Molecules"; "Babaev's Hyperperiodic System, and Its Prospects"; "What Presuppositions are Revealed by the Creation of Chemical Periodic Systems?"; and "What Problems Arise in Testing Against Data and Predicting New Data with Chemical Systems?" is composed of the presuppositions, problems, and prospects for the other class of molecular classifications, i.e., chemical systems—but in a different order. The prospects occupy the sections on "Prospects for Chemical Periodic Systems of Molecules" and "Babaev's Hyperperiodic System, and Its Prospects"; the presuppositions are the subject of the section "What Presuppositions are Revealed by the Creation of Chemical Periodic Systems?"; the problems are found in the section "What Problems Arise in Testing Against Data and Predicting New Data with Chemical Systems?" The two streams of narrative converge in section on "What are the General Prospects for Molecular Periodic Systems?" where the general prospects for the whole endeavor are considered.

Expressions such as "physical periodic system of molecules" occur very often; the use of notations like PPSM does not seem appropriate.

WHY WOULD ONE CONSTRUCT OR USE A MOLECULAR PERIODIC SYSTEM?

A periodic system of molecules is the mapping of molecular symbols onto a space that has axes related to some property or properties of the constituent atoms, or related to some enumeration of the atomic constituents. Such spaces are "chemical spaces," not to be confused with "chemical property space" as used in the study of chemical similarity.

There have been three reasons for wanting to *create* a periodic system of some or all molecules. The first, to organize vast numbers of species into some elegant form (yes, like stamp collecting). The second, to provide a graphic representation of trends among molecular properties for the educational enterprise (with the added hope that some of the trends may surprise the research establishment). The third, simply to satisfy curiosity. These three reasons are all in the spirit of "pure" science.

There have been two reasons for wanting to *use* a periodic system of molecules for predictive purposes. They are in the spirit of "applied" science. The first has been to obtain numerical data needed, even requested, by other scientists. Indeed, there is a great need for data in order to study atmospheres of the earth, the planets, and the stars; the interstellar medium; and nebulae and other celestial objects. There also is a serious need for data in studies of combustion in engines, furnaces, and incinerators, and in research on such varied topics as toxicity, pharmacology, cosmetology, and paint technology. There are other, very active, fields in chemistry whose practitioners are attempting to fill the same needs. Among them are QSPR (the search for quantitative structure and property relationships) and the search for a numerical index unique to each molecular graph. Work in these and related fields is difficult, and so the parallel effort to construct periodic systems has taken root. The second reason for wanting to make predictions is the realization by the architect that his or her periodic system will gain credibility only if it results in predictions of molecular data. A similar list of reasons, but under the headings of "philosophical, psychological, educational, and pragmatic needs," is given by Babaev and Hefferlin (1996, 62).

WHAT PRESUPPOSITIONS ARE COMMON TO WORK ON ALL PERIODIC SYSTEMS?

This section is based on discussions and correspondence, during over two decades, with the seven of the other living investigators who have created molecular periodic systems (and of course on the author's own experience). All of them share the belief that the universe of discourse is comprehensible and that our modes of thinking and communicating about it are adequate.

One might suppose that the investigators, being motivated by one or more of the reasons listed in the section on "Why Would One Construct or Use a Molecular Periodic System?" or coming across some previously existing periodic system, were struck by a rough idea of how a better system might look. Next, that they made an immediate literature search. Finally, that they tentatively or confidently filled in the details of the system, comparing against data in articles or critical tables. This scenario completely fails to take into account how powerful an experience it is to be struck by the notion of creating a periodic system of *molecules*. Each person was so captivated by the notion that the filling in of the details proceeded immediately without a literature search. In five cases, there was previous familiarity only with the Mendeleev chart; in a sixth case, there was acquaintance also with the Grimm hydrogen-displacement law; and in the seventh and eighth cases, there was knowledge of related work such as QSPR or the search for graphical indices. Of the first

Table 1: First publication, author, and distinguishing features of molecular periodic systems.

Published	Author	Distinguishing features	N	Axes	Class
1862	Newlands	Organic molecules arranged in table	Any	2	Chemical
1907	Morozov	Alkanes and elements in juxtaposition	Any	2	Chemical
1929	Grimm	Hydrogen-displacement law; semblance of outer product	To 5	2	Chemical
1935	Clark	All but hydrides related to unique originating chart	2	2	Physical
1936	Clark	Hydrides in a chart resembling his originating chart	2	2	Physical
1969	Syrkin	Dimers, hydrides, oxides, fluorides	2	2	Physical
1971	Gorski	Core charge, shells, redox, and acid-base tendencies	Any	5	Chemical
1976	SAU collaboration	Dimers, hydrides, oxides, fluorides	2	2	Physical
1979	SAU collaboration	All species arranged by (C_1, C_2); R_1 and R_2 combined	2	3	Physical
1984	Monyakin	All species arranged by (R_1, R_2); C_1 and C_2 combined	2	3	Physical
1980	Hall	Electron donor versus electron acceptor ability	Any	2	Chemical
1982	Dias	Sextet benzenoids	Formula	2	Chemical
1982	Haas	Functional groups; bootstrapping; follows Grimm	Any	2	Chemical
1982	Kong	Main-group diatomics; period sum versus group sum	2	2	Physical
1983	Kaslin	All diatomics except rare-earth molecules	2	2	Physical
1983	SAU collaboration	Outer product for N atomic molecules	2	4	Physical
1983	SAU collaboration	Reducible multiplets of $SO(3) \times SU(2)$	2	4	Physical
1984	SAU collaboration	More of first SAU 1983 entry; also isoelectronic and ionized species	N	$2N$	Physical
1984	Huang	All triatomic molecules	3	2	Physical
1989	Kong	Main-group triatomic molecules	3	3	Physical
1990	Voskresenski	Non-hydrogen atoms, valence, lone pairs, MO, etc.	Any	6	Chemical
1991	Dias	Totally resonant sextet benzenoids	Formula	2	Chemical
1994	Boldyrev and Simon	Molecules on ($C_1 + C_2$) axis, in ($R_1 + R_2$) order	2	2	Physical
1992	Dias	Fluoranoid/fluoranthenoid hydrocarbons	Formula	2	Chemical
1992	Dias	Indacenoid hydrocarbons	Formula	2	Chemical
1992	SAU collaboration	Irreducible representations of $SO(3) \times SU(2)$	2, 3	4	Physical
1994	SAU collaboration	Irreducible representations of $SO(2, 1)$	2, 3	4	Physical
1994	SAU collaboration	Irreducible representations of $SU(\nu)$	Any	1	Physical
1996	Babaev	Sketch of global general system for molecules	Any	2	Chemical

five investigators, two were struck by the creative notion upon learning of a previous periodic system of molecules, and the procedures for creating their own new systems emerged clearly, at once. One was struck by the idea, without seeing another system, while under a lucky but tedious house arrest during the Cultural Revolution in China.

All five of the first investigators at once supposed it possible to insert all possible molecular symbols into the conceived system. This supposition is tantamount to assuming the existence of all molecules for which they wrote or imagined the symbols. To put it even more bluntly, the supposition implies the *prediction* that these thousands of molecules will be discovered; characterized experimentally; or have their properties computed. It does not imply the prediction that these species will all be stable under standard conditions. They may exist only under the most unusual conditions and even then the molecules or molecular states may be transitory.

Of the sixth and seventh investigators, one realized that it might be possible to extend the hydrogen-displacement law to make a classification of functional groups. The other suspected that there might be a molecular periodicity (beyond that shown by the elements) and sought to expand the chart of the elements into other dimensions containing charts of molecules and in so doing to include "hyperperiodicity."

HOW ADVANCED IS THE FIELD OF MOLECULAR PERIODIC SYSTEMS?

Table 1 presents a chronological listing of proposed periodic systems. Table 1 lists features that distinguish each system, the numbers of atoms in the molecules included, and the dimensionalities of the spaces in which the systems are built.

There is no entry for Mendeleev's chart, because he did not as much as sketch a chart specifically for molecular species. The beautiful group-dynamic treatment of carbon tetra-hydrogen-halides by Komarov and Lyakhovskii (1986) is omitted because it did not result in a graphical representation. Allen's (1992) triangular diagram is left out because it includes only a few molecules and they are shown only to illustrate various kinds and degrees of bonding. There are no entries in Table 1 for classifications that, even if called "periodic systems," do not have specific locations for individual molecules. An example is the elegant collection of reduced potential curves presented by Jenč (1996). "SAU" refers to Southern Adventist University.

Table 1 indicates membership in one of the two classes of molecular periodic systems. Those in the first class utilize either the period numbers or the group numbers of their atoms in the chart of the elements, or some function of them. Periodic systems in this class ["physical periodic systems," as defined by Hefferlin and Burdick (1994)] always involve molecules with just one fixed number, N, of atoms. In principle, N may be arbitrarily large; however, it will soon be seen that in practice N is probably less than 5.

Systems in the second class contain molecules having various atom counts. They include a set of molecules for which the investigator has many known reliable data. A striking example is the periodic chart of functional groups proposed by Haas (1982, 1983, 1984), which can easily be used to classify halogen-containing species with any number of atoms (section on "Haas's Periodic System of Functional Groups, and Its Prospects"). Periodic systems in this second class ["chemical periodic systems," Hefferlin and Burdick (1994)] frequently do not employ any of the coordinates of the chart of the elements, though in one case the sum of the atomic group numbers is one independent variable.

It is not a purpose of this chapter to repeat the reviews given by Hefferlin (1989a, Chapter 12, 1994) and by Hefferlin and Burdick (1994) of these periodic systems. A few systems are explained to the extent necessary for the unique assumptions, obstacles, or future directions of research associated with them. At this point, it is relevant to note how some contributions specifically influenced later work. For instance, the 1979 SAU three-dimensional system (Hefferlin et al. 1979) evolved into the 1994 outer matrix product system (Hefferlin 1994). It also evolved, on a parallel path, into systems based on group dynamics (Zhuvikin and Hefferlin 1983; Hefferlin et al. 1984; Zhuvikin and Hefferlin 1992, 1994; Carlson, Hefferlin, and Zhuvikin 1995; Carlson et al. 1996). Grimm's hydrogen-displacement law (Grimm 1925) led directly to Haas's periodic table of functional groups. The superb classifications of main-group triatomic molecules by Kong [in Hefferlin (1989a) Chapter 11] led directly its broadening to include transition-metal molecules by Huang (unpublished).

Scientists consider it imperative that theory be tested by experiment and should expect others (the listening audience, the referees, or the community of researchers) to insist on a demonstration that a periodic system agrees with the data. The demonstration in this case consists of plotting the data (Figures 1–6).

The demonstration has also been done by means of least-squares smoothing or by neural-network training and validation. Both of these methods have led to large numbers of forecasted data of moderate accuracy (Carlson et al. 1997; Wohlers et al. 1998; Hefferlin, Davis, and Ileto 2003).

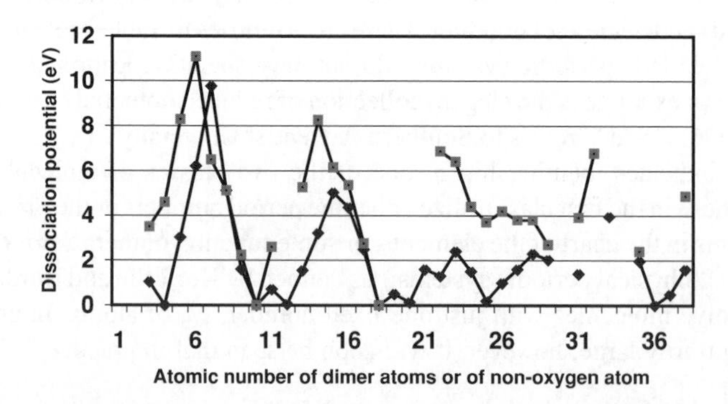

Figure 1: Dissociation potentials for homonuclear diatomic molecules plotted against either atomic number (lower graph, diamonds), and for monoxides plotted against oxidized atom (upper graph, squares). The displaced peaks are due to iselectronicity.

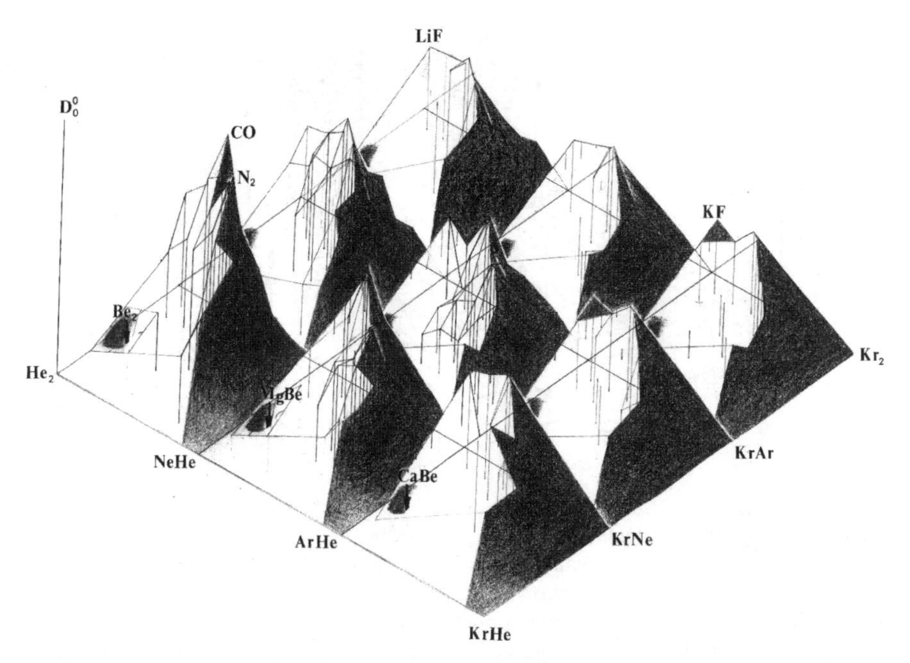

Figure 2: Dissociation potential "vertically," and Z_1 and Z_2 from 2 to 36 "horizontally," in almost isometric projection. Transition-metal and rare-earth molecules have been cut out, and the resulting pieces of the surface have been slid and joined to pieces at smaller Z. Homonuclear molecules are on the left-right diagonal, and the terrain is symmetrical with respect to a vertical plane through that diagonal. Any series of isoelectronic molecules has addresses going horizontally and normal to that plane. The scale is established by the value for N_2 (9.79 eV), which lies between the peaks for CO and OC. The figure is constructed from stick graphs by drawing lines of least descent from each peak, by "draping" a surface from those lines down to the 0 eV valleys of rare-gas molecules. In addition, an attempt is made to indicate craters in the bottoms of which are found the alkaline-earth pairs.

WHAT PRESUPPOSITIONS ARE REVEALED BY PHYSICAL PERIODIC SYSTEMS?

Similarity to the periodic chart

It is commonly believed that quantum mechanics is sufficient to recreate the chart of the elements. Even if this were so, quantum mechanics gives no justification for the presupposition that a molecular system should resemble the chart of the elements. Thus, the idea that the construction of physical periodic system requires some similarity with the chart of the elements qualifies as an assumption. Consider, for example, diatomic systems. The architectures of Kong (Kong 1982, 1989) have located molecules using two coordinates—a group axis based on the sums of the atomic group numbers, and a period axis based on the sums of the atomic period numbers. Another architecture has the group numbers of the two atoms separate and has their period

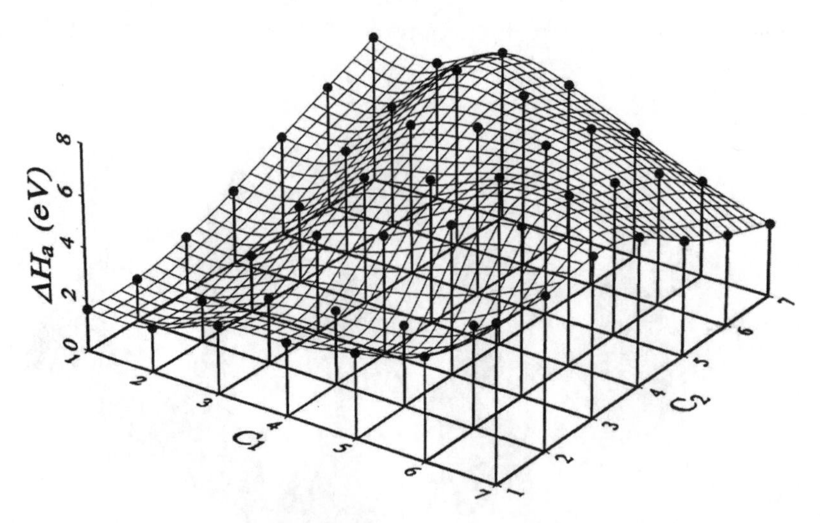

Figure 3: Dissociation potential in electron volts vertically, and C_1 and C_2 (where C_i is the group number of atom i) horizontally, in isometric projection. Li_2 is at the extreme left and F_2 is at the extreme right, i.e., the data pertain to molecules with period numbers $(R_1, R_2) = (2, 2)$. The points represent initial neural-network predictions (unpublished) rather than tabulated data. The surface has a tendency for its highest values to lie along a line near the isoelectronic series with molecules having 10 valence electrons. It also has a hint of the depression at Be_2,only a hint because very few alkaline-earth pair data were in the training set. Very similar surfaces, but with decreasing heights, pertain to the neural-network predictions for the main-group molecules with increasing values of (R_1,R_2), a phenomenon evident in Figure 1. The traditional symbol for dissociation potential appears as ΔH_a (enthalpy of atomization). The surface does not extend to the valleys where data for rare-gas molecules lie.

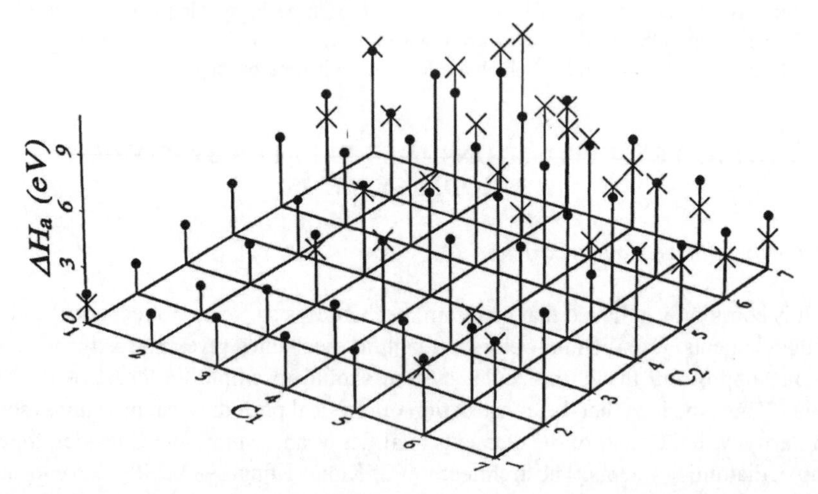

Figure 4: Same as Figure 2 except that the mesh has been removed, leaving only the points, and the tabulated data are shown (X). There is no tabulated datum for Be_2, however the neural network did have inputs for other alkaline-earth pairs.

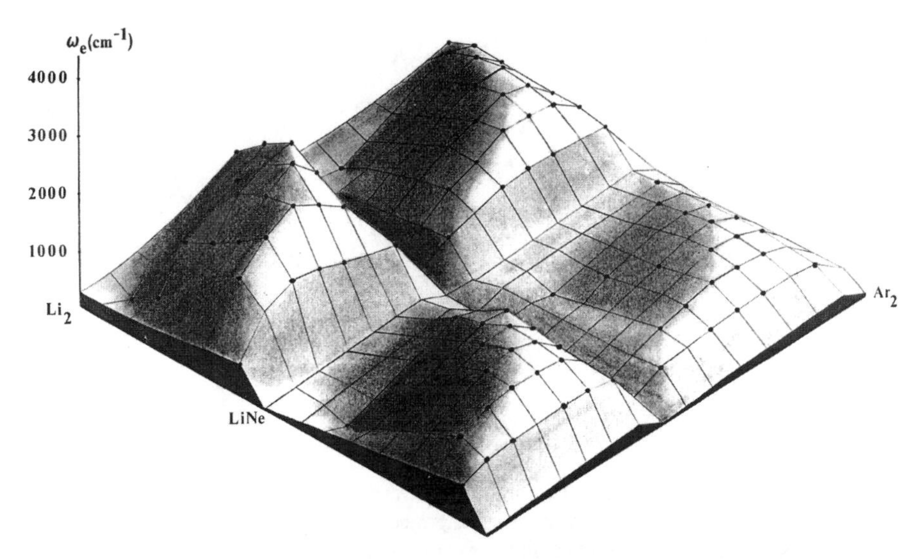

Figure 5: Vibration frequency in cm^{-1} vertically, and Z_1 and Z_2 from 2 to 18 "horizontally," in the same projection as Figure 2. The figure was constructed from stick graphs by drawing by "draping" a surface over the tabulated data (dots) into the valleys of rare-gas molecules. The depths of the valleys were estimated based on the few existing data. Depressions are clearly visible for data at the addresses of the alkaline-earth pairs. This figure is taken, by permission, from *Periodic Systems and Their Relation to the Systematic Analysis of Molecular Data*, The Edwin Mellen Press, Winter Springs, Florida, USA. Plate 6.

numbers combined (e.g., multiplied) to form a total of three axes (Hefferlin 1989a, Chapter 12, 1994). Molecules with various combinations of atoms (*ss*, *sp*, etc.) may be organized into separated blocks. Conversely, the architecture of Monyakin (Hefferlin 1989a, Chapter 12, 1994) has the period numbers of the two atoms separate and has the group numbers combined (i.e., subtracted) to form a total of three axes. In all cases, however, the chart of the elements has provided the starting point.

Molecular groups and periods

The assumption of resemblance reveals a second, subtler, presupposition. The periodic chart places elements in columns, or groups, based on the numbers of their valence electrons. Thus, nitrogen is placed in group 5 (15 in the IUPAC scheme) even though it frequently expresses a valence of three. Fixed-period molecules with the same total number of atomic valence-shell electrons ("isoelectronic," "horizontally isoelectronic," or "isosteric" molecules such as N_2 and CO) usually have properties more similar than do molecules selected at random. Molecules whose atoms come from different periods but have the same numbers of valence electrons ("vertically isoelectronic" or "isovalent" molecules such as the salts LiF, NaI, and CsCl), often have somewhat similar properties. So, the sum of the atomic valence electron counts, i.e., the sum of the atomic group numbers, is important. Thus, it appears that using

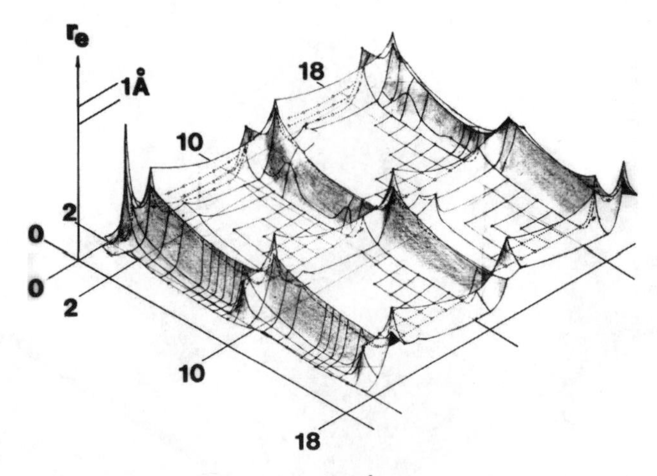

Figure 6: Ground-state inter-nuclear separation in Å vertically, and Z_1 and Z_2 from $-1/2$ (unipositive hydrides) to 19 horizontally in the same projection as Figure 1. Homonuclear species from H_2 to K_2 are on the left-right diagonal. •, tabulated data; o, tabulated data hidden from view by terrain. The figure was made in the same manner as was Figure 2, with particular attention to the "monkey saddles" located at the addresses of inert-gas pairs. Diatomic-molecular ions have been incorporated into the surface. It is shown in Chapter 8 (Hefferlin 1989a; Hefferlin et al. 1984) that if the r_e datum for a once-ionized molecule is moved $-1/2$ unit in each "horizontal" direction, then the datum fits the neutral-molecular surface quite well. [The same result usually pertains also to electron configurations and to dissociation potentials (Hefferlin 1989a, 581–596).] The resulting half-integer address lies between the addresses of neutral molecules. In contrast, a doubly ionized molecule such as He_2^{++} is shifted to the address of an existing neutral molecule (in this case H_2; the r_e values are 0.704 and 0.7414 Å, respectively). He_2 is at the center of the monkey saddle between the two high peaks for HeH. The uninegative molecular ion $LiCl^-$ (2.18Å) is shifted to the address of the unipositive molecular ion $BeAr^+$ (2.085Å). $BeAr^+$ is on the flank of the ridge at the far end of the figure, and $LiCl^-$ is on the rear of that same ridge. The close agreements of these data are remarkable, considering the jagged nature of the surface on which they lie.

the chart of the elements as a template is not inconsistent with the presupposition that molecular isoelectronicity is important.

Is it necessarily true that some combination of the atomic *period* numbers has significance? The highest principle quantum number of the electron shells that are being filled or have just been filled determines the row, or period, of the atomic chart in which an atom is placed. Thus, K through Kr are all in period 4. There is no simple combination of atomic principle quantum numbers which necessarily describes a principle quantum number of the molecular configuration. For instance, boron is in period 2, but the configuration of B_2 is $(\sigma_g 1s)^2(\sigma_u 1s)^2(\sigma_g 2s)^2(\sigma_u 2s)^2(\pi_u 2p)^2$, or in another notation $[He_2]$ $(\sigma_g 2s)^2(\sigma_u 2s)^2(\pi_u 2p)^2$; since this configuration is on the way to becoming that of Ne_2, one could argue that the molecular period number is also 2. The assumption that some combination of atomic principle quantum numbers defines some principle quantum number of the molecular configuration has not been used in a consistent manner.

The originating chart of the elements

There are many two- and three-dimensional versions of the periodic chart of the elements (Van Spronsen 1969; Mazurs 1974). There are "short" charts, "long" charts, and charts based on the symmetry considerations of group dynamics (Barut 1972; Rumer and Fet 1972). Given that the chart of the elements is to be a template for the molecular periodic system, it follows that the choice of the former will greatly influence the appearance of the latter. The third assumption made by a designer of physical periodic systems, then, has to be that one certain two-dimensional chart is the best template for his or her molecular system.

The organization of only N-atomic molecules with a given structure

A fourth assumption has been that one should start with a periodic system for diatomic molecules. Then, if the results turn out to be promising and if circumstances allow it, the process would go on to use the same approach for linear and cyclic triatomic molecules, for the several structures of tetra-atomic molecules, and possibly for yet larger molecules. This assumption was clearly visible in the successive publications, concerning diatomic and then triatomic periodic systems, by Kong and by Hefferlin and his group.

Dimensionality

The fifth assumption concerns the numbers of dimensions that the system should have for maximum visualizability and usefulness. The dimensionality has been often taken as two to make the systems ideal for publication and ready reading, and sometimes to evoke the chart of the elements (Table 1). Three-dimensional periodic systems allow for more resolution in molecular classification but at an almost total loss of anyone's capability for finding a given molecule. The Southern Adventist University (SAU) collaboration has found that $2N$ dimensions are necessary to construct the most general periodic systems for N-atom molecules (Hefferlin et al. 1984; Hefferlin 1994) and assumes that clever means of visualization will eventually make it possible to find the molecules.

Beginning the classification with main-group molecules

Finally, it was usually taken as obvious that initial construction of small-molecule systems should begin with main-group molecules and only later include molecules with transition-metal atoms ("transition-metal molecules") and rare-earth atoms ("rare-earth molecules"). The presupposition comes naturally upon considering that data for main-group molecules are relatively more plentiful. It is exemplified in the works of Kong (section on "Kong's Periodic System and Its Prospects") and of Hefferlin and his group (section on "The SAU System and Its Prospects"). Hydrides have often been treated separately, beginning with Clark (1935), as their properties tend to be again quite different.

PROSPECTS FOR THE BEST *N*-ATOMIC PHYSICAL MOLECULAR PERIODIC SYSTEMS

Kong's periodic system and its prospects

Kong's periodic systems of diatomic and triatomic molecules (Kong 1982, 1989) are truly beautiful. They are the silver lining of the dark cloud of his long-term house arrest related to the Cultural Revolution in the People's Republic of China. His systems place molecules in the compartments of a flat chart with approximately the same shape as the chart of the elements. The period and group numbers are the sums of the atoms' period and group numbers.

There is a molecular number that proceeds from upper left to lower right, much as does the charge number of the atomic chart. This number is piecewise continuous and within each piece is linear with respect to (but is not equal to) the sum of the atomic numbers. The discontinuities exist in part because numbers have been reserved for transition-metal molecules that are not present in Kong's systems.

The construction of Kong's periodic table, using row and column sums, is in harmony with the additivity or near-additivity of some molecular properties. Atomic numbers and electron counts are obviously additive; atomic weights sum to molecular weights except for the effects of binding energies. Inter-nuclear separations of diatomic molecules are (usually by definition) additive when compared with covalent radii of atoms or ionic radii of charged species (Campbell 1970). There are other, less well-known additive properties (Hefferlin 1989a, Chapter 13).

Many of the compartments in Kong's tables contain several molecules; indeed, the paper that introduces his diatomic-molecular system comes with a lexicon listing the molecules contained in each compartment. The list is two-dimensional (arranged according to the differences of the atoms' period and group numbers), which suggests that Kong's table is in fact a projection onto two-dimensional space of a four-dimensional architecture.

One prospect for Kong's elegant construction is that it may be extended to tetra-atomic molecules; a proof-of-principle data analysis has already been made. Whether the same can be accomplished for larger molecules remains to be seen. Another possible prospect is the preparation of tables including all molecules in the compartments of Kong's periodic systems of triatomic and tetra-atomic molecules; they should require four and six atom row and column differences, respectively, if the same protocol is used as for diatomic molecules.

The periodic system of Boldyrev, Gonzales, and Simons and its prospects

The periodic system of Boldyrev, Gonzales, and Simons (1994) provides useful data at a glance. Diatomic molecules are arranged horizontally according to electronic configurations and placed vertically, with no gaps, according to the sum of the atomic numbers. Since there are no gaps and since there are so many more molecules in group-sums near the center of the configuration axis than at the extremes, the atomic-number sum varies across a given horizontal line. In other words, a vertical atomic-number-sum axis has been partially collapsed for sake of compactness. Subsequently, three wall charts (Boldyrev and Simons 1997) have been prepared—one for main-group molecules, one for molecules with one main-group atom and one transition-metal atom, and one for molecules formed from two transition-metal atoms. The prospects for this useful table include the publication of a book and CD-ROM diskettes containing much more information about each molecule.

The SAU system and its prospects

The SAU collaboration of Hefferlin, undergraduate students at various places, and (intermittently) colleagues around the world have produced what is the most general, $2N$ dimensional, physical periodic system of molecules. This molecular periodic system can, with suitable choices of originating tables of the elements, reproduce all but a few *ad hoc* details of all the other physical periodic systems in Table 1 and can lead to the creation of others. For example, Clark's (1935) and Kaslin's (1983a, b, 1985) systems are quilts consisting of slices of the general four-dimensional architecture. Kong's systems (1982, 1989) are two-dimensional projections of the four-dimensional architecture. Monyakin's (Hefferlin 1989a, Chapter 12) proposal is a three-dimensional collapsing of its four-dimensional architecture, formed by combining two of the four basis vectors.

The SAU collaboration system has an *ad hoc* theoretical formulation. An originating atomic chart is considered as a two-subscript null matrix; it is formed by replacing appropriate zeros of the matrix by symbols of the elements. The outer product of this matrix with itself is taken once to create the periodic system for diatomic molecules, twice for triatomic molecules (acyclic or cyclic), and so on. The result is a four, six, or more subscript matrix for those molecules.

A four-subscript matrix can be imagined as a four-dimensional array of symbols; a six-subscript matrix can be imagined as a six-dimensional array, and so on; these arrays are the periodic system (Hefferlin and Kuhlman 1980; Hefferlin 1989a, Chapter 10). In general, the outer product is taken $N - 1$ times to create the $2N$-dimensional periodic system for N atomic molecules.

The eight-element rows of the periodic chart have been likened to musical octaves; to the same extent that this analogy holds, these periodic systems allow two-, three-, and poly-phonic harmonies.

Some molecular properties are nearly additive (section on "Kong's Periodic System and Its Prospects"). This additivity can be reconciled with the multiplication just described by considering molecular data as exponents of some base, for instance the natural number e.

The main *scientific* prospect for the SAU systems is the publication of atlases, the first of which contains forecasted inter-nuclear separations for diatomic molecules (Hefferlin, Davis and Ileto 2003) and the second of which presents vibration frequencies (Davis and Hefferlin, in preparation). The prospect for *public* acceptance of this very general periodic system could improve if the public develops a taste for multiple dimensions. One way in which this might occur is by exposure to written (Abbott 1952; Burger 1983) or filmed expositions (Banchoff and Strauss Productions 1979), or science fiction tales, which involve objects in a world with four spatial dimensions (hypercubes, hyperspheres, etc.). For those with a passion for algebra, there are quite convincing mathematical expressions for various measures of the point, line, square, cube, and hypercube; of the point, circle, sphere, and hypersphere; and the like in spaces of progressive dimensionalities. For those with a love of fine art, Salvador Dali's impressive painting, "Corpus Hypercubus" shows the instrument of crucifixion as a hypercube unfolded into three-dimensional space.

Group-dynamic systems

Quantum mechanics is sufficient to recreate the general outlines of the chart of the elements by doing computations for one atom at a time. It is also possible to reconstruct the outlines of the chart of the elements from a completely different starting point, namely group dynamics. Group dynamics practically demands that a symmetry-based chart of the elements be extended to higher spaces. Anyone somewhat familiar with the ideas of fundamental-particle physics will recognize the parallel between this line of reasoning and that which resulted in the construction of the "periodic charts" of hadrons from the "periodic chart" of the three light quarks in $SU(3)$ symmetry (Hefferlin 1989a, 557–561; Young and Freedman 1996). Group dynamics, using *reducible* representations, has been used as a theoretical basis for the SAU molecular classification (Hefferlin et al. 1984). The use of *irreducible* representations results in some compartments being filled with linear superpositions of molecular symbols (Zhuvikin and Hefferlin 1983; Hefferlin et al. 1984; Zhuvikin and Hefferlin 1992, 1994; Carlson, Hefferlin, and Zhuvikin 1995; Carlson et al. 1996). Current research is concentrated on group chains that result in individual molecular symbols occupying each compartment.

WHAT PROBLEMS ARISE IN TESTING AGAINST DATA AND IN PREDICTING NEW DATA?

Technical problems in testing the physical periodic system against data

The first technical problem is encountered immediately upon beginning to test how well a periodic system concept agrees with the data. The problem consists of the surprisingly large errors that often accompany data in journal articles (occasionally as much as 20% for dissociation potentials and 100% for oscillator strengths of diatomic molecules). The errors are far less serious in quality critical tables.

A second technical problem has been encountered in regard to the periodicity that is basic to physical periodic systems. That periodicity is pronounced for main-group molecules (Figures 1–6). The expectation has been that periodicity would be similarly visible in transition-metal molecules. For instance, it was expected by the author that transition-metal molecules having as one atom Zn, Cd, or Hg would mark the end of a period by having low dissociation potentials. This expectation was rewarded (Hefferlin 1989a, Chapter 5). The expectation that lanthanoid molecules having Lu as one atom would do the same was not rewarded. Even the main-group periodicity seen in diatomic species (Figures 1–6) is been found to fade slightly for heavy molecules. For example, whereas NeN and NeO have magnitudes near zero, XeN and XeO have dissociation potentials with magnitudes close to the average value for all other molecules (Krasnov 1979).

Problems in using the physical periodic system to forecast new data

This section takes up problems that result when new data are forecasted by physical systems. The first problem is that the scientific community is (in general) confident of quantum theorists' ability to compute, or of experimentalists' skill in measuring, the needed data to any degree of precision. Much of the community fails to realize that neither quantum theorists nor experimentalists have the interest or the resources to do the work rapidly enough. The second problem is that the scientific community largely considers any semi-empirical (heuristic) forecasts as undependable. Much of the community is unaware of the many successful forecasts that have been made in this field of research as well as in others. It is the author's experience that these two problems of perception are more severe in the "West" than elsewhere.

The third problem has arisen upon attempting global predictions and is due to the relative shortage of data even for neutral, ground-state, species. There are, for example, a few hundred data for many spectroscopic and thermodynamic properties of diatomic molecules; there are a bit more than half that many for the analogous properties of acyclic triatomic molecules (and very few for the cyclic species). There are yet fewer for tetra-atomic molecules, and those data have to be rationed out among four-atom species with several structural-isomeric forms. The fraction of molecules for which data can be found for a given structural form becomes vanishingly small by the time one passes four-atomic molecules. This situation is due to the vast numbers of molecules and their isomers that can exist with a given number of atoms.

Thus, it seems that the physical periodic systems can be used for predictions of massive numbers of molecules only for diatomic and acyclic triatomic molecules and possibly for one or another structural form of tetra-atomic molecules. This limitation is not due to the method of construction of the periodic system, but to the scarcity of data with which to set up the least-squares or neural-network computing.

This problem has been somewhat ameliorated by collapsing coordinates, i.e., using functions that combine coordinates. For example, Kong has plotted molecular data on the total numbers of valence electrons with great success. For another example, the SAU three-dimensional system for acyclic triatomic molecules (Carlson et al. 1997; Wohlers et al. 1998) uses axes $(R_1R_2 + R_2R_3)$, $(C_1 + C_2 + C_3)$, and C_2, thus collapsing coordinates of both rows and columns and producing a three-dimensional system from one of six dimensions.

All coordinate-collapsing methods cause there to be more than one molecule at a given address. The tabulated data for all these molecules are combined in any analysis, and any predictions made for them will be identical. (A precedent appears in the chart of the elements, where more than one isotope may occupy an element's compartment.) It is clear that a loss of accuracy will take place using collapsed coordinates.

PROSPECTS FOR CHEMICAL PERIODIC SYSTEMS OF MOLECULES

The previous section took up the problem of data shortages and the use of collapsed coordinates as a "work-around." There exists the second class of periodic systems that goes to where the data are instead of waiting for the data to come to them. These are the "chemical" periodic systems, which include only molecules for which the investigator has special expertise and a comprehensive database. Now, a few chemical periodic systems will be presented along with the prospects foreseen for them.

Dias's periodic table of polycyclic aromatic hydrocarbons, and its prospects

From 1982 and to the present time, Dias has worked out periodic tables for polycyclic aromatic hydrocarbons—for sextet benzenoids (1982, 1994), for a subset of these, namely total resonant sextet benzenoids (1994), for fluoranoid/fluoranthenoid hydrocarbons (1992a), and for indacenoid hydrocarbons (1992b). Dias has created tables for additional species and has inter-related all of these periodic tables (1996). This work illustrates how a very narrow class of molecules can have a huge population and can have many interesting features, including isomer counts and topologies (1991).

It is impossible to exaggerate how thoroughly Dias has explored these species in his many publications. The prospects can only be good for his discovering further vistas of great value.

Haas's periodic system of functional groups, and its prospects

Haas published a periodic system of functional groups (Haas 1982, 1983, 1984). This system is an imaginative extension of Grimm's hydrogen-displacement principle. It contains, in principle, an infinite number of members of a limited class of molecules (perfluorinated functional groups). First, Haas substitutes any halide for the hydrogen in Grimm's hydrogen-displacement principle, so that the hydrogen-displacement principle becomes the "element displacement principle" and the groups are "paraelements." The process can be iterated, resulting in ever more complicated paraelements. Then, Haas repeats the entire process beginning with atoms from rows three and more.

The prospects for this periodic system have already been realized. It is easy to write formulae for paraelements using the system; Haas and his group, world-renown experts on perfluorinated compounds, have synthesized the compounds thus suggested.

BABAEV'S HYPERPERIODIC SYSTEM AND ITS PROSPECTS

The hyperperiodic chart

Babaev has sketched a periodic system that, in principle, includes all possible main-group molecules, with even numbers of electrons, and their isomers (Babaev and

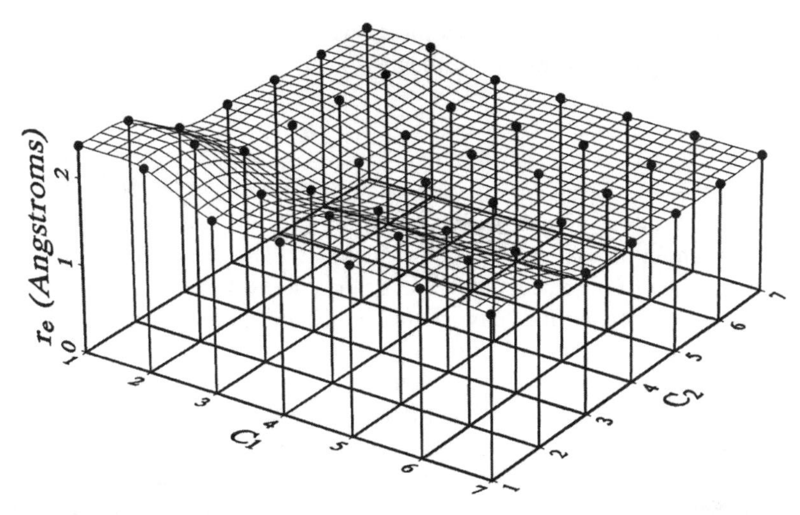

Figure 7: A portion of Babaev's periodic system. Each compartment $\chi > 0$ contains more than one molecule due to the fact that hydrogen-containing species may be included and also ($\chi > 1$) due to proton "shifting" (e.g., from Be_2 to LiC). These different species could be shown if additional dimensions were available. Molecules with odd numbers of electrons are not shown. All entries are from the second period of the chart of the elements; entries including atoms from other periods could be displayed if other additional dimensions were invoked. Various kinds of isomers could be exhibited if yet more dimensions were utilized. The right-hand edge of this table contains species at the limit of stability; the addition of any more electrons results in instability. However, if atoms from higher periods are allowed, then some species with Euler characteristic >2 can be stable. Some of them are mentioned in the text (Adapted from Babaev and Hefferlin 1996).

Hefferlin 1996). It features a periodicity completely unrelated to atomic periodicity; hence the name "hyperperiodicity."

Figure 7 shows a schematic of this hyperperiodic table. This table appears in the figure to be two-dimensional. It has the number of non-hydrogen atoms as the vertical axis and a parameter related to the shapes of the molecules as the horizontal axis. The hyperperiodicity is related to these molecular shapes, as will be shown by examples in the following section. The figure favors, at least in the upper portions, molecules with row-two atoms.

Many molecules can exist in one compartment, even if the restriction to row-two atoms is maintained; for instance, CO and BF are in the same compartment as N_2, and FNO_2, O_4, and N_2F_2 are in the same compartment as BF_3. The schematic does not show the additional axes necessary to distinguish between these horizontally isoelectronic molecules; such axes are shown explicitly in the lexicon for Kong's periodic system for diatomics. The schematic also does not indicate the additional axes necessary to separate vertically isoelectronic molecules like O_2, SO, S_2, SeO, ... There must be N of these additional axes to enumerate the periods of the atoms in molecules of a row containing N non-hydrogen atoms. Finally, the schematic lacks an indication of the axes necessary to resolve isomers. The definition of these isomer-enumerating axes will be extremely difficult.

Hyperperiodicity

It is convenient to begin with the diagonal fluorine series BeF_2, BF_3, CF_4, PF_5, and SF_6. The first three of these species come from row-two atoms; the significance of the deviation for the last two will be taken up later. The shapes are defined by the lone-pair electrons of the parent atoms and by the fluorine ligands. They are a line, a triangle, a tetra hedron, a trigonal bipyramid, and an octahedron, respectively. These shapes have one, two, three, and four cycles (as determined by finding the minimum number of edges that, when removed, eliminate all cyclic paths along the edges). It follows that these molecules have Euler topological characteristics $\chi = -2, 0, 2, 4,$ and 6 (Babaev and Hefferlin 1996). The same result follows when looking along the horizontal direction. The last few members of the hyper-rows beginning with three of the molecules just named have species with the same geometrical sequences as those above. Thus, BeF_2, CF_2 (isoelectronic with FNO), SF_2 (isovalent with F_2O), and XeF_2 have shapes from line to trigonal bipyramid; BF_3, NF_3, and ClF_3 have forms from triangle to trigonal bipyramid; and CF_4, SF_4, and XeF_4 are tetra hedron to octahedron.

Corroboration is provided by sets of neutral, same-row, isoelectronic species: AlF_3, $SiOF_2$, PO_2F, and SO_3 ($\chi = 0$) are all triangles; SiF_4, POF_3, SO_2F_2, and ClO_3F ($\chi = 2$) are all tetra hedra, etc.

Finally, looking along the vertical direction reveals that members of hypercolumns all have the same shapes. F_2O, NF_3, and CF_4 ($\chi = 2$) are all tetra-hedral; XeF_2, ClF_3, SF_4, and PF_5 ($\chi = 4$) are all trigonal bipyramids; XeF_4, IF_5, and SF_6 ($\chi = 6$) are all octahedra. These shapes are all defined by the lone electron pairs belonging to the parent atoms and by the ligands. Thus, it is clear that the hypercolumns are defined by the molecular shapes in the portion of the system where these species lie.

Although there is a steady increase in valence-shell electron count along each hyper-row, the count varies in a hypercolumn. In fact, it increases by six from one hyper-row to the next. The reason for this shift is explained in the next paragraph. The increase along each row and the increment of six in going down a column of the figure confirm the locations of the atoms and ions (point shapes), the dimers (line shapes), and the species further to the left in the table.

Each compartment in Figure 7 shows only one representative entry. However, it also contains other molecules with the same number of non-hydrogen (row-two) atoms and the same number of valence electrons. For example, along with Ne in $\chi = 2$ are FH, OH_2, NH_3, and CH_4 (the last column in Grimm's hydrogen-displacement diagram). Along with F_2 are OFH, H_2O, NFH_2, CFH_3, NOH_3, ..., and C_2H_6. And along with F_2O are many molecules with progressively more hydrogen atoms until C_3H_8 terminates the list. Thus, the most hydrogenated molecule in this hypercolumn is always an alkane (C_nH_{2n+2}). Alkanes have the greatest possible number of hydrogen atoms that can bond with a given number of atoms. The addition of one more creates an unstable species. Similarly, the other species in a $\chi = 2$ compartment are such that the substitution of any row-two atom with more valence electrons results in an unstable combination. There indeed are stable molecules in $\chi > 2$, but they involve atoms from other periods and result from more complex bonding considerations.

The molecules with the greatest numbers of hydrogen atoms in $\chi = 0$ are alkenes or a cyclo-alkanes (C_nH_{2n}), e.g., FNO and C_3H_6; those in $\chi = -2$ are alkynes (C_nH_{2n-2}), e.g., FN_3 and C_4H_6.

Much more is involved in the hypercolumns than has been explained here (Babaev 1988; Babaev and Hefferlin 1996).

The system subsumes other periodic systems of molecules

Babaev's periodic system appears capable of reproducing Morozov's (1907) system when only the hydrocarbons, and their derivatives, in each compartment are shown (Babaev and Hefferlin 1996). Parenthetically, it is of human interest that Morozov created his classification while in the Schlisselburg island prison due to his political activities, just as Kong did his creative work while in house arrest. Babaev's system yields Haas's system when the odd-numbered valences are included and only paraelements are displayed.

If the row containing atoms were to be extended normal to the plane of the chart and enumerated with an axis for the periods and if the odd elements were to be supplied, then the chart of the elements would be produced. If the row containing diatomic molecules were twice to be extended normal to the plane of the chart and enumerated with two independent axes for periods, and if radicals were to be introduced, then the four-dimensional physical periodic system emerges. Analogous proliferation of axes for other rows of the figure would produce outer-product periodic systems for these N-atom molecules.

There is work that could be done on this general hyperperiodic system. While it is in principle completely general, it does not now have any group-dynamic or other foundation and does not, at this time, have a means of iterating to larger molecules such as is inherent in the periodic system of Haas. It should be fleshed out much more fully and checked against many atomic and molecular properties.

Emergence

Two beautiful examples of emergence are seen in the presentations made so far: the increase in the numbers of molecular structures (isomers) as N increases, and the appearance of hyperperiodicity.

WHAT PRESUPPOSITIONS ARE REVEALED BY THE CREATION OF CHEMICAL PERIODIC SYSTEMS?

The designers of these schemes had such a thorough knowledge of the molecules in their periodic systems that they report not having checked the designs with additional literature values. The prediction of new data has not been featured in their publications. The periodic systems have usually been presented as two-dimensional even if they are multi-dimensional, i.e., no attempt has been made to display the additional dimensions. This presentation style may be related to the fact that the designers are usually chemists in areas that do not require thinking in multi-dimensional spaces.

WHAT PROBLEMS ARISE IN TESTING AGAINST DATA AND PREDICTING NEW DATA WITH CHEMICAL SYSTEMS?

The very motivation and methods of construction of chemical periodic systems guarantee that there will be data for at least a few molecules with many more than four or five atoms. Unfortunately, the fact remains that there will eventually be, as the numbers of atoms in the molecules increases, an intolerable shortage of data compared with the numbers of species in compartments of any system.

This problem is not due to any lack of effort in measuring or computing data, since over 20 million species have been characterized and accessioned by Chemical Abstracts Service. It is difficult to imagine the amount of effort that has been expended in obtaining this vast quantity of information, and the construction of periodic systems to organize the data can only be considered as homage to the effort.

WHAT ARE THE GENERAL PROSPECTS FOR MOLECULAR PERIODIC SYSTEMS?

Molecular similarity, mathematical chemistry, and combinatorial chemistries

There are related fields of activity such as the search for molecular similarities (Johnson and Maggiora 1990), and for quantitative structure and property/activity relationships, where the insights gained from molecular classifications based on periodicity may be helpful.

One object of mathematical chemistry is to associate a unique index to each skeleton (i.e., no hydrogen atoms) molecular diagram (Balaban 1973; King 1977; Randić 1992). The index may be a number, a set of numbers, or a matrix. Hydrogen and other atoms with various valences have since been included and three-dimensional or knotted structures have been treated. In the meantime, however, combinatorial chemistries have accumulated so rapidly that the prospect may be for them to outpace the progress in correlating data to one or another index unless a breakthrough occurs in the latter.

The classroom and the scientifically literate public

In 1967, Sanderson published a textbook with a unique chart of the elements, and related charts for hydrogen, nitrogen, oxygen, sulfur, halide, methyl, ethyl, and phenyl compounds. Shchukarev (1970, 1974) published scores of elaborate graphs for classes of molecules (such as oxides of nitrogen) showing how values of molecular properties depend on some significant measure (such as the oxidation state of the nitrogen atom). The effort devoted to this research at Leningrad State University is beyond imagination and deserves more attention in the West (Latysheva and Hefferlin 2004). The effort by these authors to demonstrate molecular periodicity is very evident. It is interesting that Mendeleev, Morozov, Sanderson, and Shchukarev all presented their concepts of atomic or molecular periodicity in textbooks. Hopefully, the charts by Boldyrev and Simon or by Kong may eventually be seen on classroom walls, in textbooks, and on wallet cards.

New Scientist (Rouvray 1994), Scientific American (Scerri 1998), and probably other magazines for the same reading audience have carried articles making some mention of periodic systems of molecules. The 1989 McGraw-Hill Encyclopedia Yearbook carries a short section featuring some physical molecular systems (Hefferlin 1989b), as does the 1997 McMillan's encyclopedia of Chemistry (Scerri 1997). Babaev has dramatized periodic charts and systems on programs for Russian and British television.

ACKNOWLEDGMENTS

The author is deeply indebted to Dr. Chris Hansen (SAU), Dr. Evgenii Babaev (Moscow State University), and Dr. Gary Burdick (Andrews University) for help with this chapter or for ideas expressed therein.

REFERENCES

Abbott, E.A. 1952. *Flatland*. New York: Dover Publications.

Allen, L.C. 1992. Extension and Completion of the Periodic Table. *Journal of the American Chemical Society* 114: 1510–1511.

Babaev, E.V. 1988. Vozmozhno li Periodicheskaya Sistema Molekul? In *Istoria i Metodologiya Estestvennykh Nauk. Filosofskie Problemy Khimii* 35. Rudenko A.P., Ed. Moscow, Moscow University Press, pp. 121–140.

Babaev, E.V. and Hefferlin, R. 1996. The Concepts of Periodicity and Hyperperiodicity: From Atoms to Molecules. In: Rouvray, D.H. and Kirby, E.C. (eds.), *Concepts in Chemistry*. Taunton, Somerset, England: Research Studies Press Limited.

Balaban, A.T. 1973. Chemical graphs. XVIII. Graphs of degrees four or less, isomers of annulenes, and nomenclature of bridged polycyclic structures. *Rev. Roum. Chim.* 18: 635–653.

Banchoff and Strauss Productions. 1979. The Hypercube: Projections and Slicing (film). Providence, Rhode Island.

Barut, A.O. 1972. *On the Group Structure of the Periodic Table of the Elements* in Wybourne, B. Structure of Matter (Proceedings of the Rutherford Centennary Symposium, 1971). Canterbury: University of Canterbury Press.

Boldyrev, A.I., Gonzales, N., and Simons, J. 1994. Periodicity and peculiarity in 120 First- and Second-Row Diatomic Molecules. *Journal of Physical Chemistry* 98: 9931–9944.

Boldyrev, A.I. and Simons, J. 1997. *Periodic Table of Diatomic Molecules* (*Wall Charts A, B, and C*). Hoboken, New Jersey: John Wiley & Sons.

Burger, D. 1983. *Sphereland*. New York: Barnes & Noble.

Campbell, J.A. 1970. *Chemical Systems: Energetics, Dynamics, Structure.* New York: W.H. Freeman and Company.

Carlson, C.M., Cavanaugh, R.J., Zhuvikin, G.V., and Hefferlin, R.A. 1996. Periodic Systems of Molecular States from the Boson Group Dynamics of $SO(3) \times SU(2)_S$. *Journal of Chemical Information Computer Science* 36: 396–389.

Carlson, C.M., Gilkeson, J., Linderman, K., LeBlanc, S., and Hefferlin, R. 1997. Global Forecasting of Data Using Least-Squares Methods and Molecular Databases: A Feasibility Study Using Triatomic Molecules. *Croatica Chemica Acta* 70: 479–508.

Carlson, C.M., Hefferlin, R., and Zhuvikin, G.V. 1995. Analysis of Group Theoretical Periodic Systems of Molecules Using Tabulated Data. *Joint Report No. 2*. Physics Departments of Southern College, Collegedale, TN, USA; St. Petersburg University, St. Petersburg, Russia (March).

Clark, C.H.D. 1935. The Periodic Groups of Non-hydride Di-atoms. *Transactions of the Faraday Society* 3: 1017–1036.

Davis, W.B. and Hefferlin, R. An Atlas of Forecasted Molecular Data II: Vibration Frequencies of Main-Group and Transition-Metal Neutral Gas-Phase Diatomic Molecules in the Ground State (in preparation).

Dias, J.R. 1982. A Periodic Table of Polycyclic Aromatic Hydrocarbons. Isomer Enumeration of Fused Polycyclic Aromatic Hydrocarbons. *Journal of Chemical Information and Computer Sciences* 22: 15–22.

Dias, J.R. 1991. Benzenoid Series Having a Constant Number of Isomers. Part 3. *Journal of Chemical Information Computer Science* 31: 89–96.

Dias, J.R. 1992a. Studies in Deciphering the Information Content of Chemical Formulas: A Comprehensive Study of Fluorenes and Fluoranthenes. *Journal of Chemical Information Computer Science* 32: 2–11.

Dias, J.R. 1992b. Deciphering the Information Content of Chemical Formulas: Chemical and Structural Characteristics and Enumeration of Indacenes. *Journal of Chemical Information Computer Science* 32: 203–209.

Dias, J.R. 1994. Setting the Benzenoids to Order. *Chemistry in Britain* (May): 384–386.

Dias, J.R. 1996. Formula Periodic Tables—Their Construction and Related Symmetries. *Journal of Chemical Information Computer Science* 36: 361–366.

Grimm, H.G. 1925. Zur Systematik der Chemischen Verbindunger von Standpunkt der Atomforschung, Zugleich Uber Einige Aufgabe der Experimentalchemie. *Zeitschrift Fur Elektrochemie* 31: 474–480.

Haas, A. 1982. A New Classification Principle: The Periodic System of Functional Groups. *Chemicker-Zeitung* 106: 239–248.

Haas, A. 1983. Novyi Printzip Klassifikatzii Funktzional'nykh Grupp v Svyazi c Periodicheskoi Sistemoi. *Zhurnal Vse-soyuznovo Khimicheskovo Obshchestva* 28: 545–655.

Haas, A. 1984. The Element Displacement Principle: A New Guide in P-block Element Chemistry. *Advances in Inorganic Chemistry and Radiochemistry* 28: 167–202.

Hefferlin, R. 1989a. *Periodic Systems and Their Relation to the Systematic Analysis of Molecular Data.* Lewiston, NY: Edwin Mellen Press.

Hefferlin, R. 1989b. Molecule. *McGraw-Hill Yearbook of Science and Technology* 224–228.

Hefferlin, R. 1994. Matrix-Product Periodic Systems of Molecules. *Journal of Chemical Information Computer Science* 34: 313–317.

Hefferlin, R. and Burdick, G.W. 1994. Fizicheskie i Khimicheskie Periodicheskie Sistemy Molekul. *Zhurnal Obshchei Khimii* 64: 1870–1885. (English translation: 1994. Periodic Systems of Molecules: Physical and Chemical. *Russian Journal General Chemistry* 64: 1659–1674.)

Hefferlin, R., Campbell, G.D., Kuhlman, H., and Cayton, T. 1979. The Periodic Table of Diatomic Molecules—I. An Algorithm for Retrieval and Predication of Spectrophysical Properties. *Journal of Quantitative Spectroscopy Radiation Transfer* 21: 315–336.

Hefferlin, R., Davis, W.B., and Ileto, J. 2003. An Atlas of Forecasted Molecular Data I: Internuclear Separations of Main-Group and Transition-Metal Neutral Gas-Phase Diatomic Molecules in the Ground State. *Journal of Chemical Information Computer Science* 43: 622–628.

Hefferlin, R. and Kuhlman, H. 1980. The Periodic System for Free Diatomic Molecules-III. Theoretical Articulation. *Journal of Quantitative Spectroscopy Radiation Transfer* 24: 379–383.

Hefferlin, R., Zhuvikin, G.V., Caviness, K.E., and Duerksen, P.J. 1984. Periodic Systems of N-Atom Molecules. *Journal of Quantitative Spectroscopy Radiation Transfer* 32: 257–268.

Jenč, F. 1996. The Reduced Potential Curve (RPC) Method and Its Applications. *International Review of Physiology Chemistry* 15: 467–523.

Johnson, M.A. and Maggiora, G.M. 1990. *Concepts and Applications of Molecular Similarity*. Hoboken, New Jersy, USA: John Wiley.

Kaslin, V.M. 1983a. *Tables of Force Constants k_e and Vibrations Constants ω_e of Ground Electronic States of Diatomic Molecules, Composed of Atoms with Composition of s and p Shells (Atom from the Chemical Groups of Lithium, Beryllium, Boron and Carbon). Preprint 302.* Moscow: Optics Laboratory, Optics and Spectroscopy Department, Physical Institute.

Kaslin, V.M. 1983b. *Tables of Force Constants k_e Vibration Constants ω_e of Ground Electronic States of Diatomic Molecules, Composed of Atoms with Composition of s and p Shells (Atom from Chemical Groups of Nitrogen, Oxygen, Fluorine, and Neon). Preprint 303.* Moscow: Optics Laboratory, Optics and Spectroscopy Department, Physical Institute.

Kaslin, V.M. 1985. O Svyazi Spektroskopicheskikh Konstant Dvukhatomnykh Molekul c Parametrami Atomov. *Optika i Spektroskopiya* 59: 667–770 (text and graphics without tables).

King, R.B. 1977. The Elplacarnet Tree: A Complement to the Periodic Table for the Organometallic Chemist. *Journal of Chemistry* 15: 181–188.

Komarov, V.S. and Lyakhovskii, V.D. 1986. Unitarnaya Simmetriya Molekul. *Khimicheskaya Physica* 5: 914–924. (English translation: 1986. Unitary Symmetry of Molecules. *Soviet Journal of Chemistry and Physics* 5: 1501–1519.)

Kong, F.A. 1982. The Periodicity of Diatomic Molecules. *Journal of Molecular Structure* 90: 17–28.

Kong, F.A. 1989. In: Hefferlin, R. 1989a, Chapter 11.

Latysheva, V.A. and Hefferlin, R. 2004. Periodic Systems of Molecules as Elements of Shchukarev's "Supermatrix", i.e. Chemical Element Periodic System. *Journal of Chemical Information Computer Science* 44: 1202–1209.

Mazurs, E.G. 1974. *Graphic Representations of the Periodic System During One Hundred Years*. Tuscaloosa, Alabama, USA: University of Alabama Press.

Morozov, N. 1907. *Stroeniya Veshchestva*. Moscow: I. D. Sytina Publication.

Randić, M. 1992. Chemical Structure—What is She? *Journal of Chemical Education* 69: 713–718.

Rouvray, D. 1994. Elementary, My Dear Mendeleev. *New Scientist* (Issue # 1912): 36–39.

Rumer, Yu.B. and Fet, A.I. 1972. The Group *spin*-4 and the Mendeleev System. *Theory of Mathematical Physics* 9: 1081–1085.

Sanderson, R.T. 1967. *Inorganic Chemistry*. New York: Reinhold.

Scerri, E. 1997. Periodicity, Chemical. *McMillan's Encyclopedia of Chemistry* III: 22–32.

Scerri, E. 1998. The Evolution of the Periodic System. *Scientific American* 279: 78–81.

Shchukarev, S.A. 1970. *Neorganicheskaya Khimiya*, Vol. 1. Moscow: Vysshaya Shkola.

Shchukarev, S.A. 1974. *Neorganicheskaya Khimiya*, Vol. 2. Moscow: Vysshaya Shkola.

Van Spronsen, J.W. 1969. *The Periodic System of Chemical Elements.* Amsterdam: Elsevier.

Wohlers, J., Laing, W.B., Hefferlin, R., and Davis, W.B. 1998. Least-Squares and Neural-Network Forecasting from Critical Data: Diatomic Molecular r_e and Triatomic ΔH_a and *IP*. In: Carbo-Dorca, R. and Mezey, P.G. (eds.), *Advances in Molecular Similarity.* Stamford, CT: JAI Press, 265–287.

Young, H.D. and Freedman, R.A. 1996. *University Physics.* Reading, Massachusetts: Addison-Wesley.

Zhuvikin, G.V. and Hefferlin, R. 1983. Periodicheskaya Sistema Dvukhatomnykh Molekul: Teoretiko-Gruppovoi Podkhod. *Vestnik Leningradskovo Universiteta* 16: 10–16.

Zhuvikin, G.V. and Hefferlin, R. 1992. Bosonic Symmetry and Periodic Systems of Molecules, in M.A. del Olmo, M. Santander, and J.M. Guilarte, Group Theoretical Methods in Physics. *Proceedings of the XIX International Colloquium*, Salamanca, Spain, Vol. II. Anales de Física, Monographias 1, Real Sociedad Espanola de Física, 358–361.

Zhuvikin, G.V. and Hefferlin, R. 1994. Symmetry Principles for Periodic Systems of Molecules. *Joint Report No. 1.* Physics Departments of Southern College, Collegedale, TN, USA; St. Petersburg University, St. Petersburg, Russia (September 11). Library of Congress number QD467.Z48.

CHAPTER 13

A NEW PARADIGM FOR SCHRÖDINGER AND KOHN

JACK R. WOODYARD
Department of Community Services, Colby Community College, 1255 South Range, Colby, KS 67701, USA

INTRODUCTION

An alternative to Hilbert space in non-relativistic quantum mechanics gives Schrödinger back his "matter waves," and passes the "duck test"[1] for the Kohn Sham equations. According to Wylie (1999), "... it cannot be assumed that the sciences presuppose an orderly world, that they are united by the goal of systematically describing and explaining this order, or that they rely on distinctively scientific methodologies which, properly applied, produce domain-specific results that converge on a single coherent and comprehensive system of knowledge." Discovery is nearly always chaotic. The development of a new view, particularly one bridging concepts between chemistry and physics, with potential applications to biology and medicine elicits the skeptic in not only the scientific community, but in the individual developer as well. There must be some driving force to make the effort worthwhile. Looking at the literature and conversations with both philosophers and scientists show all is not as well as one would like to assume (Tegmark and Wheeler 2001). Even technically, it is possible to show quantum chemical calculations converging to answers often uncomfortably removed from experimental values, and supposed improvements actually making the situation worse (Feller and Peterson 1998). Many staples of teaching, particularly to undergraduates and non-scientists fail both logically and efficaciously, and call into question current practices. At the heart of the matter is what really constitutes an "explanation." At issue is, "How does interpretation affect the use and results of scientific observations or calculations?" The matter finally culminates in the statement, "Quantum mechanics is non-intuitive" (Pesic 2002). Why? Can a new approach do something about it? Review of the original work and early development of, in this case quantum mechanics and its application to both physics and chemistry, shows some assumptions, their displacement by current knowledge, and the natural resistance to recognition of such displacement.[2] In particular, some very simple modifications to Schrödinger's work allow much greater intuitivity and make atomic and molecular systems able to be visualized. Some specific mathematical results of general validity are necessary to this exposition, as is the discussion of the use of Hilbert space in atomic and molecular problems. Once the traditional method is understood, one may

D. Baird et al. (eds.), Philosophy of Chemistry, 245–269.

approach the new view in which Hilbert space is no longer useful, and the relevant equations, and their solutions are retained in complex 3-space. The electron, proton, indeed, all charged "particles" are described as waves at the onset, with a particular archetype chosen to match the theory to experiment. The pertinent interactions are seen as the spectral interference of these waves. The results are related to traditional methods, including density functional theory[3,4], and show both calculation and descriptive efficacy. Examples from an early feasibility study are given, and compared to results from a current calculation. The full theory appears robust with respect to both calculational efficiency—values have a standard error of 0.0009% of experiment[5] for the systems tested using only one configuration, and zero iterations, and the work has internal consistency.[6] This chapter should illuminate the problems associated with developing a new theory, which challenges current dogma, and illustrate some of the contortions necessary to make it work. Unfortunately, acceptance by the scientific community hinges on factors not necessarily associated with the competence of the logic or the value of the results—challenges for the development of a coherent philosophy of chemistry, or any other science. Some of these factors are exposed in this work.

PLANCK'S LAMENT

According to urban legend, Max Planck once said that new ideas do not gain acceptance—it is just that their opponents die out. Anyone who has attempted to do something new has had experiences similar to having a reviewer state, "Electronic structure theory is a mature field and has no need of naïve theories." Even though a brief scanning of the literature shows this particular statement to be short sighted at best.[7] They may even have had the experience of a supposedly competent scientist looking them straight in the eye and refusing to even look at their mathematics, despite their personal credentials. This is all part of the very human and social endeavor we call science (Shermer 2001). Of equal interest, however, is the internal dialog of the same nature. One is steeped in the knowledge of the time. Frequently, this knowledge is inconsistent, but it appears consistent until deeply analyzed. Little things get in the way—comments by old professors, or mis-statements or errors in old books have to be rooted out and subjected to the acid test. At best progress is slow, and fraught with self-doubt and potential error. Each new revelation must be tested, tested, and tested again. If the method is accurate and efficacious, as apparent from Tables 1 and 2,[8] and from the method's explanatory power, one would hope it will eventually be recognized and used. History indicates this could be through the persistent efforts of the author, or may not happen until a savant with the right political connections proposes it again. Of some concern, however is the technological age in which we live. Many current results are achieved simply by applying the monstrous power of modern computers. The elegance of a new theory may be lost in the shuffling of numbers. All in all it is much more an adventure than a scientific expedition.

Table 1: Feasibility studies with simplified models.

The crystal approximation–lay out the atomic or molecular system so that the electrons are as far apart as possible consistent with the presence of the nuclei. Effective nuclear charges were determined by using the Virial theorem. Experimental values are from Moore's tables or from the literature. These values were obtained for one configuration and zero iterations.

Ion	Calc.	Exp.	Ion	Calc.	Exp.	Ion	Calc.	Exp.
H^-	−0.5174	−0.5277	He-I	−2.8797	−2.9049	Li-II	−7.1976	−7.2837
Be-III	−13.4962	−13.6637	B-IV	−21.7838	−22.0463	C-V	−32.0643	−32.4329
N-VI	−44.3400	−44.8248	O-VII	−58.6119	−59.2228			

The restricted density functional approximation—restrict the one-electron functions to superposition of the ϕs and determine an effective nuclear charge on the nuclei through the S-integral. Test cases were prepared with one configuration and zero iterations.

Ion	% of exp.	Ion	% of exp.	Ion	% of exp.	Ion	% of exp.	Ion	% of exp.
H^-	99.34	He-I	102.13	Li-II	102.99	Be-III	102.47	B-IV	102.06
C-V	101.75	N-VI	101.51	O-VII	101.31				

The non-holonomic constraint approximation—eliminate the electron–electron interaction from the Hamiltonian and replace it with a set of one-dimensional Schrödinger equations modeling the "motion" of the electrons as they are allowed to move freely within the crystal approximation. Quantizing this motion provides a "zero point energy" that approximates the energy of the system. Tests were prepared for two- and three-electron atomic cases in ground and excited states, diatomic molecules, and simple crystals. Data below are for one configuration and zero iterations. The atomic number is used as the effective nuclear charge. Ions designated with "*" followed by a number are in excited states, e.g., *1 = first excited state.

				Energy (atomic)					
Ion	% of exp.	Ion	% of exp.	Ion	% of exp.	Ion	% of exp.	Ion	% of exp.
H^-	100.4	He-I	103.4	Li-II	102.8	Be-III	102.5	B-IV	102.1
C-V	101.9	N-VI	101.6	O-VII	101.5	Li-I	101.1	He-I*1	101.2
Li-II*1	101.3	Be-III*1	101.1	B-IV*1	100.9	C-V*1	100.8	N-VI*1	100.6
O-VII*1	100.5	He I*2	101.9	He I*3	100.1	Li-II*3	101.2	Be-III*3	101.3
B-IV*3	101.2	C-V*3	101.1	N-VI*3	101.0	O-VII*3	100.9		

(cont.)

Table 1: *(Continued)*

Mean ionic radii (preliminary calculations, frozen cores)									
Ion	% of exp.	Ion	% of exp.	Ion	% of exp.	Ion	% of exp.	Ion	% of exp.
H^-	101	He I	108	Li-II	107	C-V	97	N-VI	99
Bond lengths (selected diatomic systems, preliminary calculations)									
Molecule	% of exp.	Molecule	% of exp.	Molecule	% of exp.				
H_2	106	Li_2	99.5	LiH	110				
In this next test the effective nuclear charges were determined for the non-holonomic constraint method using the Virial theorem.									
Energy (atomic)									
Ion	% of exp.	Ion	% of exp.	Ion	% of exp.	Ion	% of exp.	Ion	% of exp.
H^-	100.4	He-I	101.1	Li-II	101.7	Be-III	101.7	B-IV	101.5
C-V	101.4	N-VI	101.3	O-VII	101.2				

For an advanced calculation of the hydrogen molecule, $E = 1.0024\ E_{exp}$; $\delta = 0.9838\ \delta_{exp}$; where δ is the mean nuclear separation—Crystal structures were correctly predicted for LiH, metallic hydrogen, and crystalline helium using the mean charge distributions calculated with this method. Molecular structures of carbon dioxide and water and the approximate density difference between liquid water and ice were also correctly predicted. (Models of the charge distribution at the mean value of $r(\theta, \phi)$ were constructed from modeling clay and assembled like three-dimensional puzzle pieces to make these predictions. A full analysis would require the energy to be relaxed as a function of the geometry.)

"... OF SHOES, OF SHIPS, OF SEALING WAX ..."

Authors such as Wylie (1999) and Cartwright (1999) point out the limits of modern scientific endeavor, but working scientists tend to get captured by the system. It is very easy to follow the concepts of a discipline, such as quantum chemistry on a *pro forma* basis, questioning little in the rush to achieve one more decimal point in the target answer, or the next large government or industrial grant. Quantum mechanics has been a powerful tool for many years, but still, much needs to be explored and examined.[9] (1) From a purely calculational point of view, it is fully possible to improve both the theory and the number of terms in the series solution and come out with a less correct answer.[10] (2) We can calculate the chemical bond rather well, but we still do not know exactly what it is (Scerri 1994). We really cannot explain it very well except as a calculational procedure. (3) The periodic table pays lip-service to quantum mechanics, but in order to determine a configuration, all possible configurations must be calculated and the one with the lowest energy chosen Scerri (1997; 1998a, b, c). Frequently, our calculations are not very accurate, and elements like chromium are still a mystery. (4) We even have trouble explaining how elements, such as carbon can have so many valence states, when many of the valence electrons are paired. (5) Prediction of crystal structure, bonding angles, or the energy of molecules, glasses, or other materials requires calculation of all possible structures and taking the one with the least energy. (6) Simple principles, such as the Pauli exclusion principle, Hundt's rule, and the wave–particle duality are still laws of nature. (7) From another direction, philosophers ask each other why we teach orbitals Scerri (1991). After all, orbitals have no physical meaning, and, in all but hydrogen, exist only in a multi-dimensional, Hilbert space. (8) There is no way to visualize atomic structure, or even the charge distribution around an atom, or group of atoms, without very messy calculations. (9) The closest thing to an intuitive method is density functional theory, and it is founded on two major flaws (Dreizler and Gross 1990, op. cit.): (a) the derivation only works for the ground state—excited states are a band aid (Nagy and Adachi 2000); (b) only model potentials exist for the exchange-correlation energy—there is no *a priori* method of determining this potential.

This state of affairs is something of a historical accident. Schrödinger and Einstein wanted something gentler than what eventually evolved into quantum mechanics as we know it (Scott 1967). Schrödinger wanted matter waves and got wavefunctions.[11] In a demonstration of proof-by-intimidation, Bohr and friends pushed the Hilbert space representation quite successfully[12]. At issue was how to combine hydrogen-like wavefunctions to represent multi-electron atoms. Schrödinger's attempts at simple superposition did not work. The form of the electron–electron interaction term seemed to dictate the single-electron wavefunctions should be combined in the form of a product. This, of course, all evolved into the configuration–interaction scheme of totally antisymmetric product wavefunctions familiar today. One of the really rude things about this is, Hilbert space is very strange. Each electron exists in its own three-dimensional space, independently, and all of these three-dimensional spaces, with spin, are superposed. Only the electron–electron operator, which bridges the "gaps" in these superposed spaces can remedy the situation. One should also note, as written in the Schrödinger equation, the electron is a particle until the equation is solved. Only then can some of the wave nature be manifest. The old issue of the wave–particle duality is a mystery, and "Quantum mechanics is non-intuitive."

This chapter takes the approach in which charged "particles" are actually waves in complex 3-space, and rewrites the Schrödinger equation to accommodate this idea. The square magnitude of these waves, including their spectral interference, provides the mean charge distributions, positive and negative, for the system. High quality results are presented for calculations developed from only one configuration and zero iterations, and many phenomena are explained as the result of the concept of electron interference. Hartree's atomic units are used throughout (see Bethe and Salpeter 1957).

SCHRÖDINGER'S FOLLY[13]

Schrödinger constructed his wave equation in non-relativistic form, but he also brought Coulomb's law directly into the mix—the functional form of the potential energy between the electron and the nucleus, between nuclei, or between electrons, is exactly Coulomb's law for two point particles. There was no guarantee it would work, except for the great success of the one-electron atomic system calculations. The kicker may have entered into the electron–electron calculations.

Unbeknownst to Schrödinger, there is a singularity imposed when the one-electron functions are superposed in a concentric manner. Should one treat the one-electron functions as standing waves with centers displaced from each other by an amount ε, see Figure 1, and take the square magnitude of the resultant spectral interference as the physical quantity, the result as $\varepsilon \to 0$ does not necessarily equal the result at $\varepsilon = 0$. This had a very deleterious effect on his calculations.

Figure 1 is a calculated contour graph of two interacting 1s electrons. The centers-of-mass are distributed symmetrically about the nucleus in the center. We have chosen a displacement of the centers-of-mass of 0.4 hartree for clarity. We first show the trial function wave, Ψ, i.e., the resultant of our model, in segments "a" and "c". Segment "a" is for paired electrons while segment "c" is for antipaired electrons. We then show the square magnitude, $\Psi^*\Psi$, which we style as the charge distribution. Again, segment "b" is for paired electrons and segment "d" is for antipaired electrons. Note in segment "b," even with a separation of 0.4 hartree, the two-electron waves have merged into one distinctive wave. This is the spin-zero boson of superconductivity fame. Note in segment "c" how the two-electron waves appear to be pushing each other apart. This is even more pronounced in segment "d," the antipaired charge distribution. This causes a marked increase in the calculated energy even if the range parameter is taken as just the atomic number. Such atomic systems cannot exist. Only for quite large ε, or separate association with multiple nuclei, can the two electrons coexist in a bound system.

Another issue Schrödinger could not have been expected to know was these same one-electron functions must change mean radii as they approach each other. This is required by the Virial theorem[14], and relates to propagation of the wave packet. In

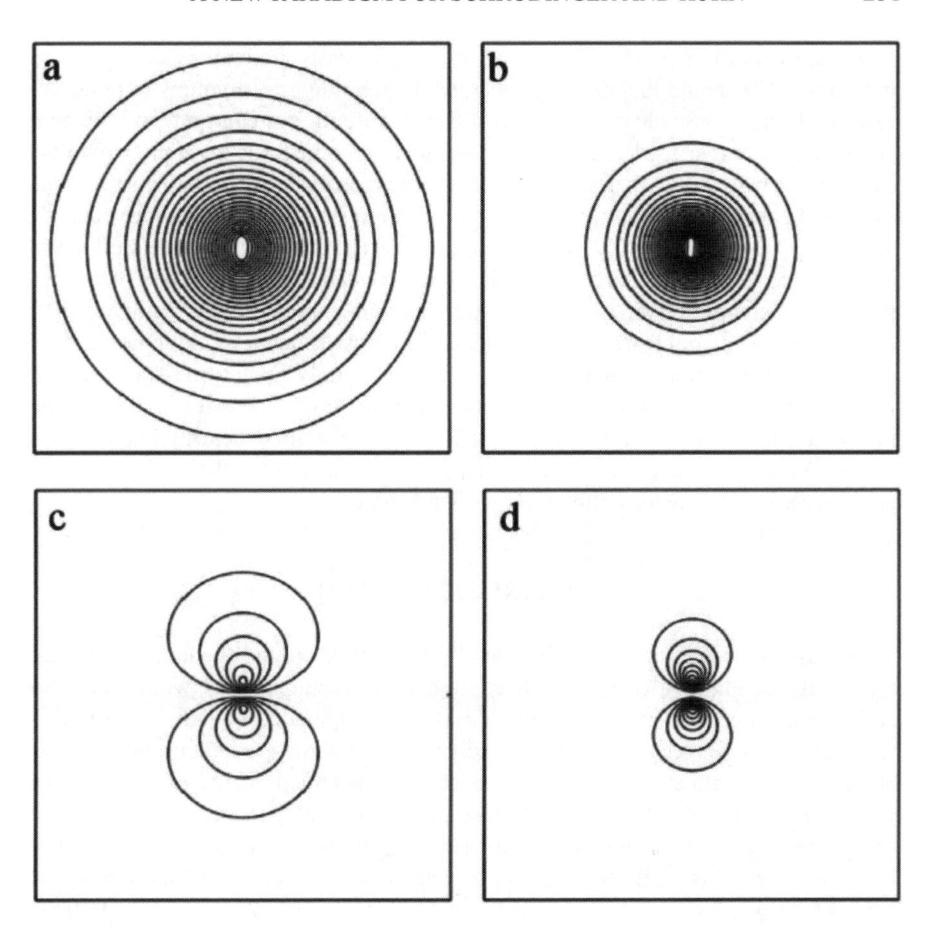

Figure 1: One-electron functions separated by 0.4 Hartree superposed to represent two electrons in an atomic system: (a) $1s^2$ $1S^0$ electron wave; (b) $1s^2$ $^1S_0$ charge distribution; (c) $1s^2$ ^3P electron wave; (d) $1s^2$ ^3P charge distribution. Graphs are of the x–z cross-section and are 4 hartree on a side.

fact, one discovers a general tendency of the one-electron waves to compress as they approach a nucleus, so the normally expected wave packet spreading is reversed, at least, in bounded systems.

The early practitioners and developers of density functional theory encountered yet another problem related to the extended nature of the number density primitive—self-interaction (Dreizler and Gross 1990, op. cit.). Specialized correction procedures were developed quite successfully and are in use in all modern calculations. In our case, again, because of the extended nature of our representation of the charged "particle," there is a potential problem with what amounts to self-interference which needs to be examined. The potential self-interference corrections are always finite, and generally small which is a great advantage over the situation in quantum electrodynamics.

Armed with these pieces of information, we are nearly ready to rebuild quantum mechanics. We should first note, however, while Schrödinger's attempts were unsuccessful, Hyllaras was able to calculate two-electron atomic systems quite well without going into Hilbert space.[15] As another issue, there is some basis for approaching the electron as a wave rather than a particle, or, at least as an extended body. This was done early in the development of quantum electrodynamics by Fermi (Schweber 2002)[16], and string theory (Woit 2002) is approaching the issue from a similar direction in modern times. Our plan is to rewrite Schrödinger's equation from the point of view of the electron and proton as standing waves rather than point particles. While we shall end up with an integral equation, rather than a differential equation, and some required changes in interpretation, the effect is very close to simply using a non-standard trial function. This trial function is wholly contained in complex 3-space, and takes the interesting form of a "totally antisymmetric sum" of non-concentric, one-electron waves. Fortunately the extremely large mass difference between the nucleus and the electron makes the nucleus effectively a point source.

THE VIRIAL THEOREM

One of those powerful results which holds both in classical and quantum mechanics, the Virial theorem is vastly underutilized. It gives the relationship between the space-and-time averaged kinetic energy and Virial for any bounded system. The Virial is related to the dot-product between the radius vector and the gradient of the potential energy. In this manner, only for a central field, the potential energy is -2 times the kinetic energy. For all other systems the Virial is considerably more complicated. The derivation of the Virial theorem is very general and, in quantum mechanics, takes the form of analysis of the commutator between the Hamiltonian and the symmetric form of the radius vector dotted with the momentum operator for each particle, after time and space averaging. The Virial theorem applies to such diverse things as planetary motion, a collection of Fermions, and Bose–Einstein condensates as long as the system is bounded. We will use the Virial theorem to probe the validity of Coulomb's law between a proton and an electron down to subatomic sizes when dealing with the quantities of quantum mechanics. This is necessary as the only measurable values are the mean values of the dynamic variables. We shall also use the Virial theorem to determine the parameter, called the range parameter, which defines the mean radius of the electron or proton wave.

THE HAMILTONIAN

Using Schrödinger's postulates, the Hamiltonian operator, H, for each atom in the system may be written as,

$$H = T + V = \left[-\frac{1}{2} \sum_{i=1}^{n} \nabla_i^2 \right] + \left[-\sum_{i=1}^{n} \frac{Z}{r_i} + \sum_{i=1}^{n-1} \sum_{j=i+1}^{n} \frac{1}{r_{ij}} \right] \tag{1}$$

where T is the kinetic energy operator associated with the term containing the sum of the Laplacians and V is the potential energy operator associated with the second term. The second term is further separated into the (negative) potential energy of interaction between the nucleus and each electron and the (positive) potential energy of interaction between the electrons, pair wise. There are "n" electrons in the system. The total Hamiltonian operator for, e.g., a molecule is the sum of all of the atomic Hamiltonians plus the corresponding forms for the nucleus–nucleus interactions and nuclear kinetic energies. To form Schrödinger's equation, this Hamiltonian operator is applied to a trial wavefunction, Ψ, and is set equal to the energy operator applied to the same trial wavefunction. For stable atoms and molecules, this result is just an eigenenergy times the same wavefunction if the solution has been found. Traditionally, Ψ is the sum of totally antisymmetric product form wavefunctions and resides in Hilbert space. This means each of the Laplacians is effective only in its own, ith subspace. That is,

$$\nabla_i^2 = \frac{\partial^2}{\partial x_i^2} + \frac{\partial^2}{\partial y_i^2} + \frac{\partial^2}{\partial z_i^2} \tag{2}$$

in Cartesian coordinates. If we retain the Schrödinger equation in complex 3-space, along with its trial function Ψ, the meaning is entirely different. There is only one Laplacian in such a system, and it is invariant over the entire space. The terms ∇_i^2 represent only a change in the locus of the origin and the Schrödinger equation may be written as

$$\left[-\frac{n}{2}\nabla^2\Psi\right] + \left[-\sum_{i=1}^{n}\frac{Z}{r_i} + \sum_{i=1}^{n-1}\sum_{j=i+1}^{n}\frac{1}{r_{ij}}\right]\Psi = E\Psi \tag{3}$$

where the eigenvalue of the system is E. The proper form of Ψ no longer involves the product of one-electron functions.

To be useful with electrons and protons as waves, the Hamiltonian of Eq. (3) must be integrated over all space to form the mean energy and the extended form of Schrödinger's equation.

THE ARCHETYPE

Any complete set of complex, three-dimensional functions would do to represent the charge-waves, but certain boundary conditions are required: (1) we need the wave to extend through all space, but have a finite mean radius, (2) we need the wave to behave as a particle under specific circumstances and as a wave in others, (3) we need the proper interaction between waves in the form of spectral interference, (4) the waves need to exist in complex 3-space with the square magnitude representative of some physical reality. In short, we need the representation to be true to our experimental knowledge. As it is our theory, we are free to choose what works as long as it is internally consistent and agrees with nature. Any theory will have a definite range of validity. We just want the range of validity as large as possible.

We choose, as an archetype, the one-electron function of Schrödinger, ψ. Using Condon and Shortley (1967) phase, this is expressed in somewhat modified form as,

$$\mu\psi_{n,\ell,m_\ell} = \mu \frac{-(n-\ell-1)!}{(n+\ell)!^{\frac{3}{2}}\,(2n)^{\frac{1}{2}}} \left(\frac{2m\beta}{n}\right)^{\frac{3}{2}} \mathrm{e}^{-\frac{m\beta r}{n}} \left(\frac{2m\beta r}{n}\right) L_{n+\ell}^{2\ell+1}\left(\frac{2m\beta r}{n}\right) Y_\ell^m(\theta,\phi) \tag{4}$$

In Eq. (4), we have been a bit audacious. We are multiplying the spatial part, ψ, by, not the spin, but the product of the spin with the magnetic moment, μ. We have also included the mass of the "particle," m, even though we are using hartree units. This is because we wish also to represent the proton, or nucleus, with this same scheme. The μ-factor helps relate the magnetic interaction between the nucleus and the electron at the correct magnitude. Except for β, which requires more explanation, the rest of the notation is standard as found in, e.g., Bethe and Salpter, with L the associated Legerre polynomial, and Y the spherical harmonic. The cofactor β is normally associated with the nuclear charge in the one-electron atomic system, or, for theories using screening, is the effective nuclear charge. We wish to represent free-standing electrons or nuclei, and so this association is not appropriate. We shall call β the range parameter as it serves primarily to determine the mean radius of the wave.

To be sure, any "particle" represented by this archetype, will, in general, not be in a specific state. Therefore, the actual charge-wave will be an infinite sum over the complete set of ψs, but with the same spin-magnetic moment. This sum will superpose these ψs concentrically, as they are representing only one charge. The boundary conditions determine the weights of the various components.

SPECTRAL INTERFERENCE

To be useful, the trial function Ψ must be formed in a particular way to carefully reproduce experimental knowledge. Waves interfere by either adding or subtracting, which translates as constructive or destructive interference, but we have some specific requirements. As is well known experimentally, and put into some specific words by Pauli, atomic systems do not exist in which any two electrons have the same quantum numbers. To model this with interfering waves, we must require the resultant, Ψ for two-electron waves, ψ_1, and ψ_2, take the form

$$\Psi = \frac{1}{\sqrt{\gamma}}(\psi_1 - \mathrm{o}\psi_2) \tag{5}$$

where γ is a normalization factor, and o is the sign of the product of the electron spins. This form is antisymmetric under exchange of indices, as needed for the Pauli exclusion principle. If the quantity with physical meaning is the square magnitude, $\Psi^*\Psi$, then there will be cross-terms. For paired electrons, same spatial quantum numbers but antiparallel spins, the cross-terms are positive and enhance the square magnitude between the two charges. For all quantum numbers the same, the cross-terms ensure vanishing of the square magnitude at the midpoint between the centers of

ψ_1 and ψ_2. For paired electrons, interference is constructive. For antipaired electrons, interference is destructive. Figure 1 gives an example for two, 1s electrons. For all cases other than paired or antipaired electrons, interference is a mixture of constructive and destructive interference as the spatial parts of the waves may be either positive or negative over a given range.

Equation 5 is generalized by: (1) using μ to allow ψ_i, to represent protons as well as electrons, and forming an antisymmetrization operator so the cross-terms always come out negative times the sign of the spin, or in this case, μ, product. This allows the proper form of interference in all cases, including those without all spatial quantum numbers the same. Incidentally, this "antisymmetric sum" form of the trial function has an additional benefit. Since the Laplacian is invariant throughout the complex 3-space, one may transform to the center-of-mass of each one-electron function in turn. Under these circumstances, Schrödinger's equation applies individually to each ψ_i. This greatly simplifies the evaluation of the Laplacian operating on the trial function. This is demonstrated in the Appendix.

THE NUCLEAR ATOM

The peculiar choice of Eq. (4) has a number of salutary effects. One of the best is the relationship between the nucleus and the electron. Because the mass term, m, is so large for the nucleus with respect to the electron, the cross-terms all vanish except at nuclear distances. Because the magnetic moment of the proton and neutron is so much smaller than the magnetic moment of the electron, the magnetic interactions are small enough to be discarded—as desired for this particular study. The net result is the Schrödinger equation for a collection of nuclei and electrons can treat all of the nuclei as point sources, without the spectral interference cross-terms. Finally, the resultant equation looks a great deal like the traditional equation, except for existing in complex 3-space, rather than Hilbert space.[17]

THE INTERPRETATION OF $\Psi^*\Psi$

Bohr, and friends, would have us interpret $\Psi^*\Psi$ as the probability density for the locus of the electron. In Hilbert space, this interpretation is not too helpful. There are an infinite number of Hilbert space geometries, which will give the same projection into 3-space. It is only to this projection which we have any hope of attaching meaningful measurements. Only for the one-electron atomic system does the probability density concept really have any potential physical meaning. Since we have retained the new Schrödinger equation in complex 3-space, we may analyze the one-electron atomic case for this meaning. Remember, on the quantum scale, because of the Heisenberg uncertainty principle, we may not follow the evolution of a system in detail. It is only the mean values, averaged over space and time, which have physical significance. Only some of these mean values are constants of the motion, i.e., their operators commute with the Hamiltonian. In complex 3-space, multiplying $\Psi^*\Psi$ by the total electronic charge makes it a charge density as well as a probability density. The proof is as follows: Take a cell in the probability density at time t. Either the electron is there or not. If there, then Coulomb's law applies to the dynamics of the system. If not, Coulomb's law applies by default as the null result. Therefore, each cell in the probability density behaves as if it were an infinitesimal charge with value equal to the product of the probability density at that point and the total electronic charge of the system. This is only possible as long as the equation and its solutions remain in complex 3-space.

THE VIRIAL AND COULOMB'S LAW

We have no guarantee that Coulomb's law works on the atomic, or sub atomic level without testing the hypothesis. Fortunately, this is readily done using the Virial theorem. We shall begin by accepting Schrödinger's postulates, but we shall cast a wary eye on the potential energy term. We shall look at a system involving a proton and an electron, as this is easily generalized to a nucleus and an electron, and a bounded system is required to avail ourselves of the Virial Theorem. We shall assume the functional form of the electron–electron repulsion should be somewhat analogous to the attractive, proton–electron interaction, and we are dealing with a two-body force. Our trial function will involve the antisymmetric sum of a "heavy" proton function and a "light" electron function with appropriate magnetic moments. We form the mean value of the kinetic energy operator by operating on the trial function, multiplying on the left by the complex conjugate of the trial function and integrating the whole over all space. By the Virial theorem, since we are expecting a central field, we may expect the mean potential energy to be just -2 times the mean kinetic energy. When one evaluates the Laplacian using the one-electron Schrödinger equation, and integrates over all space, one discovers: (1) the cross-terms vanish for all but nuclear distances; (2) the spin–spin interactions are very small and may be ignored; (3) the final form of the expression fits the form required by mathematically determining the mean value of Coulomb's law for this case using the trial function, *only at* $\varepsilon = 0$. For $\varepsilon > 0$, a complicated formula relates the potential energy to the parameters of the system. If we retain the known form of Coulomb's law, we may use the Virial Theorem for each value of ε to determine the value of β, which will satisfy the required dependence. We select this as the simplest approach for now. It appears to give us good comparison to experiment, and minimally affects the theoretical basis of our work. Although the numbers come out right, this exposition is somewhat speculative. To fully describe the Coulomb interaction, the system should be made fully covariant in a relativistic formalism. This is beyond the scope of this chapter.

DENSITY FUNCTIONAL THEORY

By requiring Schrödinger's equation and its solutions to remain in complex 3-space, we have done something interesting. The trial functions are actually the sum, albeit the antisymmetric sum, of one-electron functions. With only small manipulation of the resultant equation, and suggesting the one-electron functions may be re-arranged into individual one-electron representations, one attains a set of simultaneous equations having the same functional form as the Kohn–Sham equations, but: (1) the physical interpretation of the symbols is different; (2) the derivation works for all states—ground and excited; and (3) there is a definite functional form required of the exchange-correlation energy. Still *the functional form is exactly the same* once the system is generalized. It is here suggested, once subjected to the "duck" test, the system would pass in a quack. Sample calculations of simple systems indicate results accurate to within a few percent of experiment, even with some simplifying assumptions. Some of these results are presented in Table 1, but the actual derivation is beyond the scope of this particular chapter.

EXPLAINING THINGS—SOME CONJECTURE

What does it mean to explain something? It seems fairly comfortable to distrust simply displaying a number, i.e., calculating all possible configurations and picking the one with the lowest energy. This, however, seems to have some popularity among, particularly physicists, but other scientists as well.[18]

We may explain something using information outside of the discipline of discourse, such as using radio-carbon dating to explain the age of an artifact. We also tend to make explanations of poorly understood items by referencing concepts we feel are better understood. To what extent does such an exercise become circular?

We have already started explaining things in a non-traditional manner with Figure 1. There is much more. Let us take it step-by-step from the introductory sections and add a few items of interest along the way. Some of these explanations are partially conjectural, however, and require further calculational study to be verified. These will be specifically noted in the text below.

The issue of the internal consistency brought up by Fellers and Peterson, of course can never be truly laid to rest (Scerri 1998a, b, c), but note—while conventional systems require literally hundreds of configurations for success in *ab initio* calculations (see Kinghorn and Adamowicz 1997), and many, many iterations in density functional calculations, even the simple models listed below give agreement with experiment to within a few percent—with only one configuration and zero iterations. The calculations listed in the Appendix, and Table 2, agree with experiment to an astonishing degree; again for just one configuration and zero iterations. Anyone familiar with series solutions knows that the series converges to the solution of the equation in question. If the answer is wrong, i.e., does not agree with experiment, then the equation is wrong. In our case, almost all of the experimental result is recovered in only one configuration, and zero iterations. Such tight agreement with nature seems highly significant.[19]

The chemical bond has never been adequately explained by traditional quantum mechanics. We wave our hands and talk about valence as opposed to paired electrons, but choke when we see the valence electrons of, e.g., the Group II elements

Table 2: Discovery oriented calculations.

Element	H^-	He-I	Li-II	Be-III	B-IV	C-V	N-VI	O-VII
Energy								
SIC	−0.5724	−3.1037	−7.6349	−14.1662	−22.6974	−33.2287	−45.7599	−60.2912
SICSE	−0.5295	−2.9039	−7.2816	−13.6627	−22.0471	−32.4350	−44.8262	−59.2208
Full	−0.4482	−2.8232	−7.1982	−13.5732	−21.9482	−32.3232	−44.6982	−59.0732
FullSE	−0.5275	−2.9050	−7.2836	−13.6639	−22.0466	−32.4329	−44.8242	−59.2231
Exp.	−0.5277	−2.9049	−7.2836	−13.6637	−22.0463	−32.4329	−44.8248	−59.2228

Note: Experimental values are from Moore. Round-off error for numerical integration estimated at approximately ±0.0005. SIC, calculated with self-interference correction; SICSE, calculated with SIC and a semi-empirical correction of the form $0.98429 - 0.161599\beta + 0.001690\beta^2$; Full, calculated with no correction; FullSE, calculated with semi-empirical correction based on the formula $0.075 + 0.003\exp(0.403\beta)$.

are paired. Many other problems exist. With the new model, we see all is a function of the spectral interference between the electron waves. Ionic bonds clearly show in the calculations one, or more electrons physically transferred from one atom to another. The $\Psi^*\Psi$ calculation clearly demonstrates this in the process of the calculation itself. Van der Waals bonds clearly show the interaction of the quadrupole moments of the participating atoms, which are dictated by the interaction between the electron–electron repulsion, and the spectral interference between the interacting electrons in each atom. Metallic bonds clearly show the electron waves permeating all of the metal space due to the spectral interference, and becoming what amounts to a somewhat lumpy soup of indistinguishable, constructively and destructively interfering waves. The covalent bond is more exciting, however. In the new model, the covalent bond is actually broken into two separate forms. The "common" form is in which the electrons are actually shared between the atoms, as well as bonding through constructive interference. Examples are the hydrogen molecule and carbon dioxide. In such cases, the electrons from the interacting atoms are of the same principal and angular quantum numbers and are totally interchangeable. The mean locus of the electrons tends to hover between the participating atoms, rather than be imbedded in one or the other, but is still highly localized. The "interference" form is actually the result of constructive interference between electrons in differing quantum states, obtained from the two or more atoms. Examples include water and other compounds in which electrons from one atom are in a different quantum state than electrons from the other. The chemical bond is the result of the constructive interference between the electron waves, and the Coulomb repulsion is neutralized by the presence of the attractive nuclei. The electrons tend to be more closely associated with their "own" atoms, merely shaking hands, as it were, to form the bond.

Claiming the periodic table is explained by traditional quantum mechanics is flat wrong (Scerri 1998a, b, c).[20] The new model, however, has much to say. One problem is the issue of the Group II elements. By all rights they should be chemically inert. The valence electrons are paired. A rather subtle result from the new model is the issue of the spin-zero boson. If paired electrons approach closely enough, as in the core of heavier atoms, or, even as simple a system as helium, the constructive interference is so complete that the electrons loose all identity and effectively form a spin-zero boson (see Figure 1b). This is the same spin-zero boson famous from the BCS theory of superconductivity, but also exists in the atomic systems as well. An interesting conjecture occurs in the case of the Group II elements, the nuclear charge is small enough and the core electrons are tightly packed enough, so the outer valence electrons are forced too far apart and the spin-zero boson is broken into two vulnerable electrons, which are still loosely paired. The net result—Group II atoms interacting chemically with others tend to shed both electrons simultaneously. Similarly, Group II electrons bound together as a metal tend to donate both electrons simultaneously to the electron soup of the metallic bond. The structured nature of the pseudo-pairs in this soup tend to make such combinations rather poor metals. This notion has been partially verified by some of the preliminary calculations found in Table 1.

Perhaps a more dramatic, although somewhat conjectural, example is the explanation of Hundt's rule. We shall use the 2p electron system as the example, but the demonstration applies to all electron states. Refer to Figure 2 for a stylized picture of this argument. (The pictures are actual contour plots of the x–y sections of the charge

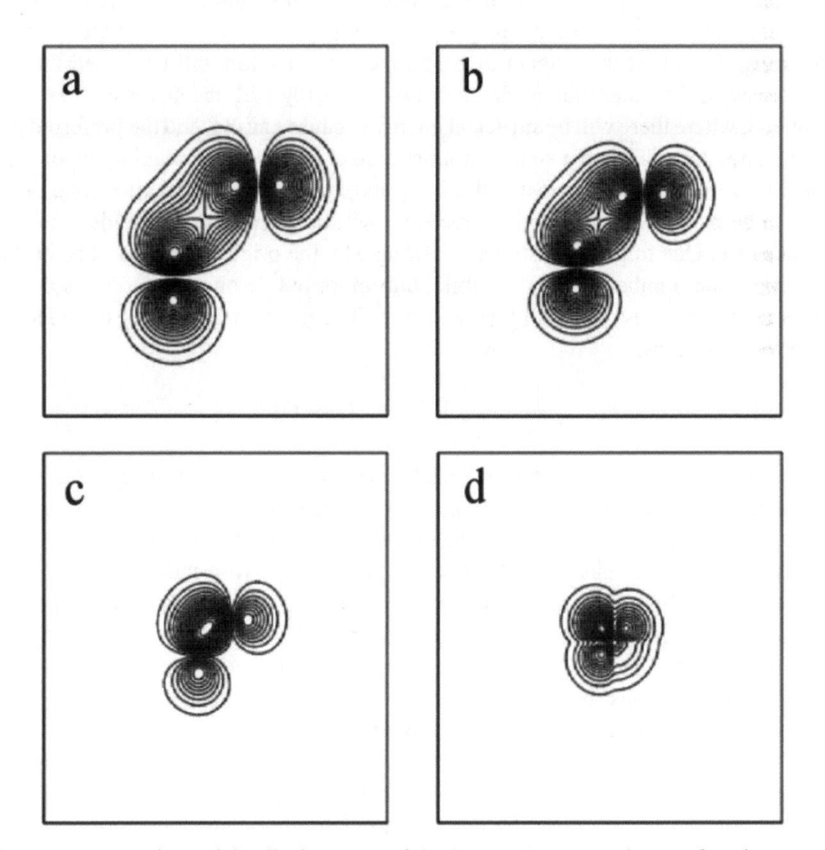

Figure 2: Comparison of the displacement of the 2p-x + 2p-y one-electron functions, same spin, for carbon through fluorine: (a) carbon; (b) nitrogen; (c) oxygen; (d) fluorine.

distributions from the interfering electrons, but the β, ε, and overlap parameters are estimated. The interfering electrons are actually what is normally termed the 2p-x and 2p-y functions, which are a linear combination of the 2p states.) Assume we are walking through the periodic table, one atomic number at a time. The present configuration is $1s^2 2s^2$, and we move to atomic number 5. Energy dictates, the next open space is in the 2p subshell, but there is no preference for the next electron. In the new model, this electron will be listed as a superposition of all of the possible 2p states which we will label $2p_{-1}\uparrow, 2p_0\uparrow, 2p_{+1}\uparrow, 2p_{-1}\downarrow, 2p_0\downarrow, 2p_{+1}\downarrow$; where we have represented the spin state by an up, or a down arrow. We have no way of knowing which of these, or what combination of these states the system is in unless we do some kind of experiment, such as reacting the system with another atom, but, no matter. Let us walk a little farther. Moving to atomic number 6, we may add another electron, again in the 2p subshell, but the model now makes some demands. The $2p_{-1}$, and $2p_{+1}$ substates interfere constructively if they have the same spins, as one has a positive spatial part, where the other has a negative spatial part. This gives a small, but definite energy preference for the configuration $1s^2 2s^2 2p_{-1} 2p_{+1}$ with the outer spins aligned. Now, taking one step further, atomic number 7, we now have a situation in which there is a preferred direction in space due to the magnetic alignment of the previous two spins. To reduce the magnetic energy, the next electron will be forced into the $2p_0$ state with the same spin as the other two. It is only with the next step, to atomic number 8, where there will be sufficient energy to cause pairing, and the preferred pair will be $2p_0\uparrow 2p_0\downarrow$. The next step to atomic number 9 will force breaking the original constructive interference between the $2p_{-1}$ and $2p_{+1}$ sub states to allow one or the other to be paired. Finally, the last free state will be paired in neon. This is exactly Hundt's rule. One might be justified in asking why the original pair could be broken only at atomic number 9. This is subtle, but important. If one were to consider the old screening models one would be led astray. The mean radius of any one-electron function is given by,

$$\langle r \rangle = \frac{1}{2\beta}\left[3n^2 - \ell(\ell + 1)\right] \tag{6}$$

where, in this case, β is the effective nuclear charge. A glance at the periodic table, however, indicates all atoms are approximately the same mean radius. Therefore, each electron wave must see nearly all of the nuclear charge, only slightly modified by the presence of the other electrons. In this new model, no two-electron waves may be concentric, but the mean separation between their centers-of-mass is determined by the nuclear charge. Further, look at the geometrical forms associated with the 2p functions. Figure 2 indicates the picture of the x–y cross-section of the 2p-x + 2p-y one-electron functions, both with the same spin, for: (a) the carbon atom, (b) the nitrogen atom, (c) the oxygen atom, and (d) the fluorine atom.

For oxygen, the nuclear attraction is sufficiently weak to allow the two waves to interfere strongly at the overlapping end. For fluorine, the two waves are pulled into a nearly concentric form and the overlap is much less—thus allowing the pairing overlap to take precedence. This is the reason for the particular bonding angle of water, and a similar effect is responsible for the peculiar configuration of chromium.

Even though the model predicts spherical symmetry until some experiment is performed, a particular combination of the subshell functions is then chosen to accommodate the environmental requirements.[21] Note—the separation parameter is not a dynamic variable, but serves only as an optimization parameter to find the minimum energy of the system.[22] The addition of another electron, a perturbing field, or bonding with another atom may draw out any potential asymmetry in the charge distribution.

Prediction of crystal structures may also be done using the new method. The interference requirements dictate definite forms for atoms interacting with other atoms as exhibited in the oxygen, forming water, model above. One may make clay models of such structures and put them together to form models of molecules, crystals, and even glasses. Just a brief session with these clay models of the structures of various materials may be readily understood, even by the typical undergraduate, non-major. The full calculations must be done to give final bonding angles, mean lattice parameters, and final charge distributions, but the models do quite well in qualitative predictions. These predictions have been verified by some of the preliminary calculations of Table 1, with actual models having been made for several atoms and molecules from rows 1–2 of the periodic table.[23]

One of the most interesting predictions of the new method is the issue of the Pauli exclusion principle. When one attempts to calculate the energy of an antipair one immediately discovers the system is not bounded—the Virial ratio does not work. More work is needed, but once continuum electrons are brought into the structure, it should be possible to directly calculate the energy deficit encountered when the Pauli exclusion principle is violated. This gives a meaningful explanation of the principle in terms of constructive and destructive interference.

FEASIBILITY STUDY

When this model was just starting to come together, it was so outrageous that some test had to be developed to determine its utility. We were going to be doing calculations very much unlike the canned ones now available off the shelf, and worse yet, we would be using a system of non-orthonormal functions. A simple approximation, called the non-holonomic constraint method, in which the electron–electron interaction was replaced by forcing the electron waves to remain as far apart as possible, and allowing them to vibrate like the vibrational modes of molecules was used for this study. Other techniques applied frozen cores, or the new version of density functional theory. The results for several simple systems are given in Table 1.

The success of these preliminary results was deemed sufficient to continue the project. In fact, some of these simple approximations may be adequate for certain commercial applications. These results frequently show too much binding, particularly for the non-holonomic constraint method. Normally, because traditional work is based on a variational principle, there is too little binding, but these models are over constrained and may loose some of the kinetic energy, or some of the electron–electron interaction potential.

The full model is still under development and yielding new information. An example is reviewed in the Appendix in which the calculated energies agree with experiment over the isoelectronic series from the negative hydrogen ion through oxygen-VI. Several interesting implications come from this calculation: (1) Previous attempts corrected for self-interference by eliminating what appeared to be obvious terms, but this new calculation indicates that the electron waves are so intimately mixed that some self-interference may be required to model reality accurately. Only those terms that result in a constant (in terms of ε) really qualify. (2) The two calculations using the spherical harmonic expansion of the electron–electron interaction, considering each element in space multiplied by $\Psi^*\Psi$, and the number of electrons as an infinitesimal charge yields relatively poor results for the lighter elements until very simple corrections are applied. The corrections apply to the total mean energy, but not to the Virial, indicating that there is a here-to-fore missed component of the interaction that is perpendicular to the radius vector. (3) The correction as determined applies only to the minimum in the mean-energy–ε curve, and so a functional form for the complete correction must be found. The author's conjecture is that the spin should be treated as the charge, i.e., distributed over the entire electron wave. Traditionally, paired spins would add zero magnetic component to the energy, but a spin density allows for the paired spins to provide some attractive energy just like two magnets set side-by-side with opposite poles next to each other. The bulk of the additional energy would be repulsive, however, from the nearest neighbors that are above and below, lending weight to using the partial self-interference correction mentioned above. The size of the correction should be approximately right as the magnetic interactions are about an order of magnitude smaller than the electrostatic interactions. Significantly, there is a component in the dipole–dipole interaction that is perpendicular to the radius vector—and thus affects the potential, but not the Virial as required by the current observations.

FUTURE

We have attempted to illustrate the problems associated with a major change in the view and interpretation of an accepted scientific theory. The new approach has been illustrated in the broad brush—the author apologizes for the technical detail necessary to describe it, but, particularly in this case, the devil is in the details.

Work continues on this new approach. Many problems are still there, due in part to the thorough indoctrination of the author into the traditional methods of quantum mechanics, and lead to the necessity of finding and blocking blind alleys of research. The current, corrected model interaction is shown in the Appendix.

Of first priority is the proof or disproof of the conjecture about the spin density. It is hoped in the not-to-distant future, several significant systems of multi-electron atoms in the ground and excited states, through polyatomic molecules will be calculated and used as evidence of the efficacy of this method. Eventually, a fully relativistically covariant model will be developed. Beside philosophical interest, and fundamental scientific interest, there is potential commercial interest (Eberhart 1994)[24]. The process of calculation is greatly speeded-up by only having to do one configuration and zero iterations, but many of the calculations are functionally easier. Being able to casually predict the molecular, or material structure of a system has been almost fantasy up until now. Such improved properties have great potential interest to the designers and manufacturers of both new medications and new materials. Unfortunately, unless the new approach can gain academic acceptance, no such applications are likely to materialize. Under any circumstances, many questions relating to the philosophy of chemistry have been exposed. It will require great insight to answer them adequately.

APPENDIX

Electron waves in the $1S^2\ {}^1S_0$ *configuration—discovery oriented research*

The development of this simple, but important system illustrates the mathematics of the new method, but also shows some of the difficulties in the development of a new idea. The traditional approach would be to form the Hamiltonian operator, and then apply it to the trial function to form the Schrödinger equation. In order to compensate for the extended nature of the electron wave, we must deal with the mean values of the operators at the start. Figure 3 shows the graphic layout used for this development.

In Figure 3, the two-electron waves are illustrated as simulated 1s functions, one with its center-of-mass on the positive z-axis a distance ε_1 from the nucleus at the origin, the other on the negative z-axis a distance ε_2 from the origin. An arbitrary infinitesimal volume is represented by the small cube residing at the end of the radius vectors: r_1 from the center of the first wave; r from the nucleus; and r_2 from the second wave. The figure is the cross-section in the x–z plane.

Using the fact of invariance of the Laplacian in 3-space, the mean kinetic energy, T′, is given by

$$T' = -\int_{\text{all}}^{\text{space}} \Psi^*\nabla^2\Psi\,\mathrm{d}\tau = -\frac{1}{\gamma}\int \psi_1^*\nabla^2\psi_1 + \psi_2^*\nabla^2\psi_2 - \mathrm{o}\left(\psi_1^*\nabla^2\psi_2 + \psi_2^*\nabla^2\psi_1\right)\mathrm{d}\tau \tag{6}$$

Where we have used the trial function, Ψ, and summed over the two electrons. The equation expands as shown through Eq. (5). Since Ψ is normalized to 1, the normalization factor, γ, has the form,

$$\gamma = 2(1 - \mathrm{o}\,\mathrm{Re}\,(S_{1,2})) \tag{7}$$

where o is the sign of the product of the spins, and we take the real part of the overlap integral, $S_{1,2}$. The overlap integral and the U-integral are defined by

$$S_{i,j} = \int_{\text{all}}^{\text{space}} \psi_i^*\psi_j\,\mathrm{d}\tau \qquad U_{i,j,k} = \int_{\text{all}}^{\text{space}} \frac{\psi_i^*\psi_j}{r_k}\,\mathrm{d}\tau \tag{8}$$

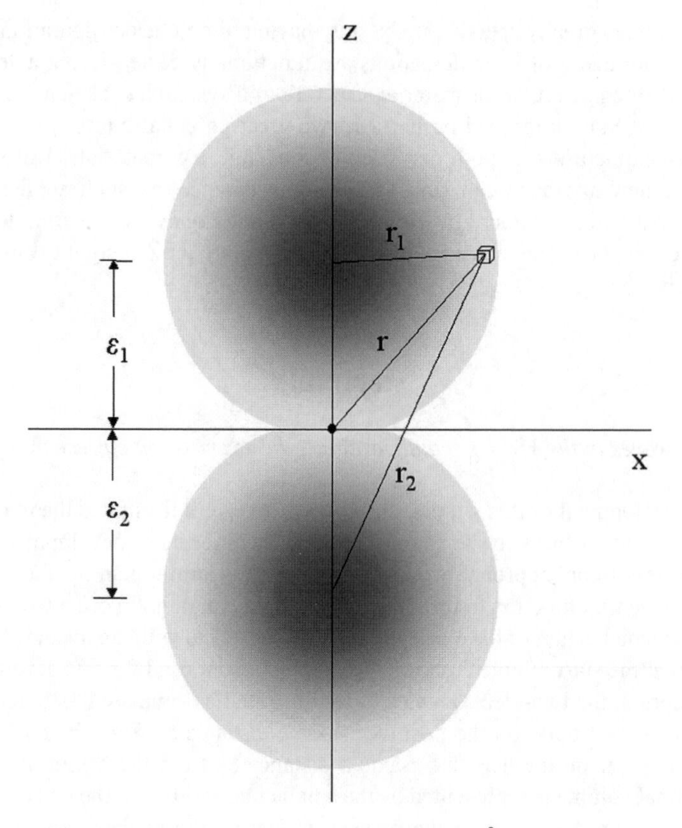

Figure 3: Two-electron atomic system in the $1s^2$ configuration.

The ψ_i are orthonormal only when the separation between their centers-of-mass is zero. By the invariance of the Laplacian, we may use the relationships derived from the solution of the one-electron atomic system case

$$\nabla^2\psi_i = \left(\frac{\beta_i^2}{n_i^2} - \frac{2\beta_i}{r_i}\right)\psi_i \quad \int_{\text{all}}^{\text{space}} \psi_i^* \nabla^2 \psi_i \, \mathrm{d}\tau = -\frac{\beta_i^2}{n_i^2} \quad \int_{\text{all}}^{\text{space}} \frac{\psi_i^* \psi_i}{r_i} \mathrm{d}\tau = \frac{\beta_i}{n_i^2} \tag{9}$$

so equation (6) becomes

$$T' = \frac{\left(\beta_1^2 (1 + \mathrm{o} S_{2,1})/n_1^2\right) + \left(\beta_2^2 (1 + \mathrm{o} S_{1,2})/n_2^2\right) - 2\mathrm{o}\,(\beta_1 U_{2,1,1} + \beta_2 U_{1,2,2})}{2\,(1 - \mathrm{o}\,\mathrm{Re}\,(S_{1,2}))} \tag{10}$$

The mean potential energy of interaction between the resultant electron wave and the nucleus is

$$V' = -Z_{\mathrm{a}} \int_{\text{all}}^{\text{space}} \frac{\Psi^*\Psi}{r} \, \mathrm{d}\tau = -Z_{\mathrm{a}} \frac{U_{1,1,0} + U_{2,2,0} - \mathrm{o}\,(U_{1,2,0} + U_{2,1,0})}{1 - \mathrm{o}\,\mathrm{Re}\,(S_{1,2})} \tag{11}$$

where we have taken advantage of the homogeneity of the resultant wave—there is no necessity of interacting each one-electron wave independently. This, by the way, is true indistinguishability of the electrons—a natural consequence of this model, and only forced onto the system in traditional quantum mechanics.

The next step is one of Schrödinger's old bug-a-boo's, the determination of the electron interaction, v', without going to Hilbert space. The functional form is deceptively simple

$$v' = \int_{\text{all}}^{\text{space}} \frac{\Psi^*\Psi}{r_{\text{a,b}}} \, \mathrm{d}\tau \tag{12}$$

where, for the traditional theory, $r_{\text{a,b}}$ is the distance between the electrons, and must be expanded in a sum of spherical harmonics as is done in classical electromagnetic theory. We can approach it in the same manner with the new method, but the results at first appear less than desirable. The calculated energy for the negative hydrogen ion recovers only 85% of the experimental value, and that for He-I, only 97%. Things get steadily better, and for Li-II the calculated energy recovers 99% of the experimental value. The author has stumbled around several ways of doing this by attempting to maintain at least some contact with traditional methods. For example, one would expect that some self-interference correction should be made.

The ancient Greeks worked up their geometrical ideas from wooden or clay models and drawings in the sand. Then and only then did they return to mathematics to prove their new theorems. This self-same process of discovery has driven this particular work, only the models have taken the form of new equations tested against experiment, and using semi-empirical adjustments to fine-tune them. If such an adjustment was robust under changing from one system, configuration, etc., to another then some theoretical reason for the adjustment had to be found and incorporated into the fundamental body of the theory. If it proved non-robust, it was discarded and another avenue tried. This final example illustrates this practice.

The latest method relies on the notion of $\Psi^*\Psi$ times the number of electrons giving the charge density of the system. This works very well in the electron–nucleus interaction and quite clearly gives the value of V' found in Eq. (11)[25]. In this manner, we look at the interaction of one charge element in the resultant wave with another, and integrate over all space to find the total interaction. This takes the form of Eq. (13).

$$v' = \frac{1}{2} \int_{\text{all}}^{\text{space}} \Psi_{\text{a}}^*\Psi_{\text{a}} \int_{\text{all}}^{\text{space}} \frac{\Psi_{\text{b}}^*\Psi_{\text{b}}}{r_{\text{ab}}} \, \mathrm{d}\tau_{\text{b}} \, \mathrm{d}\tau_{\text{a}} \tag{13}$$

The expansion of this proves quite messy and will not be reproduced here. The factor of 1/2 is included to prevent double counting. Equation 13 was first expanded in general, keeping all terms, and then a separate calculation was made including a self-interference correction which amounted to eliminating those terms in which a component interacted with itself in such a way that the result was independent of ε.

We now must determine the Virial, which is simply the dot-product between the radius vector and the force on the electron at each element in the "other" electron wave. Some vector analysis yields,

$$\varpi' = \frac{\upsilon'}{2} \qquad \varsigma' = V' + \varpi' \tag{14}$$

where ϖ' is the electron–electron portion of the Virial; ς' is the total Virial, and the Virial ratio, P, is given by

$$P = \frac{\varsigma'}{T'} = -2 \tag{15}$$

In practice, o, Z_a, and ε are chosen, and Eq. (15) is solved for β. This value is used in the energy equation to find the eigenenergy as a function of ε. The minimum of this curve gives the eigenenergy of the system for use in all calculations of atomic and molecular parameters. For the $1s^2$ configurations, many symmetries apply greatly reducing the calculational difficulty. As mentioned, the raw results are relatively poor, but analysis of the absolute error shows that very simple corrections bring the calculated data into alignment with the experimental data. Strangely, this correction is to be applied to υ', but not to ϖ'.

Table 2 compares calculated values with experimental values in hartree. There are entries for calculations containing a self-interference correction, the full expansion of υ', and the semi-empirical numerical corrections.

This example highlights one of the main problems in the development of a new theory, i.e., each revelation opens a whole can of worms! In this case, whence commeth the corrections? Much is yet to be explained. Such is both the beauty and the frustration discovery oriented research. Both the preliminary studies, and the full theory give results which should not be ignored—results which appear to remain consistent with exploration of ever more complicated systems.[26] Unfortunately, the concepts are far enough away from tradition so the perpetrator is constantly in doubt as to fully understanding the method or its implications. As a conjecture, the author suspects that the spin should be included in the theory in the same manner as the charge, yielding a spin density as well as a charge density from $\Psi^*\Psi$. The magnitude is about right, and such a modification would include a portion that affects the electron–electron interaction, but not the Virial.

ACKNOWLEDGMENTS

The author wishes to thank the Kansas Technology Enterprise Corporation (KTEC) for grant #150134, and Mr. Clyde Engert, Former Vice President of Innovation and Marketing, for his personal interest and efforts to find proponents for this work.

NOTES

1. In one of several forms, "If it looks like a duck, walks like a duck, and quacks like a duck, it is a duck." Throughout the history of science there are examples in which the form stays the same, but the

interpretation changes with the times. An example is the Lorentz contraction. The form was known long before Einstein developed special relativity.

2. See the interplay between the publication of some experimental work on "d-orbitals" and the reaction to it as exemplified in Scerri's Editorial 4. By traditional methods Scerri is right. Other "strange" experiments are ongoing relating to the splitting of a single electron, etc. This new method presented here allows such experimental evidence to happen! Scerri (2000), and references there in.
3. An excellent summary of modern methods is found in Bernholc (1999).
4. The development of density functional theory is given in Dreizler and Gross (1990).
5. As an example, for the negative hydrogen ion the calculated value is -0.527463 compared to the experimental value of -0.527731, but the absolute value of the round-off error is ~ 0.0005 based on calculating some of the integrals in reverse order.
6. By internal consistency, I mean that a correction applied to one instance, say the negative hydrogen ion, applies to all instances of the same group, in this case, He-I through O-VI, but also, the theory is developed in general, with no special instances where any constants must be adjusted individually, or *ad hoc* additions are necessary.
7. This experience is not infrequent as evidenced by a letter to the editor from Brian K. Ridley, *Physics Today* 80, September 1999.
8. Experimental energies are adapted from Moore (1971), with the exception of the negative hydrogen ion, which are adapted from Kinghorn and Adamowicz (1997). Other data are from Lide (1999), and Herzberg (1989).
9. Some of this examination gives credence to the necessity of a separate philosophy of Chemistry. See Scerri and McIntyre (1997).
10. Feller and Peterson, op. cit. They state, "Straightforward increases in the size of the one-particle basis set of Gaussian functions or the use of more sophisticated levels of theory can frequently lead to a seemingly paradoxical deterioration in the agreement with experiment." Hardly a characteristic for a "mature" theory.
11. Max Born's criticism of some of Schrödinger's original papers examined in Scott (1967, op. cit.).
12. See a comment by Carver Mead in an article about him and his book, *Collective Electrodynamics*, Karlgaard (2002).
13. Much of the historical material is covered in detail in Scott (1967, op. cit.).
14. For an excellent discussion of the Virial theorem, see Wannier (1966).
15. Hyllaras was unable to meaningfully extend his method to multiple electron systems.
16. "Fermi, on investigating this same problem of the electrons self-energy, recognized that the divergences resulted from the point-like character of the charges" (Schweber 2002).
17. This new method study reduces to the one-electron Schrödinger equation for one nucleus and one electron.
18. There may be some justification, in certain circumstances, for listing a formal procedure for determining some value or outcome as an explanation, particularly when dealing with something mechanical.
19. There is another issue which should be mentioned. All traditional calculations are based on a variational principle, and thus can give only an upper bound on the energy of the system in question. Several forms of the new method, particularly the semi-traditional method with self-interference correction, mentioned in the appendix always give energies too low, i.e., too much binding. Therefore, the new method produces a lower bound of the system in question. This is a new capability and of considerable interest as mentioned by Scerri in several of his papers.
20. Some intriguing attempts have been made, however. See Melrose and Scerri (1996).
21. Mathematically, the coefficients of the terms in the resultant are determined by the boundary values.
22. Attempts to base the full theory on a kinetic energy term associated with the separation parameter generally fail. Only in the non-holonomic constraint method has any success been achieved with a dynamic interpretation of the separation parameter, and it falls apart with the one-electron atomic system.
23. Unlike traditional molecular modeling systems, these clay models fit together like puzzle pieces. For example, the primitive model of the negative hydrogen ion is somewhat like a dog-bone. They fit together leaving chambers appropriate for inserting the positively charged lithium ion so that lithium hydride is

projected to have the sodium chloride structure. Final relaxation, however, blends the electron waves together in such a manner that each nucleus appears to be surrounded with a spherically symmetrical cloud of electrons, much like the conventional model. In the conventional model it is the relative sizes of the ions that causes the structure. Another clever model is the ice–water transformation. When the hydrogen bonds are active the molecule array has a very specific, three-dimensional pattern, full of holes. In the liquid state, the molecules are free to dove-tail together into a much more compact form. This is a great illustration to the undergraduate as to why ice is less dense than water.

24. A multi-million dollar corporation, Pharmacopia, has been marketing software for this purpose using traditional methods for several years now.
25. It should be mentioned, when this construction is applied to the one-electron atomic case it reduces precisely to Schrödinger's equation for the one-electron atomic case at the limit as ε approaches zero, i.e., at the minimum of the energy–ε curve.
26. See the three-electron cases, excited states, and simple molecules listed in Table 1.

REFERENCES

Bernholc, J. 1999. Computational Materials Science: The Era of Applied Quantum Mechanics. *Physics Today* (September): 30–35.

Bethe, H.A. and Salpeter, E.E. 1957. *Quantum Mechanics of One- and Two-Electron Atoms*. Berlin: Springer.

Cartwright, N. 1999. The Limits of Exact Science, From Economics to Physics. *Perspectives on Science* 7: 318–336.

Condon, E.U. and Shortley, G.H. 1967. *The Theory of Atomic Spectra*. New York: Cambridge University Press.

Dreizler, R.M. and Gross, E.K.U. 1990. *Density Functional Theory*. Berlin: Springer.

Eberhart, M. 1994. Computational Metallurgy. *Science* 265: 332–333.

Feller, D. and Peterson, K.A. 1998. An Examination of Intrinsic Errors in Electronic Structure Methods Using the Environmental Molecular Sciences Laboratory Computational Results Database and the Gaussian-2 Set. *Journal of Chemical Physics* 108: 154–176.

Herzberg, G. 1989. *Molecular Spectra and Molecular Structure 1. Spectra of Diatomic Molecules*. Malabar, Florida: Krieger.

Karlgaard, R. 2002. Incomparable Carver. *Forbes* (March).

Kinghorn, D.B. and Adamowicz, L. 1997. Electron Affinity of Hydrogen, Deuterium, and Tritium: A Nonadiabatic Variational Calculation Using Explicitly Correlated Gaussian Basis Functions. *Journal of Chemical Physics* 106: 4589–4595.

Lide, D.R. (ed.). 1999. *CRC Handbook of Chemistry and Physics*, 79th ed. Boca Raton: CRC Press.

Melrose, M.P. and Scerri, E.R. 1996. Why the 4s Orbital is Occupied Before the 3d. *Journal of Chemical Education* 73: 498–503.

Moore, C.E. 1971. *Atomic Energy Levels*, Vol. 1. National Technical Information Service.

Nagy, A. and Adachi, H. 2000. Total Energy Versus One-Electron Energy Differences in the Excited-State Density Functional Theory. *Journal of Physics B-Atomic Molecular and Optical Physics* 33: L585–L589.

Pesic, P. 2002. Quantum Identity. *American Scientist* 90: 262–267.

Scerri, E.R. 1991. The Electronic Configuration Model, Quantum Mechanics and Reduction. *British Journal for the Philosophy of Science* 42: 309–325.

Scerri, E.R. 1994. Has Chemistry Been at Least Approximately Reduced to Quantum Mechanics? *PSA* 1: 160–170.

Scerri, E.R. 1997. The Periodic Table and the Electron. *American Scientist* 85: 546–553.

Scerri, E.R. 1998a. The Evolution of the Periodic System. *Scientific American* (September): 78–83.

Scerri, E.R. 1998b. How Good is the Quantum Mechanical Explanation of the Periodic System? *Journal of Chemical Education* 75: 1384–1385.

Scerri, E.R. 1998c. Popper's Naturalized Approach to the Reduction of Chemistry. *International Studies in the Philosophy of Science* 12: 33–44.

Scerri, E.R. 2000. Editorial 4 (Experimental Display of "d" Orbitals). *Foundations of Chemistry* 2: 1–4.

Scerri, E.R. and McIntyre, L. 1997. The Case for the Philosophy of Chemistry. *Synthese* 111: 213–232.

Schweber, S.S. 2002. Enrico Fermi and Quantum Electrodynamics 1929–32. *Physics Today* (June): 31–36.

Scott, W.T. 1967. *Erwin Schrödinger, An Introduction to His Writings*. Amherst, MA: University of Massachusetts Press.

Shermer, M. 2001. I Was Wrong. *Scientific American* (October): 30.

Tegmark, M. and Wheeler, J.A. 2001. 100 Years of Quantum Mysteries. *Scientific American* (February): 68–75.

Wannier, G.H. 1966. *Statistical Physics*. New York: Wiley.

Woit, P. 2002. Is String Theory Even Wrong. *American Scientist* 90: 110–120.

Wylie, A. 1999. Rethinking Unity as a "Working Hypothesis" for Philosophy of Science: How Archaeologists Exploit the Disunities of Science. *Perspectives on Science* 7: 293–317.

SECTION VI

CHEMISTRY AND ITS TOOLS OF REPRESENTATION

CHAPTER 14

VIRTUAL TOOLS

The Epistemological and Social Issues of Computer-Aided Chemical Process Design

ANN JOHNSON

Departments of History and Philosophy, University of South Carolina

To compute or not to compute.
That is the question.
Whether 'tis nobler in the mind to suffer
The slings and arrows of stochastic fortune
Or, to take algorithms to a sea of problems
And by heuristic solve them.
(Malpas 1989, 599)

INTRODUCTION

It is very striking to contrast the work practices of chemical engineers in the 1950s with those of the present day. While chemical engineers from both periods may give similar vague descriptions of their work—such as designing the processes that "transform raw materials into desired products" (Felder and Rousseau 1986, 3)—the ways they go about doing this work are quite different. Today, many chemical engineers begin using computer-aided design software from the first freshman class they ever take in the subject. Nearly, all use computation software like MATLAB in introductory classes. While some professors instruct them in the "hand" methods that they were taught in the pre-PC era, few chemical engineering majors actually intend to solve problems without the aid of a computer. As a result of the ubiquitous personal computer, the work practices of the chemical process designer have completely changed—to the extent that a hypothetical chemical engineer awakening from a deep sleep begun in 1960 would be no more able to work today than a recent chemical engineering graduate could be transported back to the slide rule era and quickly take up a place in a design team. I will argue here that this constitutes a kind of incommensurability—one focused on the day-to-day world of practice, rather than the rarified and esoteric world of theory. By "practices," I mean the day-to-day or even hour-to-hour activities of the chemical engineer engaged in process design. If a chemical engineer identifies her work as process design, this tells an outsider nothing about what she does for 8 or more hours a day. In thinking about this, I am reminded of several television segments I have seen where young children explain what their parents do for a living. While the child of a chemical engineer might know to say their

D. Baird et al. (eds.), Philosophy of Chemistry, 273–291.

Mommy works on chemicals, it would be most interesting to hear her interpolation of the actual daily business of process design. I would venture that many of today's children of chemical engineers would identify their parent's activity with working on a computer, something none would have thought of in the 1950s and very few in the 1960s. There is a serious lesson here for those of us who study science and engineering—design can be usefully discussed at a meta-level, but this is disconnected from the work scientists and engineers do. This distinction between theory and work practices is particularly crucial in studying science and engineering in industry, where doing counts for everything and new theories are often by-products. Since chemistry has always had very close ties to industry, and a vast majority of chemists and chemical engineers work in industry, focusing on doing is particularly apropos in the philosophy of chemistry. As a result, what I am interested in here is how computer-aided design tools and techniques were developed, and how they have transformed the work of process design in such a way that the today's work is incommensurable with that of the age before the ubiquitous computer.

While arguing that computers have fostered significant and important changes in engineers' work practices, I do not want to convey the notion that an external technological force called "the computer" acted upon passive chemical engineers. While the use of the computer did change the way chemical engineers work, as they used the computer they also changed chemical engineering itself, what computers could do, and, most profoundly, participated in a broader redefinition of what the computer was.[1] This feedback cycle demonstrates a way to eliminate the misleading determinism that tends to creep into any story about the ways computers and society have interacted.

The story I want to tell about the computer and chemical process design involves both the computer and chemical engineering practices mutually transforming one another in the hands of human agents. To some extent, this is the traditional narrative of the development of the computer—looking at the computer's development often requires looking at the ballistics tables ENIAC was designed to produce, or looking at the census figures UNIVAC was designed to crunch, or the physics problems behind MANIAC. However, when the focus shifts to the applications, rather than the development, of the computer, a more deterministic voice seems to appear. I want to lay my cards on the table right away—I disavow this deterministic attitude in all forms.[2] That said, this chapter will not be a symmetrical account of the co-development of the computer and its software with chemical engineering design practices, but will instead focus more heavily on the incorporation of new computer-aided methods of doing design work. The history of computing and, particularly, of software—which is more germane to the questions I want to ask—constitutes a complex project and one that would take me too far astray from chemical process design.

EPISTEMOLOGY AND ENGINEERING COMMUNITIES

In trying to understand what changed with the integration of computer-aided techniques, it is important to explore the relationship between engineering epistemology and engineering communities. While I will discuss this relationship from the perspective of chemical engineering, a similar dynamic holds in various forms for many different kinds of engineering and science. First of all, the basic social structure of engineers is the problem-oriented research group, or, what I refer to as "the community," an intimate group with a common *very specific* research interest. Engineering communities of this nature range from, at the smallest, a dozen or so members to the 100s. The critical defining aspects are the group's collective definition of their research subject, and the fact that they all know each other, sometimes through reputation and intermediaries, but more often through face-to-face encounters. Rarely are there any more than two degrees of separation between any two members of a community. So, a discipline like chemical engineering does not represent a problem-oriented community, nor does a subfield such as process design. The problem-oriented community I focus on here is the group of engineers interested in the utilization of the computer as a tool for process design. They asked how they could make the computer a central tool in chemical engineering, and in doing so, increase the speed, accuracy, and complexity of their designs. Once engineers had accomplished this, they moved on to other problems—how to solve more and bigger problems with new computers. Often as the problem evolved, so did the community. At some point in this evolutionary process, the community ceased to be the same group as it started out—it had many new members and they were asking different questions. In the case of computer-aided design, the initial community then splintered into many more communities—generally divided by the types of processes being designed. That is, the initial problem connecting the members of the community ceased to be a problem; therefore, new problems were pinpointed and new communities formed. As can be seen, the community is a very fluid arrangement—ever changing in both epistemological focus and in its membership. For this reason, it demands historical examination for seeing it fixed in time means losing its most essential quality.

While a research community can reside completely with a discipline, more often problem-oriented communities are highly inter- and multi-disciplinary. Because the practitioners are all acquainted with each another, disciplinary boundaries are broken down. When a community begins to exceed this highly personal size, it often begins to splinter. Splintering generates new communities and spells the end of the original group.[3] In the 20th century, communities can and often do span national and company lines.[4]

THE EVOLUTION OF ENGINEERING COMMUNITIES

Engineering communities evolve because engineering knowledge evolves. Therefore, separating the social arrangements from the knowledge around which they are formed often results in inaccurate depictions of both the social and epistemological aspects. As a result, I would like first to present schematically the evolution of an engineering community, then subsequently detail a case study involving the computer-aided process design as the exemplary community. The case study will demonstrate the intertwined problems of social and epistemological structures within engineering communities and flesh these schematic stages out. Ultimately, I want to show how incommensurable process design practices developed in the course of community and epistemological development.

Perhaps it is obvious, but a problem-oriented research group must begin with a problem. For engineering communities, the initial problem is usually of the form "How do you make . . . " Historically speaking, many engineers in different institutions are often asking such questions simultaneously for several possible reasons. Perhaps, a new market demand has been presented, for example, from changes in legislation.[5] In other cases, a device from another field becomes available for adaptation. This is the case for the application of the computer—though as I insist in the Introduction, I do not believe that the computer is a completely exogenous introduction. A third motive for a new problem becoming central to community formation is a new promise of solving an existing issue—for example, a new scientific understanding of a disease as an impetus for pharmaceutical development, or a new phenomenon in the case of instrument development. Clearly, understanding the motives for the formation of a certain community is important. However, since they are also peculiar to each case study, a long list of possible motives would be useless in a schematic sense.

Suffice to say that for various reasons over the course of a few years, groups of engineers at different institutions begin to investigate a set of promising issues, which all swirl around a single definable problem—how to make, determine, prevent, clean-up, improve, cheapen, and so on. At conferences and workshops of a broader professional nature—for example, the annual meeting of the American Institute of Chemical Engineers (AIChE)—these engineers meet to discover their common interests. While these initial meetings can be in formal settings, such as at paper sessions, they are more commonly less formal—informal conversation through personal contacts, for example. At these informal meetings, information, or new knowledge, is exchanged. The exchange of information in these informal settings is a critical dimension of community formation, since the kind of information a potential community member has and is willing to share will play a role in his or her status within the nascent community. I liken this stage to a poker game, in which an "ante" of information is required. Then, a game begins—each participant wondering what knowledge the others may have and be willing to share. Offering one's knowledge to the group is a gamble—will the information received from others justify the knowledge given up? These questions are more than simply about increasing one's individual status. Because most engineers work in the private sector, giving up information has potentially serious economic consequences. Still, not participating in knowledge sharing can have even more dire consequences. It is a moral economy of sorts, but it would be naïve to dismiss the gamesmanship of the moment.

As engineers recognize who is interested in a particular problem or set of related problems, they also use professional resources and organize formal sessions on their problem at more broadly organized conferences. This allows them to introduce themselves to one another and recruit new members from related fields and problems. In these formal sessions, they often hone the problem and begin to delimit solution sets—what will count as solution and what solutions may be unacceptable for any number of reasons, including feasibility within a given production plant, cost, legislation, environmental impact, material usage, and so on. From these sessions at very large conferences, members plan smaller workshops devoted to their problem. Once small workshops and conferences are held that focus on the problem, one can safely acknowledge that a community exists and from participant lists one can discern who belongs to this community. In the process of organizing these meetings, some individuals take a leadership role in the new community, and depending on personalities, these people are often the strongest producers of new knowledge applicable to the problem.[6]

Ironically, just at this moment of coherent community formation, resistance to the community often increases. Members of larger professional organizations may deny either the validity of the community's problem or that it belongs to the larger discipline. For example, at nearly the same time that computer-aided chemical engineers began to hold their own meetings, their articles failed to address more general problems of chemical engineering. They discovered they needed their own journals—which had different standards and, in this particular case, would publish algorithms, programs, and new applications of commercially available programs, instead of traditional methodological articles.

Once, a community has its own meetings and publishes its own journal or special interest annuals, it achieves a mature, although still dynamic, state. Ironically, in many cases, this is the beginning of the end. One of two situations develops that spells the end of this particular community. In many cases, the problem around which the community formed is solved. One marker of a solution is a product being introduced to the market and becoming successful.[7] Often, a one-company's version of the product becomes dominant, that is "closure" around a specific artifact is achieved (Law 1987, 111–114; Misa 1992, 109–111). Engineers working for other companies often refocus their own work to compete more directly with the leading product. Because, product competition is so keen in these last steps, any orientation toward an intellectual community suffers—less and less is to be gained from sharing knowledge of any sort. Patents take over from informal conversations and drawings on cocktail napkins. Sometimes, instead of competition killing the community, the group grows so large as to cease to function as a knowledge "co-op"—then splintering occurs and smaller communities form focused on what were subproblems in the original community. The cycle of community development then begins again. However, it is crucial to note that the central focus of a community is its knowledge—epistemological and social issues are not completely distinct.

THE PECULIAR PROBLEMS OF CHEMICAL ENGINEERING

Chemistry and chemical engineering share many qualities. Perhaps then, it should not surprise anyone that the history and philosophy of these two disciplines also share a dominant characteristic, which is that they are far less studied than other disciplines. While there are far more chemists than physicists in the world, the history and philosophy of physics dwarfs the history and philosophy of chemistry. Chemical engineering holds a similar, but even more ignored, position. There are more graduates of chemical engineering programs than any other engineering discipline, yet there is very little attention paid to the history of chemical engineering in comparison with the other engineering disciplines—civil, mechanical, electrical, even aerospace.[8] Given the lack of attention to the history of chemical engineering, a few words about chemical engineering and its historical relationship with chemistry may be useful before moving to my discussion of the development of computer-aided process design.

Chemical engineering is the youngest of the four main engineering disciplines, and it has the closest ties to a scientific discipline. As a result, it is a good case study in how one discipline was self-consciously constructed to be distinct from another. Much of the history of chemical engineering has focused on this early period of discipline formation. However, looking at this period does not lead to a useful characterization of the practices of chemical engineering, because in its early, nascent form, common practices were only beginning to develop—consensus had not yet been reached. By the 1920s, a consensus had formed as to the nature of chemical engineering, and common generalizable practices were being used in industry and taught in the universities. These discipline-defining techniques usually fall under the category of unit operations, whereby chemical production was looked at as a series of physical processes, such as distillation, heating and cooling, crystallization, drying, filtration, and so on.[9] Under development largely at MIT, this conceptualization of chemical engineering was disseminated in a textbook by Walker, Lewis, and MacAdams titled, *The Principles of Chemical Engineering*, which was published in 1923. At MIT, the focus of chemical engineering was on design for chemical industry—from plant design to process design to production control to problems of scale-up. From the beginning, chemical engineering was unabashedly commercial—distinguishable from chemistry by its complete orientation toward large-scale chemical production.[10]

Chemical engineering at MIT in the 1920s had a strong, although perhaps unsophisticated, mathematical side to it, with Walker, Lewis, and MacAdams writing, "so far as it is possible, the treatment is mathematically quantitative as well as qualitatively descriptive" (Freshwater 1989, 20). In the 1930s, applied thermodynamics began to change the focus of chemical engineering and was the next step in making the mathematics of chemical engineering more sophisticated. At Purdue University in the 1930s, R.N. Shreve began to depart from a physical conception of processes and move toward a chemical conception, by focusing on unit processes in addition to unit operations. He argued that industrial chemistry could be presented as a series of unit processes such as alkylation, nitration, oxidation, and sulfuration, and so on (Peppas 1989, 129). However, Shreve was not able to further the mathematization of the discipline using unit processes at this point, and Shreve was criticized for the largely descriptive approach he took to the subject (141). In chemical engineering, as in many other scientific and engineering disciplines in the early 20th century, mathematization became the single most important dimension of advancement.

THE 1950S: MAKING NEW MODELS

Chemical engineering became unwaveringly mathematical in the 1950s, led by Neal R. Amundson at the University of Minnesota. Amundson, who actually earned his Ph.D. in mathematics, became head of the chemical engineering program at

Minnesota in 1949, when he was just 33 years old. Amundson's unusual mathematics background led him to do groundbreaking work in the transformation of chemical engineering from an experimental to an analytical discipline. Amundson explained the difficulty in getting his master's thesis on an analytical solution to the McCabe-Thiele method in a 1990 interview: "the engineers who were on the committee really did not understand what I was doing, because chemical engineering was very non-mathematical. It was very primitive from a mathematical point of view. That is one of the things I changed later" (Amundson 1990, 14). Over the next two decades, he transformed the program at Minnesota into the leading institution for the mathematization of chemical engineering (Amundson 1990, 22). In the 1960s, often led by graduates of the Minnesota department, other programs followed suit and prioritized the construction of new increasingly complicated mathematical models. As the sophistication of the chemical engineer's mathematical tools and models increased, so did the likelihood that computers could facilitate mathematical solutions. Similar phenomena can be seen in other engineering disciplines in the mid-1950s, as iterative solutions and matrix methods became more common. Consequently, the mathematical methods of chemical engineering were maturing just as the IBM 704 and its FORTRAN programing language were being introduced. Amundson became one of the key proponents of using the computer in chemical engineering, largely because it advanced the mathematical sophistication that marks his whole career. The computer was a mathematical tool; therefore, it was of interest to Amundson, his students, and his followers (Amundson 1990, 30).

Mathematization affected the university curriculum in chemical engineering very quickly. Increasingly in the 1960s, the introductory course in American chemical engineering was a course in mass and energy balances. This course was unrelentingly mathematical, invariably requiring knowledge of calculus as a pre-requisite. The course also put process design front and center in the training of chemical engineering, thus elevating the status of this subspecialty, which had been at the center of "theory-building" in chemical engineering, but not the core of the chemical engineering curriculum in the pre-WWII period. In doing so, it also highlighted the mathematical character of chemical engineering.

THE TROUBLE WITH PDEs

In his doctoral thesis, Amundson investigated the process of diffusion analytically, using non-linear, parabolic, partial differential equations (PDEs). As was often the case, Amundson's work opened a new research field, and by the early 1950s Amundson was pioneering the use of computers, such as the IBM 605 and 650, in chemical engineering. Through Amundson's consulting activities at Remington Rand, the University of Minnesota procured a Univac ERA 1103 in 1956. Amundson claimed that the machine had two sets of users: his chemical engineering students and "a fellow in the physics department"[11] (Amundson 1990, 30). Under Amundson's direction, process design lent itself particularly well to computer applications. As processes were modeled mathematically, computers were useful in solving the complex equation sets that characterized process design, such as mass and energy balance problems. Amundson's characteristic use of PDEs lent itself to the application of the computer, since PDEs with more than two variables had to be "solved" using iterative approximations.[12]

The iterative calculations of a computer could approximate PDEs very accurately, since a computer could iterate an equation 100s of times, whereas engineers worked them dozens of times. The more iterations, the closer the approximations became. Iterative methods did not solve PDEs in an analytical way, but they could give engineers a great deal of confidence in their approximations. Still, outside Minnesota, access to computers for engineers was very limited—they were few in number and at most institutions, often their use was prioritized for defense-related physics research. Only a few chemical engineers at major research institutions gained any access to computers. Those, like Amundson, who did, publicized the potential of the computer as a machine for making PDE approximations useful instead of doubtful.

For years, engineers had avoided building mathematical models dependent on solutions to PDEs because they would never move into common practice—models that did not permit numerical solution were useless. Rather, suddenly in the mid-1950s, intractable mathematical models did not look quite as intractable. In the second half of the 1950s, chemical engineers oriented toward mathematical modeling reworked models to take advantage of the computer's ability to perform iterative approximations. They were aided by the fact that often when physics departments got new, more powerful computers, engineering programs scrambled to get the old computers as "hand-me-downs." While this stage of academic research was necessary, it can hardly be characterized as changing the practice of chemical engineering, since the community of engineers working on computer-aided chemical engineering of any nature numbered less than 20 in 1955.

PROFESSIONAL ISSUES

When chemical engineers first tried to use the computer to solve equations, they often approached it as they would a slide rule—as a tool to help with computation. This was particularly true of any problems that required iterative approximations—the computer simply replaced a human computer. However, the analog computers of 1955 were so significantly different from today's computers, it is easy to underestimate the difficulty of using the computer as a simple calculator. Programing an IBM 701 or even a 704, which used FORTRAN, was a specialist's job–that is, a programer's job. Even writing short routines for solving matrix algebra problems could be taxing. Therefore, writing a program to solve a matrix was not trivial—when engineers wrote them they often became subjects of journal articles. Using the computer often meant dealing with programers. While a few chemical engineers took it upon themselves to learn how to program before the diffusion of FORTRAN into the undergraduate engineering curriculum, senior engineers rarely wanted to spend their time learning to program—programing was perceived as clerical work.[13] Programing was more trouble than it was worth—so programing was either hired out to computer specialists or more commonly, farmed out as graduate student work.[14] Consequently, programers played an increasingly central role in the development of computer-aided chemical engineering, because they often advised engineers on how to construct mathematical models for maximum solvability. Despite the often-dismissive attitudes of some engineers, programers constituted a central part of the nascent computer-aided chemical engineering community. By the end of the 1960s, programing became a much more common skill for Ph.D. level students; until then, questions of how to program remained central to the computer-aided process design community.

Engineers working on computer applications in chemical engineering encountered resistance from other engineers who worried that the computer might transfer the real "thinking" of engineering to the machine or its tender, the programer. Some chemical engineers feared that the computer would make them obsolete, especially as automated design was discussed as the "holy grail" of computer-aided engineering.[15] Ultimately, chemical engineers wondered how they could retain ownership of their discipline—and they faced three options. First, they could deny the potential of the computer, but the 1950s were simply not a historical moment in which the potential of the computer could be denied. Physicists had already ensured the scientific importance of computer applications, even of simulations. All science and engineering disciplines were moving toward computer applications—for the AIChE to sanction hesitation in researching computer applications would be to deny the profession's status as scientific. Second, chemical engineers could try to incorporate programers into chemical engineering. Large chemical companies often approached the problem this way and tried to fashion their computer service departments as co-operative partners—more technical than clerical. The third option appeared to be the most complicated and required the most change within the field of chemical engineering—changes in the curriculum could make chemical engineers into programers. As FORTRAN became the general programing language of engineering, it became more and more common to require that graduate students, then undergraduates, learn FORTRAN as quickly as possible. Eventually, FORTRAN became a *de facto* degree requirement for chemical engineers. By the 1960s, the chemical engineering profession had defined its relationship with programing—facilitated by a general language like FORTRAN—engineers *had become* programers.[16]

The result of this first phase of computer use by programers was to map the contemporary practices of the chemical engineer onto the computer. The mapping process was colored by the fact that the early proponents of the computer were mathematically oriented, so the practices mapped were often those at the mathematical cutting edge, not necessarily the general practices of the profession. Furthermore, because the capabilities of the computer were being determined at the same time, engineers did not necessarily know the best way to use the machine. Often the engineers took their pre-existent methods and just made computers "crunch" the numbers—they had not yet figured out that they needed to reparse the tasks for optimal use of the computer.[17] Chemical engineers of the 1950s used the computer to do what they could already do without it—albeit more quickly and with greater complexity—they were not yet changing their models to get the computer to do something they could not do with their "hand" methods.

NEW COMMUNITY, NEW KNOWLEDGE

In the period from the late 1940s to the early 1960s, chemical engineers first encountered the computer as a possible ally in their construction of a more rigorous set of mathematical models. In that era, a small group of engineers, most of whom had been part of the Minnesota department, redirected their efforts from model building to computer-aided design work. In this early period, computer-aided design meant figuring out how to use the computer to produce numerical solutions to models with too many variables to iterate by hand. As they saw the potential the machine had for massive number crunching, they realized that the old rules for what constituted a solvable model were no longer valid. They could design models with much more mathematical complexity and let the computer do hundreds or even thousands of iterations to produce a solution. PDEs no longer posed insurmountable obstacles to solution—in fact, these engineers began to construct models with more and more difficult PDEs, since the computer's approximation methods were good enough to call solutions. Thus, in the 1950s, a new group of engineers came together around a new set of problems—how to best use the computer and how to reorganize mathematical models for maximum solution by machine computation. Both the process of answering these inquiries and the answers themselves constituted the new knowledge over which this new community of engineers claimed ownership.

THE 1960S: MAKING NEW PRACTICES

In the 1960s, the development of new computer-aided tools was dominated by industrial research, instead of academic. Former graduate students from highly mathematically oriented programs dominated the movement to make computer-aided chemical engineering a real-world practice. In the next two sections, I will review two major steady-state process design systems developed in the 1960s. These programs, and dozens of others like them, changed the practice of chemical engineers in a profound way from the 1950s. In order to produce these new programs, engineers first faced the problem set-up by the initial use of computers as PDE calculators. Full utilization of the potential of the computer required a different attitude toward making mathematical models. Models needed to embrace the iterative power of computing—they needed to be oriented toward matrixes and finite difference methods. In other words, chemical engineers had to move away from traditional notions of elegance and accept the brute computational force of the computer.

MONSANTO'S FLOWTRAN SYSTEM

In 1961, chemical engineers at Monsanto began programing FORTRAN computer routines to solve material and energy balance problems (Cobb 1972, 32). It took 4 years to produce enough blocks to construct a design system from these pieces. From the outset, engineers and programers designed these blocks with the capability to model entire processes in combination. In 1965, engineers began to use the system, called FLOWTRAN, to design new processes and new plants. Using this system, an engineer could specify a series of material and energy inputs, and certain sequence of unit operations. Then, the program would generate the material and energy outputs (Gallier 1972, 4). The system was physically located on a United Computing Systems CDC-6600 computer, at Monsanto's Kansas City facility. Monsanto engineers at other facilities had access to the system through a remote batch-processing mode. By 1972, FLOWTRAN was running 200 jobs a week (Cobb 1972, 36).

Unlike, the use of the computer as calculator by the academics of the 1950s, Monsanto's FLOWTRAN *simulated* chemical processes. However, simulation in 1965 was crude. C.B. Cobb, an engineer who worked on marketing FLOWTRAN to users outside Monsanto, described what that meant for use of the system, "Simulators right now are not the type of tools that you ask a question, push a button, and the answer comes out. This is simply not the case. This black box contains a mathematical model and to make it work for you, you have to know what it is about" (33). Cobb went on to say that "tuning" the model required significant experience with both the system and with process design.

Cobb claimed that simulation itself changed the practice of chemical engineering, although he clearly dismissed critics' concerns that computer-aided engineering would eliminate engineering thinking. Considering the day-to-day work practices of an engineering working on FLOWTRAN, one can see that their work does change. They must construct a flow sheet in order to use the computer, but the construction of operations and inputs in that flow sheet is completely informed by the way FLOWTRAN works. Previously, the logic of the flow sheet would not have been dictated by how processes that were defined by the programers of the blocks. The engineers had to make different choices in their design processes. Engineers needed months of formal education in FLOWTRAN in order to design processes using it. Daily practice at Monsanto in 1970 was quite different than it had been in 1960.

DIGITAL SYSTEMS CORPORATION'S PACER SYSTEM

Chemical engineers at Purdue University and Dartmouth College entered into a collaborative relationship in 1960. By 1961, using FORTRAN II, they had the beginnings of a computer-aided design program called Process and Case Evaluator Routine (PACER). It was a collection of FORTRAN modules with an executive routine with a tracing ability. By 1966, the program had been distributed to university programs in chemical engineering. In 1967, PACER's code was completely rewritten in order to become a commercial software package, called PACER 245. While the new PACER's mode of operation was similar, it was, in fact, a completely different program. Digital Systems Corporation was set up in Hanover, New Hampshire, to sell, service, and maintain PACER 245 as a commercial application.

PACER 245, as a program, had three elements (Shannon 1972, 34). First, an executive routine handled input and output and the computer's bookkeeping. While obscured from the user, this element was not trivial at all. On a programing level, this was the most complicated task. A library of standard unit operations calculations, called COMPUBLOCs, constituted the second element. COMPUBLOCs were discrete mathematical models of unit operations, often developed from the small solution routines generated in the 1950s on analog computers. PACER 245 had 165 COMPUBLOCs. Finally, PACER contained a second library of physical property data and routines.

Engineers using this system had to conceive of their designs in several different ways, in order to best use the capabilities of the system. Shannon, the system's chief spokesman and one of its designers from the beginning, warned engineers to "start simple" (34). From a simple mass balance simulation, engineers could build bigger and bigger models, chaining more and more processes together and adding more materials. Designing a plant required not a single, complicated model, but rather a series of models of increasing complexity. The engineer could then figure out what worked in isolation in the model—modules could be optimized one at a time. The engineer would know at each stage that everything up to a certain point was working—the models were functional and accurate, up to that point. Like the FLOWTRAN system, this was a somewhat different work practice from the pre-computer days. Although, if there is any common engineering "method," it is to break problems down into discrete steps and work through them individually; the computer further emphasized discretizing the problem. Furthermore, on the computer, the engineer reworked each previous step at the next juncture—a technique that would be far less common in hand methods because of the labor involved. This problem also showed how interrelated the modules of PACER really were. Each new step had to take into account the old steps—picking up the process midstream held serious potential for introducing emergent errors. While the keystrokes of PACER were completely different from FLOWTRAN, Shannon and Cobb agreed that a similar mindset was needed to use simulation in general. This attitude toward process design had to be taught, either in the corporation or preferably at university. Training on PACER took 6 months from first introduction to the point where an engineer was ready to being designing the first system (37). Shannon estimated that it would take 2 years for PACER to permeate a given company's process design department. One of the services offered by Digital Systems was training. In addition, a company could hire Digital to co-design a plant with their process design engineers, as a way to learn the system. After the collaborative design, PACER could then be leased to the client who would use it in-house.

PACER was probably the most commonly used process design software of the 1960s and early 1970s, because of its prevalence in universities and its accessibility to small companies. Still, Shannon pointed out at the 1972 conference that only a very small percentage of process designers used the computer at all—mostly those at large companies with systems designed in-house. While there were several software packages available to smaller companies who could not construct or even support their own software, using any system also required purchasing time on a mainframe computer. Computer time often exceeded the budget of a small company. In addition, PACER's cost was not necessarily trivial. Digital Systems charged a 50% surcharge on the cost of the computing time, so if computer time cost $1000, PACER cost $500. Practices were changing, but not yet for most engineers.

THE 1970S: MAKING NEW ENGINEERS

By 1970, most large chemical companies had some kind of software for process design. Most research universities were using some software package in graduate education in process design. Dozens or even hundreds of masters' theses were written to introduce either a new process design system or a new set of subroutines for an existing package. Computer-aided design had taken root in chemical engineering—computer-aided process design issues were at the center of a functioning community. As software packages became more common and practices were being set, the community began to splinter and address an array of issues that assumed the existence of functional, accurate, and useful design software.

One of the issues that splintered from the process design community early on was the new problem of how to educate chemical engineering undergraduates in a way that would prepare them for careers using computer-aided design. In 1970, undergraduates coming out of a chemical engineering program possessed limited exposure to computer-aided design. Most undergraduates in the early 1970s had no experience and of those who did, their experience was rarely sufficient for them to enter into one of the large commercial process design departments. FORTRAN was only beginning to be a requirement for undergraduate engineers. Seider (1972, 179), of the University of Pennsylvania, lamented that even when textbooks did include computer programs, "frequently the computer programs appear toward the end of the text following the subject matter which is presented traditionally; the computer program is an afterthought. It is my belief that one should present the material more logically to assure that the computer programs demonstrate principles at the right time". Computer applications had to be made integral in the chemical engineering curriculum, a process that took time and yet-to-come technological developments, like the PC.

NEW ENGINEERS, A NEW STATE OF THE ART

The chemical engineers who gathered at the first AIChE workshop on computing in chemical engineering in 1972 all agreed that computing was changing the field of chemical engineering. E.H. Blum of the Rand Corporation put it in the strongest way:

> I was very impressed and implicitly believe Professor Seider's remarks about the computer completely changing the way chemical engineers think about problems. The way you fundamentally think about stoichiometric problems is different; the way you think about equilibrium calculations is different .... The way you think about kinetics and reaction rate problems should be different .... But how do you teach people to think differently about problems? (Blum 1972, 186)

Seider replied,

> The way in which we think about a problem is affected to some extent by the way we plan to solve the problem .... Given a problem, the question is, how would you

solve it? Well, if it is in the material and energy balancing area, you begin by writing the conservation equations. Then, the question remains as to how you solve the set of equations. In the past, many problems were framed in such as way that they could be solved using shortcut methods, the methods of unit operations. Today the computer offers more flexibility in modeling and in equation solving. Hence, we can avoid the need to employ semi-empirical methods that are sometimes inconvenient and inaccurate to use (Seider 1972, 186).

The community of computer-aided chemical engineers had some answers to the questions that framed their community. They had determined an agenda for the most promising future avenues of development. However, the nature of the questions the community was asking required participation by a much larger segment of the discipline. What computer-aided engineering needed in the 1970s was a new generation of engineers, trained in a curriculum that introduced them to computer-aided engineering early and placed computers into the curriculum centrally.

Technological issues accompanied the need for more students groomed to design processes on a computer. The minicomputer was coming of age, and this promised to make computers much more accessible to both smaller university chemical engineering programs and smaller companies. Seider claimed a minicomputer capable of running process simulation software could be purchased for $20,000–$40,000 in 1972 (Seider 1972, 181). While the microcomputer of the 1980s would have a much more significant effect in terms of bring computing to the practices of every chemical engineer, the mini facilitated the spread of computer-aided engineering to a significant proportion of process design engineers by the end of the 1970s.

REDEFINING DESIGN IN CHEMICAL ENGINEERING

As Peter Galison so elegantly lays out in *Image and Logic* and addresses even more directly in his article on "Computer Simulations and the Trading Zone," the computer began its interaction with scientific communities as a tool, but quickly developed a much more complicated relationship with those communities. In fact, it is hard to come up with a term for what the computer becomes, which, in and of itself, indicates the complexity of the role it plays in the post-war development of science and engineering. Galison argues that the use of the computer forced physicists to challenge the traditional definitions and relationship between theory and experiment (Galison 1996, 119). In the case of chemical engineering, the computer played a similarly complicated role and also muddied traditional, epistemological, and social boundaries. However, it is not accurate to discuss the traditional relationships between theory and experiment, since this is not a useful dichotomy in engineering. Instead, the computer in chemical engineering challenges the relationship between mathematical modeling and design methodology.

In the 1950s, the numerical analysis of mathematical models increasingly dominated the chemical engineer's job. In order to calculate mass and energy balances, large numbers of equations had to be manipulated to find some way to solve or approximate critical values. The entry of computing into chemical engineering took the need to find numerical solutions for these equations and gave that job to a machine.

Engineers then refocused their task on two areas, how to reorganize the design process to take full advantage of the number crunching ability of the computer and how to design mathematical models that would be black-boxed into the computer's software.

The relationship between modeling and design practice is central to both the epistemology and activity of engineering. Most historians and philosophers of design have agreed that the activity of design is a, if not the, primary generator of new engineering knowledge. Design in engineering usually requires that the physical world be expressed as a set of mathematical relationships in order to provide an avenue for analysis; that is, it requires the construction of models. Consequently, the dialectic relationship between modeling and design is at the heart of engineering epistemology, in a similar way to the more traditional theory and experiment relationship in scientific knowledge. In their article on "Models as Mediating Instruments," Morgan and Morrison argue, "the capacity of mathematical/theoretical models to function as design instruments stems from the fact that they provide the kind of information that allows us to intervene in the world" (Morgan and Morrison 1999, 23). Yet, in the case of computer-aided design, another layer of analysis is required, since the computer, as well as the mathematical model, is functioning as a design instrument. Furthermore, computer-aided design not only provided an impetus for the construction of new models, but also changed the ways pre-existing models were used. The computer shifted the relationship between model and design practice by black-boxing the calculation of the solution. As Warren Seider claimed at the 1972 AIChE workshop, the reason chemical engineers in 1972 needed to think differently was that they were working differently, indicated by new methods and, particularly, new "shortcuts" using the computer (Seider 1972, 186). Designing a new process came to be understood as a different activity, requiring a changing body of knowledge. The job of the engineer, what she did on a day-to-day basis, was different with the computer as aid than with pencil and paper as aid, whether or not the algorithms themselves were new.

INCOMMENSURATE PRACTICES

The practitioners of computer-aided chemical engineering argued that by 1970 their software tools were so sophisticated that months or even years of retraining were required to use them to simulate anything but the most trivial processes. The requirement of so much specialized training showed that the old and new practices of process design were worlds apart. An engineer trained before 1960 was not able to walk into a process design department in 1975 and take over a project.[18] It is precisely the need for retraining that implies incommensurable practices. The engineers of 1960 and 1975 are not interchangeable—they do not know the same things or do the same work. One could even argue they do not understand processes in the same way because the experience of calculating is so different. This constitutes incommensurability, but in terms of work practices, not theory.

Philosophers of science who write about incommensurability typically focus on the role of theory. Thomas Kuhn, whose notion of paradigm shift is central to any discussion of incommensurability, saw new theories as the generators of incommensurability. Incommensurability resulted as new theories changed both the questions scientists asked and the answers they accepted (Kuhn 1970, 148). He further clarified that proponents of different theories speak different languages, "expressing different cognitive commitments, suitable for different worlds" (Kuhn 1977, xxiii). However, the incommensurability between pre-computer chemical engineering and computer-aided design is of a different nature, since not theory, but practices are shifting. It is difficult to apply Kuhn's ideas to this shift in engineering because theory plays no, or at least, little role. As Walter Vincenti argues in *What Engineers Know*, theories are not particularly central to engineering design. On the other hand, theoretical tools are, because tools are oriented at practice—that is, what engineers do (Vincenti 1994, 213–216).

Several philosophers of science since *Structure of Scientific Revolutions* have expanded Kuhn's look at the more social dimensions of science. This line of thought situates theory as just one dimension of many that comprise science, as both an activity and as knowledge. Along these lines, if science is constructed of multiple dimensions, then perhaps incommensurability can be the result of dimensions other than theory. Ian Hacking's phrase "topic incommensurability," while applied to theory in *Representing and Intervening*, also works for the incommensurability of engineering practices (Hacking 1983, 68). Topic incommensurability results when a new theory, or in this case practice, does "different jobs" than the old one. The new practice produces a different output, which, in engineering, may be a more epistemologically significant activity than theory-building. Hacking further captures the sense of practice incommensurability when he writes, "students of a later generation educated on T* may find T simply unintelligible until they play the role of historians and interpreters, relearning T from scratch" (Hacking 1983, 69). If one substitutes "engineering practice" for Hacking's T, then this statement conveys exactly what happened when the computer became the primary tool in chemical process design.

Ronald Giere also tackles incommensurability in ways that illuminate the shift to the computer in chemical engineering. He writes that one variety of incommensurability, drawn from Kuhn, is the incommensurability of standards, whereby Kuhnian exemplars help determine what will be accepted as significant problems and solutions (Kuhn 1970, 148; Giere 1988, 37). He explains, writing "it follows that a new solution to a problem, one appealing to a new exemplar and thus violating the standards of the older research tradition could not be recognized as a solution by that tradition" (Giere 1988, 37). Giere's notion of research tradition, with its clearly Kuhnian overtones, captures the shift in chemical engineering without recourse to theories as the primary generators of the shift.[19] In other words, the epistemological shift that occurs with the introduction of the computer into chemical process design can be adequately described by investigations into incommensurability from the philosophy of science, as long as theory is not the only vector for this shift in worldview.

CONCLUSION

In the immediate post-WWII world, academic chemical engineering, in common with most of the other engineering disciplines, focused on the construction of increasingly sophisticated mathematical models and tools. Was it just a fortuitous coincidence that at the same time, the computer was being developed by physicists, mathematicians, and electrical engineers to solve just the kind of mathematics the chemical engineers were designing into their new models and tools? I think this overstates the contingency of this history. Instead, perhaps the 1950s were a moment of synergy, where Department of Defense and military spending were driving science and engineering research down the same road. By 1952, when the IBM 701 was introduced, military spending had made the computer a workable tool. In the late 1950s, when the 704 and FORTRAN were introduced, engineering industries became a prime market for computers. While the first target for high-speed computing was the defense industry, chemical companies did not waste much time in investigating the potential of computing for their research and development work. First graduate students, but later even undergraduates, coming out of the highly mathematically oriented chemical engineering programs found significant opportunities to continue working in the community of computer-aided chemical engineering. In this way, chemical engineers from industry and universities worked together to produce a new set of work practices in process design by 1970.

NOTES

1. See Mahoney (2002).
2. By "deterministic," here I mean an argument of the form, "because of the computer, chemical engineering practices developed, or more pointedly, *had to develop* in the following way." There are two problems with arguments of this form. First, it implies that chemical engineering practices could only develop in the way they did—there is little room for contingencies in this type of argument. On the contrary, there were many ways chemical engineers responded to the existence of the computer and looking at several engineering subdisciplines one sees the variety of possible disciplinary responses. The second problem with naïve determinism is that it makes the causal agent exogenous—that is, the computer *acts on* chemical engineers. This is the problem solved by acknowledging the feedback loop between computer (especially software) development and changing design practices. For an interesting discussion of the historiographical problem of technological determinism, see Smith and Marx (1994).
3. This said, it is certainly clear that engineers can be, and usually are, members of more than one community simultaneously. In fact, I think this is a crucial dimension to the engineering profession, since new knowledge often comes from the transfer of ideas and things from other communities. Engineers must have multiple community identities in order to bring new knowledge (which includes new devices) to the table.
4. Clearly, there exist cases where the community is contained within a single company or institution, often for security reasons, though there are cases where the proprietary nature of the research effectively contains the community. Still, I think these are exceptions, and I would argue that, generally speaking, engineering communities span social institutions, such as corporations, universities, and national lines.
5. The most obvious cases for this would be either newly required safety specifications or environmental protection requirements.
6. Obviously, this strength can be defined in countless different ways.
7. Still, engineers may often continue developing a product in order to improve it or to make it more competitive with other options available on the market. However, the problems of a second and later generation product are often significantly different, involving questions about mass-production—cost, speed of manufacture, material choices—and even sale and distribution problems.

8. However, the AIChE is not the largest professional engineering society in the world; rather, the IEEE (electrical engineering) holds that distinction.
9. Of course, many historians have also seen the beginnings or antecedents of unit operations earlier than Arthur D. Little's 1915 coining of the term.
10. This is not to claim that the other engineering disciplines were not commercially oriented. However, with the possible exceptions of mining and aerospace engineering, which are much smaller and less distinct disciplines, the other engineering disciplines were not so closely identified with a single, albeit multi-faceted, industry.
11. The physicist was Nobel Prize winner William Lipscomb.
12. There is a difference between *solving* a PDE and coming up with a numerical solution. Most PDEs cannot be analyzed, and therefore, are technical unsolvable. However, one can plug in numbers for variables in a PDE and calculate the ranges in which solutions will fall—that is, they can be iterated. The more times they are iterated, the smaller the ranges are, until the ranges close in on single numerical values. These are approximate solutions. However, the number of iterations expands exponentially as variables are added, so practically speaking iterative methods are unfeasible for PDEs of more than two variables. See Forsythe and Rosenbloom (1958) for a contemporary textbook of how to use numerical methods to iterate PDEs by hand.
13. The development of programing from clerical work to profession is an important historical change, which happened right in the midst of the new engineering applications of the computer. How programing became defined both as an activity and as a nascent profession is the subject of several articles and two recent dissertations. See Ensmenger (2001) and Akera (1998).
14. Engineers' attitudes toward programing are often contradictory. While they often defended their decisions not to learn programing by claiming it was too much work, they also liked to downplay its academic legitimacy and see it as a mere service. When engineers wrote programs, they were worthy of thick description in journals, and when programers wrote programs they were just doing a good job.
15. Automated design was the concept that a computer could be fed the initial conditions or inputs of a process and could iteratively calculate the optimal combinations of materials, energy, and unit operations and processes. If fully realized (and it has not been), then the chemical engineer would become merely a data entry operator.
16. This is not to say that there were not any programers working in the chemical industries. On the contrary, there were many. However, by the 1960s, they were working with engineers with serious training and competency in programing. The engineers were determining what the programers, often called specialists, were programing. This same story can also be told from the perspective of the developing field of computer programing, see note 14.
17. I am indebted to Michael S. Mahoney for the notion of "reparsing the task." This phrase applies to many developments in the history of technology and generally means figuring out the most effective way for a machine to do something, instead of designing a machine to emulate human action. The simplest demonstration of a task being reparsed is to think about how a person puts in a screw using a screwdriver versus how a hand-held drill does the job. To design a tool to emulate human action would mean a tool that could only turn 180°, as the wrist does, instead of the 360° that a drill is capable of.
18. One could argue that the same incommensurability of practice exists between the engineer of 1975, used to time-sharing, punch cards, 80 × 80 displays, and numerical readouts and the engineer of today with discrete PCs, DVDs that hold 4.7 gigs of data, GUIs, and full-color laser graphics printouts. However, I would argue that these technologies requires less reorganization of the process of designing—the ways projects are broken down for solution—that the shift between no computer and the systems of the 1970s. Certainly, the nature of work has changed, but I think the worldview shift is less profound—the mathematics is still black boxed, and the pre-computing design analysis is rather similar. On the other hand, if this hypothetical time-traveling engineer is a designer of computing technology, that world has shifted as significantly between 75 and 95 as it did between 55 and 75.
19. This said, Giere's notion of theories as families or cluster of models may work for engineering in ways that more propositional explanations of theories do not. See Giere (1982).

REFERENCES

Akera, A. 1998. Calculating a Natural World: Scientists, Engineers and the Computer in the United States, 1937–68. Ph.D. Dissertation, University of Pennsylvania.

Amundson, N.R. 1990. Interview by James J. Bohning (October 24). Transcript, Chemical Heritage Foundation, Philadelphia, PA.

Blum, E.H. 1972. Where Lies the Future of Chemical Engineering Computing? Panel Discussion Transcript. In: *Chemical Engineering Computing*, Vol. 2. NY: American Institute for Chemical Engineering.

Cobb, C.B. 1972. Computer-Aided Chemical Process Design. Panel Discussion Transcript. In: *Chemical Engineering Computing*, Vol. 1. NY: American Institute for Chemical Engineering.

Ensmenger, N. 2001. From "Black Art" to Industrial Discipline: The Software Crisis and the Management of Programmers. Ph.D. Dissertation, University of Pennsylvania.

Felder, R. and Rousseau, R. 1986. *Elementary Principles of Chemical Processes*, 2nd ed. NY: John Wiley and Sons.

Forsythe, G.E. and Rosenbloom, P.C. 1958. *Numerical Analysis and Partial Differential Equations.* NY: John Wiley and Sons.

Freshwater, D.C. 1989. The Development of Chemical Engineering as shown by its Texts. In: Peppas, N. (ed.), *100 Years of Chemical Engineering*. Dordrecht: Kluwer.

Galison, P. 1996. Computer Simulations and the Trading Zone. In: Galison, P. and Stump, D. (eds.), *The Disunity of Science*. Stanford: Stanford University Press.

Gallier, P. 1972. Chemical Engineering Computing: Where the Action is. Panel Discussion Transcript. In: *Chemical Engineering Computing*, Vol. 1. NY: American Institute for Chemical Engineering.

Giere, R.N. 1988. *Explaining Science*. Chicago: University of Chicago Press.

Hacking, I. 1983. *Representing and Intervening*. Cambridge: Cambridge University Press.

Kuhn, T. 1970. *The Structure of Scientific Revolutions*, 2nd ed. Chicago: University of Chicago Press.

Kuhn, T. 1977. *The Essential Tension*. Chicago: University of Chicago Press.

Law, J. 1987. Technology and Heterogeneous Engineering: The Case of Portugese Expansion. In: Bijker, W., Hughes, T., and Pinch, T. (eds.), *The Social Construction of Technological Systems.* Cambridge: MIT Press.

Mahoney, M.S. 2002. In Our Own Image: Creating the Computer. In: Stamuis, I., Koetsier, T., and de Pater, K. (eds.), *The Changing Image of the Sciences*. Dordrecht: Kluwer.

Malpas, R. 1980. Invited Address to the Conference. In: *Foundations of Computer-Aided Process Design*, Vol. II. NY: Engineering Foundation.

Misa, T. 1992. Controversy and Closure in Technological Change. In: Bijker, W. and Law, J. (eds.), *Shaping Technology/Building Society.* Cambridge: MIT Press.

Morgan, M.S. and Morrison, M. 1999. *Models as Mediators: Perspectives on Natural and Social Science.* Cambridge: Cambridge University Press.

Peppas, N. 1989. Academic Connections of the Twentieth Century U.S. Chemical Engineers. In: Peppas, N. (ed.), *100 Years of Chemical Engineering*. Dordrecht: Kluwer.

Seider, W.D. 1972. Where Lies the Future of Chemical Engineering Computing? Panel Discussion Transcript. In: *Chemical Engineering Computing*, Vol. 2. NY: American Institute for Chemical Engineering.

Shannon, P.T. 1972. Computer-Aided Analysis and Design Packages. Panel Discussion Transcript. In: *Chemical Engineering Computing*, Vol. 1. NY: American Institute for Chemical Engineering.

Smith, M.R. and Marx, L. (eds.). 1994. *Does Technology Drive History: The Dilemma of Technological Determinism*. Cambridge: MIT Press.

Vincenti, W. 1994. *What Engineers Know and How They Know it*. Baltimore: Johns Hopkin University Press.

CHAPTER 15

SPACE IN MOLECULAR REPRESENTATION; OR HOW PICTURES REPRESENT OBJECTS

S.H. VOLLMER
Department of Philosophy, University of Alabama at Birmingham, Birmingham, AL 35294-1260

One of the earliest systems of representation in chemistry to show spatial relations within molecules was Dalton's system of molecular diagrams. In this system, collections of circles were used to represent molecules. The circles, themselves, represented atoms and, when there was more than one atom in a molecule, the positions of the circles relative to the other circles showed the spatial relations Dalton though might obtain between the atoms, or the geometry of the molecule. The touching of the circumferences of the circles in these diagrams represented the affinities between the atoms.

Dalton's diagrams were revolutionary in his time because of the direct spatial correspondence between the relative positions of the circles on the page and the relative positions of the atoms in the molecule represented. In this chapter, the section on Dalton's Diagrams explains Dalton's system and shows how this kind of representation differs from others, such as Berzelian chemical formulas.

The atom positions in a molecule represented in Dalton's system are all co-planar and, therefore, an account of how his molecular diagrams represent spatial relations by sharing geometries with the molecules represented is an account of the representation of two-dimensional objects (or a two-dimensional aspect of three-dimensional objects). By geometry what is meant in this chapter is the relative positions of points and not the absolute distances between them—the distances between the circles actually differ from the distances between the atoms by a staggering amount, a factor of about 10^8.

The section Representing the Third Dimension explains how the account can be extended to the representation of three-dimensional objects because, although the geometry of a three-dimensional object cannot be shared in its entirety with a planar representation, a component of it, its geometry in two dimensions, can still be shared. A further component, I show, can be represented in the same way it is on the retina of the eye in perception. My analysis of the way objects are represented by showing their geometries, then, raises the question of whether this property—sharing a geometry—is what makes a representation of an object a picture of it. In the Pictorial Signs section, an account of pictorial representation is suggested. In this account, when a representation of an object shares a geometry in two dimensions with the object, the

D. Baird et al. (eds.), Philosophy of Chemistry, 293–308.

representation is a picture of it, and when there is a third dimension to the object, it is coded in the representation—usually in the way it is coded in visual perception. There are exceptions to every rule and I note the exceptions due to deviations occurring when the object represented is either very large or very close.

My account opposes those of the semioticians who argue that pictures do not share any property with what they represent (e.g., Eco 1976) or that if they share a geometry, it is shared in a metaphorical sense (e.g., Peirce 1938). My account differs from the more recent accounts of estheticians who argue that an object and its picture resemble each other in virtue of sharing a phenomenology. For example, the two are said to be experienced as similar in shape (Peacocke 1987); or they are said to be experienced as similar in outline shape, where outline shape is defined in terms of the solid angle the marks, or object, subtend at the eye (Hopkins 2003). I suggest that a picture and the object it represents share more than a phenomenology. They share a property in two dimensions, geometry, and in the third, the property is coded in the picture, usually in the way it is coded on the retina in perception. Where earlier accounts focus on the phenomenology of pictures and suggest in virtue of what, a picture and its object are *experienced* as similar, the account given here explains in virtue of what they *are* similar.

The section on Pictures in Art explains the difficulties that arise when an account of pictorial representation is extended from a consideration of scientific objects to objects represented in art. The section on Representing Molecules returns to the representation of spatial relations between atoms showing how, in systems of molecular representation after the time of Dalton, the spatial relations between atoms were sometimes shown in ways that were, on the account given in this chapter, pictorial, as was the case with van't Hoff's representations of the geometries of carbon atoms using solid tetrahedra. Later, schematic methods of spatial representation were developed, such as the Fischer projection, which conventionalized the representation of spatial relationships.

This chapter may seem to assume, but need not (see the section on Pictures in Art), that representations of atoms refer. Arguments that they, indeed, refer include that x-ray crystallographic observations of molecules provide a kind of epistemic access that is similar to that of ordinary observation. Thus, if ordinary terms refer, atom representations and geometries can refer, too (Vollmer 2000; see also Vollmer 2003). A comparison of aspects of molecular representation besides geometry, such as the various ways in which affinities or forces between atoms can be represented, is beyond the scope of this chapter.

DALTON'S DIAGRAMS

When a system of molecular representation shows molecules as collections of representations of atoms, since there are different kinds of atoms, the presence of each of which is ascertained, in any given case, by an experimental method or methods, different signifiers for atoms are needed in the diagrams, one for each atom kind. Dalton's system assigned patterns and letters as names on a more or less arbitrary

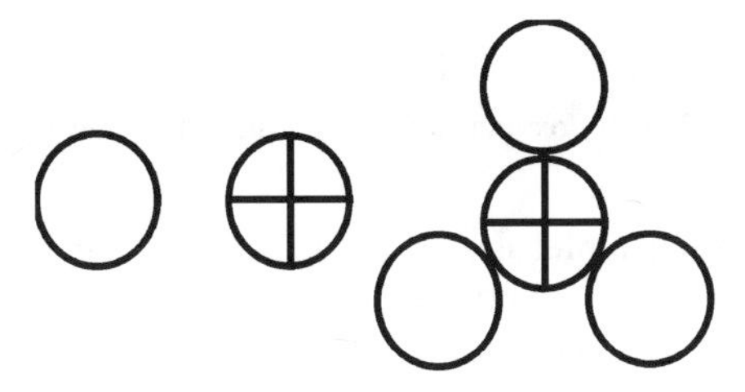

Figure 1: Dalton's symbols for an atom of oxygen, sulfur, and a molecule of sulfur trioxide, or sulfuric acid (ca. 1814).

basis. In the diagrams, he placed each signifier for an atom kind within a circle that represented an atom of that kind.

Although little was known about the relative positions of atoms in molecules in Dalton's time, he nevertheless chose to represent their relative positions; we will turn to his grounds for the particular positions he chose later, in the section Representing Molecules. He signified the positions by showing the relative locations, or geometries, of circles placed on the page. For example, the geometry of sulfur trioxide was signified by showing the relative positions of three circles placed at the vertices of an equilateral triangle, with a fourth at their center. The three circles represented oxygen and the fourth, sulfur (Figure 1).

In addition to representing the atoms in which a molecule consists, the kind to which each belongs, their affinities for other atoms, and their inter-atomic spatial relations in his diagrams, Dalton may have intended to show another property of molecules. That is, he may have intended that the circumferences of the circles represent the outer boundaries of atoms. If they did represent these boundaries, the question remains as to whether they represented the hard centers Dalton thought atoms to have or, alternatively, whether they represented the boundaries of the caloric, or heat atmospheres, of the atoms. If they represented the caloric, they did not represent its precise size. The circles were uniform in size but the caloric, on Dalton's account, differed in size according to atom kind. If each circumference, indeed, represented the outer boundary of an atom, each can be interpreted as showing a cross-section of a spherically distributed property. Alternatively, the sphere of the atom could have been represented using shaded circles to show the spheres or, less practically, actual spheres, one half extending to each side of the diagram.

Insofar as he showed them in his diagrams, Dalton's speculations about the shapes of molecules were limited to just two dimensions. Thus, the positions of the atoms of the molecules represented were, in every case, co-planar. There was, then, no need to show a third dimension, the relative positions being shown straightforwardly by the co-planarity of the circles on the page. The geometry of the positions of circles in these diagrams, then, was shared by the geometry of the positions of the atoms of the molecule represented.

In contrast to Dalton's system, Berzelius'—which predated Dalton's and is the basis for the representation of chemical compounds by their molecular formulas that

O S (S + 3 O)

Figure 2: Berzelius' symbols for oxygen, sulfur, and sulfur trioxide, or sulfuric acid (ca. 1820).

is used today—represents molecules by listing, side by side, the signifiers for atoms (Figure 2). The signifiers for the elements used in these formulas are the first letters of the Latin names for the elements. Where more than one Latin name begins with the same letter, the first letter, capitalized, was assigned to one, usually a non-metal, and the first letter followed by a lower-case letter was assigned to the other, usually a metal. The term for phosphorous, for example, was P, and that for lead, or plumbum, Pb. These signifiers differ from the ones used in earlier systems of representation; they are ordinary letters as opposed to symbols unique to chemistry. This made them, as Berzelius suggested, easier to write.

In this system, one volume of an element, later one atom, was represented by a letter or letter pair. When there was more than one atom of a given kind, the total number of atoms of each kind was shown, in the original formulas, by a number, written either preceding the term for the element or as a superscript after it, as in C^2H^5O for ethyl alcohol—at a time when the amount of hydrogen present was underestimated. Note that when Berzelius originated the formulas, the quantitative methods were far less certain than the methods for determining which atom kinds were present. A major difficulty with the former is that analytical methods involved measuring the volumes of gasses evolved and, before the time of Avogadro, there was uncertainty about whether gasses at identical temperatures and pressures all contained the same number of atoms per unit volume or whether they varied, the combustible gases, such as H_2 and O_2, containing half as many atoms per unit volume as the rest (Szabadváry 1966, 140). Not until after the acceptance of Avogadro's theories were the numbers generally reliable.

The kinds of terms in a Berzelian formula are in a one-to-one correspondence with the kinds of atoms of the substance represented. When there are no numbers in the Berzelian formula, then a different atom is signified by each term and the terms, themselves, are in a one-to-one correspondence with the atoms represented. Some have suggested that this one-to-one correspondence might make these representations pictorial, i.e., pictures of the substances represented (see Klein et al. 2003; see also Klein 2001).

However, ordinary lists are often in a one-to-one correspondence with the different parts of what is represented as, for example, in the case of a list of the states of a country or of the groceries brought home from the supermarket. Therefore, if the one-to-one correspondence displayed by a Berzelian formula were sufficient to make it a picture of the substance represented, an ordinary list (and perhaps some sentences) would be pictorial, too. But we don't usually think that an ordinary list is a picture.

Dalton's diagrams also display a one-to-one correspondence between atoms and their representations which, because the diagrams use no numbers, is always displayed. Interestingly, Dalton's diagrams seem to be unarguably pictures of the molecules they represent, whereas Berzelian formulas do not. A salient property of Dalton's diagrams, more or less absent in Berzelian formulas, is that they display the spatial relations, or geometries, of the molecules represented. Perhaps, then, showing geometry is a property necessary for a molecular diagram to be a picture. If so, to evaluate whether Berzelian formulas may be partly pictorial, we ought to ask whether Berzelian formulas can partly show geometries.

A Berzelian formula cannot show the full geometry of the atom positions of a molecule because the formula is restricted to the line on which it is written. It can, however, partially signify geometry by the positions of the element signifiers in one dimension. So, if the atoms are ordered in a formula in a way that accords, to a degree, with their spatial relations, the formula may be, on the account suggested, partly pictorial—if being a picture can be a matter of degree. Berzelian formulas are, in fact, sometimes written in multiple parts, or as lists placed side by side. In the original formulas, each of the parts showed one of the components into which the molecule represented could be transformed displaying, it has been noted, an aspect of the internal association of the substance represented (Klein et al. 2003). On the account of a picture just suggested in which showing geometry is a necessary property, if what the spatially separated lists show about the internal association of the substance is the relative spatial relations of the atoms and so their geometry—or something about their geometry—then these formulas may be partly pictorial.

REPRESENTING THE THIRD DIMENSION

As explained earlier, a diagram can straightforwardly share a property, its geometry, with the object it represents when what it represents is planar, as in the case of the relative positions of co-planar atoms.

This section shows how, in the case of three-dimensional objects as well, the geometry in two dimensions of the object can be shared with the geometry of its representation. Of course, the geometry of the third dimension of the object, corresponding to a direction perpendicular to the plane of the representation, cannot be shared, too—because the representation has no third dimension. This means that, while the first two dimensions of the object can be represented in the plane of the representation through a direct spatial correspondence, a device is needed to show the geometry of the third dimension of the object, i.e., to show some parts of the object *as if* they were closer to, and other parts farther from, the observer. For example, in an ordinary photograph, the representation of the third dimension is coded by its collapse onto the plane of the representation in the same way it is on the retina of the eye in perception.

In representations that are formed by ordinary lenses, what is displayed is the geometry of the varying reflectance of light usually at the surface of the object. To evaluate such images, consider, first, an object that is approximately planar. When the object is placed parallel to a lens, an image of the object forms from light scattered from each point (x, y) of the object, where x and y range across the object. The image forms as a consequence of changes in the direction of light caused by the lens, and the light scattered by each point (x, y) of the object converges onto a corresponding point $(-x, -y)$ in the image plane where x and y range across the image. There is, then, a one-to-one correspondence between the points of the object that scatter light and the points at which the light converges in the image. This correspondence explains the similarity between a planar object parallel to a lens and its image.

Now, replace the planar object with a three-dimensional one, or with an object that varies in the third dimension, z. Light scattered from each point (x, y, z) of the object now converges onto a corresponding point $(-x, -y)$ in the image plane and the points of scattering on the object are no longer in a one-to-one correspondence with points in the image. The z dimension, then, is collapsed, or projected onto the x, y plane.

This explains how, in photographs and the image on the retina, the z dimension of each point at which light is scattered from an object is lost in its image except insofar as it is suggested by depth cueing. The image, then, is referred to as a projection of the object or a perspective projection. It is a projection of the object as seen from a certain point, the center of projection, onto a plane. Information about the third dimension, which is retained as depth cueing, makes it possible for a perceiver like us to decode the otherwise degenerate information of the image, in the same way we decode it in visual perception, to see the geometry of the third dimension of the object that caused the image.

That two dimensions of a representation are shared with the object and the third collapsed onto the plane of the representation helps to explain how an object, such as a dinner plate, has a perspective projection that varies between being circular and elliptical as it is viewed from different angles. When the plane of the plate is parallel to the plane of the lens and the projection circular, the two dimensions of the plate sharing a geometry with the projection are described by two axes in the plane of the plate. When the plate is viewed from another angle and the projection is, for example, elliptical, the geometry of the plate along another pair of axes is shared with the geometry of the projection.

An important consequence of the collapse of the representation of the third dimension onto the plane of its image in perspective projection is that an image formed from light scattered off an image of an object can be approximately the same as the image formed from light scattered off the object, itself. This explains why the patterns of light the retina receives from a photograph or other perspective projection can be nearly identical with the patterns it receives from the object itself. It also explains why a picture of a picture can look the same as the original picture and so why, when one looks at a Wheaties box on which there is an image of someone holding a Wheaties box and on that box a similar image, and so on, the real person holding the box and each of the images can look more or less the same in all respects except size.

One might object that the experience one has of the box when seen with two eyes differs from the experience of the box when it is seen with one; however, the experience of the image does not. Indeed, in seeing an object with two eyes, the two images, one seen by each eye, differ slightly—by about 3°—in their perspectives on the object. Consequently, each carries slightly different information about the geometry of the object and together they give an experience of depth perception that is over and above that due to the depth cueing already discussed. In seeing an image, however, because the third dimension of the object is collapsed in the image, nearly all the information that can be received about the object is received, already, with just one eye and no additional information is added by the second. Therefore, when, in this chapter, it is said that an image of an object can cause an experience that is identical with the one caused by the object, what is meant is that the experience had in seeing an object with one eye can cause an experience that is identical with that had in seeing its image.

The experience of seeing an object sometimes differs from that of seeing its image for another reason. The fine structure of the object's surface is sometimes only approximated by the technique used as, for example, in the case of a relatively coarse-grained photograph, or a painting with visible brushstrokes or small points of color. This approximation places a limit on the sameness of the experience caused by seeing the object and seeing its representation. Regardless of how much the seeing experiences can differ as a result, the way the object is usually represented is, nevertheless, by showing its geometry.

As mentioned earlier, when an object is represented, what is represented is, more specifically, the geometry of the varying reflectance of light. This can be true even for extraordinarily small objects, such as molecules, when the varying reflectance, not of ordinary light, but of electromagnetic radiation of an extraordinarily small wavelength, or x-rays, is represented. The varying reflectance, then, is displayed in the image as it would appear if it had been focused by a lens—even though there is no lens that can focus x-rays. Occasionally, the geometry of an object is known in a way that does not involve the reflection of electromagnetic radiation at all but, instead, some other property, such as the electrical conductance that is measured in Scanning Tunneling Microscopy when a sharp tip approaches an atom or other object (at a distance of ca. 1 nm). From the coordinates at which the electrical conductance occurs, representations of the objects are constructed. In these representations, the geometries of the objects are often displayed *as if* the observed property were a reflectance property that had been focused by a lens, i.e., the geometry in two dimensions of each object is shared with its image and the third dimension is coded more or less in the way it is in perception.

One might object that a perspective projection of an ordinary object is not a representation of the object's geometry for what is shown is only a small part of its geometry. This is due to the fact that objects are rarely transparent. When they are transparent, as in the case of a jellyfish, we can, indeed, see the entire envelope of the object's surface from just one viewpoint, or nearly so. The geometry in two dimensions of the representation of the object, then, is the geometry of the entire object, more or less. In the usual case, though, what is represented is just the geometry of the part of the object that is visible from the point from which the object is viewed. So, when this chapter says that the geometry in two dimensions of an object is shared with its representation, what is meant is that the geometry of the part of the object that is visible is shared in two dimensions with its representation. Since there are many different points from which a representation can be made, each one more often than not showing a different geometry, there are many different geometries, each of which can represent the object by sharing a geometry with it.

There are exceptions to the rule that an object shares its geometry in two dimensions with its perspective projection. The exceptions arise because the general rule that the geometry of an object in two dimensions is the same as the geometry of a photograph or an image on the retina is, in fact, only approximate. This is because, as parallel lines recede from the observer, at infinite vanishing points they converge and the geometry becomes distorted. The general rule is a good approximation under many, if not most, conditions. It is a good approximation when the object is small enough relative to the "visual" field, or far enough away, that it subtends a limited part of that field. An object subtending $\leq 30°$ is said to sustain no distortion in its geometry (Earle 1989). It is a good approximation, also, when the object does not, like a pair of railroad tracks, wind off into the distance and, therefore, no one part of the object is represented as being closer by far to the vanishing point than any other part.

PICTORIAL SIGNS

Signs that represent objects are sometimes linguistic. "Triangle," for example, represents triangles or triangularity linguistically. This English word, like most words, has no particular properties in common with the objects it represents. Therefore, for its meaning, the word relies entirely on convention. Exceptions to this rule include the pictograms of Chinese that have the shapes of the objects they represent, and the words of the ancient language Sanskrit that, it is said, have the sounds of the objects they represent.

Pictures represent objects in a way, many think, which is not entirely by convention. Interestingly, the properties in virtue of which pictures represent objects is a matter on which there is little agreement. On some accounts, a picture and what it is a picture of have no properties in common (e.g., Eco 1976). On others, it is unclear whether they have any properties in common, and if they do it is unclear whether it is in virtue of these common properties that the representation is a picture (e.g., Goodman 1976).

On more recent accounts, the suggestion is made that what an object and its picture as presented in the visual field have in common is that they are *experienced as similar in shape* (Peacocke 1987; see also Lopes 1996); alternatively, what they have in common is the property of being *experienced as similar in outline* shape, where outline shape is defined in terms of the solid angle the marks, or object, subtend at the eye (Hopkins 2003; see also Hopkins 1998a and 1998b). These accounts, then, suggest that it is in virtue of the way it is *experienced* that a picture resembles the objects it represents. The account offered, in this chapter, differs from these accounts in that it explains that an object and its representation have more in common than a phenomenology. They share certain aspects of their geometry and it is in virtue of these shared aspects that they are experienced as similar by a perceiver like us.

A representation of a triangle that shows its sides and angles and the relationships between them is, like Dalton's diagrams, unarguably pictorial. The picture of the triangle shows the relationships that the sides and angles of the triangle bear to each other; the sides (or relative lengths of the sides) and the angles in the picture are the same as those of the triangle. Geometry, then, is a property that the picture of the triangle has in common with the triangle, i.e., the triangle and its picture share a geometry.

Geometry is a property that two-dimensional objects (or a two-dimensional aspect of three-dimensional ones), then, would seem to share with their pictures. Other examples of representations in science that share a geometry with what they represent are drawings of constellations, each of which shares a geometry with the constellation it represents (but not with the stars) and drawings of sine waves that share their geometries with the two-dimensional waves, or two-dimensional aspects of the three-dimensional waves, they represent. In each case, the fact that the representation has the same geometry as its object or objects, causes the representation to look like or cause an experience that is similar to the experience that would be caused in seeing the object itself—if the object, or just that aspect of the object, could be seen. It is plausible not only that, like drawings of triangles, these representations are pictures but also that they are pictures in virtue of showing geometries, or that showing the geometry of the object represented is sufficient for a representation to be a picture. If, as suggested in the section on Dalton's Diagrams, a representation of a two-dimensional object (or a two-dimensional aspect of a three-dimensional object) that does *not* to show a geometry is, therefore, not a picture of the object, then sharing a geometry with the object represented may be a necessary and sufficient condition for a diagram of a two-dimensional object to be a picture of it.

How, then, might one extend this account of how pictures represent to three dimensions? That is, what can one say about a picture when what it represents is, as in the usual case, three dimensions of a three-dimensional object? As shown in the section on Representing the Third Dimension, when a three-dimensional object is represented, two dimensions of its geometry can still be shared with the geometry of its representation. The geometry in the x, y plane parallel to the plane of a representation can be shared with the object, and the representation of the third dimension can be coded, usually the way it is in perception. Extending from the two-dimensional case, then, such a representation is, also, a picture of the object.

These two components—the two dimensions that share a geometry with two dimensions of the object and the third that is collapsed and represented, for example, by depth cueing—jointly determine the way the object is represented in a picture and so the way it is experienced when one looks at the picture. Returning to the dinner plate example, when the plane of the plate is parallel to the plane of the retina, the two dimensions of the plate that share a geometry with its projection are along the two axes in the plane of the plate. Each z coordinate of the outer ring of the circumference, then, is more or less identical to every other z coordinate of the circumference. Therefore, its outline shape or the solid angle subtended by the plate's circumference is circular and the geometry of the circumference is, more or less, shared in all three dimensions with the geometry in the plane of its picture, or its projected geometry. When the plate is, instead, viewed from another angle and the circumference of the plate is, for example, oblique to the plane of the lens, the outline shape and solid angle subtended are elliptical. Because the outline shape of the plate is caused by points on the plate that lie along axes oblique to the plane of the plate, they are caused by points on the plate that have z coordinates which are not identical. Therefore, the elliptical geometry in the picture is not the same as the geometry in all three dimensions of any part of the plate. It is a geometry that is the same in only two of them, the third being coded in the way it is in perception.

That pictures share a geometry in two dimensions with what they represent—and if the object is three dimensional, represent the third dimension, usually in the way it is represented on the retina of the eye—means that, on this account, pictures are in part natural. Therefore, for their meanings, they do not rely entirely on convention or cultural norms. Rather, they have a property in common with what they represent—the geometry of the first component. When the second component is, indeed, represented in the way it is on the retina, then the second component is represented in a way that, for a perceiver like us, is natural, too. It is in virtue of these two components that a picture and the object pictured, when seen by a perceiver like us, are experienced as similar.

PICTURES IN ART

Representations of scientific objects that show geometries generally show the geometries of existing objects. When representations are of classes that vary from individual to individual, such as morphological drawings of plant species, the geometries may not be the geometries of any existing individuals. They may, instead, be composites of different objects or composites of the same one at different stages in its life cycle or different seasons of the year. Many different factors, then, affect a single morphological drawing. That many different factors affect a drawing may be what is meant by the statement that morphological images are at a cross-section between an object and the pre-history of the representation of the object (Bredekamp 2000). If this statement were intended to mean that, to understand an illustration one must understand all the different factors leading up to the illustration's being drawn then it would be an overstatement. Such representations, showing complex objects, need not show them in a complex way. Rather, such drawings represent species by representing the geometries of individuals or parts of individuals. Therefore, to understand the drawings, besides needing to know some conventions, for example, those associated with the representation of seasonal and life cycle variation, the viewer needs for the most part to be able to recognize geometries. To do that, he or she does not need to know what led up to an illustration's being drawn; what the geometries signify about a species can be understood by just looking.

Representations in art are sometimes composites, as well. Perhaps the simplest kind is the composite representations that is composite in virtue of the arrangement of the objects, as when ordinary objects or buildings are, as was commonly done in the Renaissance, represented in a novel, more orderly way (Elkins 1994, 35, 126).

A more complex kind of composite in art is composite in virtue of the way it represents the spatial relations of different parts of a scene. When a representation shows not just one localized object but represents an entire scene, the scene can be shown from many different vantage points. In this kind of composite, there is no single center of projection but rather many, each slightly different from the others (see Tyler). The picture appears *as if* seen by someone at multiple places in the world at once, one for each vantage point as, for example, in Ucello's *Deluge* (Elkins 1994, 56, 131). Multiple viewpoint perspectives differ from the single viewpoint perspectives, then, in the way they represent spatial relations between the different parts of a scene. Single viewpoint perspectives, first introduced in the 1400s, were extremely popular in that century; the vantage point used was, in many cases, an inaccessible point or, as can be seen in Alberti's work, a point in the middle of the picture (Elkins 1994, 143, 145; see also Tyler).

Maps are also sometimes composites of multiple perspectives, for example, as in the case of maps of cities that represent the major civic or historical buildings, each from a vantage point that best illustrates that building, such as a point directly in front of it. The relative locations of the buildings can be shown, then, by showing the individual perspective projections of the buildings on a single-perspective projection of the city streets. There is, then, no single vantage point for the entire picture.

The composites in art discussed so far are fictional with respect to the spatial relations between the different objects represented or between the different parts of a scene. Cubist figures, such as Picasso's, that show human forms *as if* from multiple perspectives, are fictional in another respect. They show *single* objects in a way that incorporates the different outlines each object would have from a variety of different vantage points. The resulting, represented, object can be seen as an object existing in a strange, fictional, space or, alternatively, as an object seen by a fictional creature, from more than one perspective at once.

In cases in which what is represented are objects that have never existed, such as winged humans, not only just the spaces, but also the objects represented, are fictional. Pictures that show composite spaces or objects, or completely fictional ones, might seem to pose a problem for an account on which pictures represent objects by representing their geometries. That is, if no existing object or space is represented then there may, in fact, be no geometry that is represented either, and so no geometry that can be shared with its picture.

One way to avoid this difficulty is to explicitly accept the intuition that, when we think about fictional objects or spaces, our ideas about them have an "as if seeing" aspect to them. This means that the geometries, or the intended geometries, of fictional objects and spaces are implicit in our ideas about them. Admittedly imprecise, this notion is, nevertheless, intuitive. If we appeal to it, we can say that our ideas of fictional objects and spaces have intended geometries and, therefore, a fictional object or space can share its intended geometry with a picture of it.

Finally, it is often assumed that representations not just of fictional objects and spaces, but of ordinary ones, too, depend heavily on cultural norms and, therefore, cannot really be represented in as straightforward a way as I have suggested—by sharing geometries. There are two reasons this assumption is made. First, the techniques used to represent an object can vary greatly. For example, as suggested earlier, the fine detail of a picture can vary from being as fine-grained as a photograph to having visible brushstrokes, to consisting of small points of color. Furthermore, the way the third dimension is represented can also vary, for example, from being shown as in a photograph to having the exaggerated flatness of a medieval stained glass or a Japanese print. Because such differences in techniques of representation are commonplace, the fact that a significant element remains constant across the differences is commonly overlooked.

The second reason it is generally assumed that pictures cannot, really, represent in a way that is as straightforward as I suggest is that objects have an all-pervasive metaphorical content. This chapter does not deny that a consequence of this metaphorical content is that meanings, including the meanings of pictures, differ greatly from culture to culture (see Kuhn 1962). Nevertheless, the metaphorical content, for example, of a picture of a winged female, depends on a more basic representational technique, i.e., the representation of the object through a figure in the plane of the paper that refers to the object by sharing a geometry with it in two dimensions and representing its geometry in the third. Therefore, if a winged female represents victory, or great depths represent eternities metaphorically, these metaphorical interpretations rely on the capacity of pictures to share geometries with what they represent.

REPRESENTING MOLECULES

Dalton originated his method for representing the spatial relations within molecules at a time when the subject was highly speculative. The principle that an atom repels others of the same kind and therefore takes up a geometry, or "station," in which it is as distant as possible from the others was the grounds for Dalton's assignment of geometries. It was on the basis of this principle that he represented, for example, the three oxygens of the planar sulfur trioxide at the three apices of an equilateral triangle, as shown in Figure 1.

When chemists came to learn more about the arrangements of the atoms of molecules in space, they came to need ways to display these geometries in three dimensions. In response to this need, chemists made use of perspective projections. van't Hoff, for example, developed a system for representing the geometry of carbon centers through displaying perspective projections of solid tetrahedra (see Ramberg and Somsen 2001; see also Hudson 1992). In these diagrams, the center of each tetrahedron signifies the position of a central carbon atom, and the tetrahedron's free corners signify the positions of the atom's nearest neighbors. Unless labeled by a letter, each atom is, perhaps surprisingly, signified exclusively by its position—at the center or corner of a tetrahedron. In Figure 3, which displays two tetrahedra sharing a corner, the geometry of the two adjacent tetrahedral carbon centers of ethane are shown by their implied positions at the centers of the tetrahedra; the positions of the six hydrogens are signified by the six free corners of the tetrahedra. Tetrahedra are used to show the geometry of trigonal carbon centers as well, as in ethylene, by displaying two tetrahedra sharing an edge; the positions of the two trigonal carbons are implied by the two centers of the tetrahedra and the four adjacent atoms are signified by their four free corners. Linear carbon centers, as in acetylene, are shown using two face-sharing tetrahedra; the

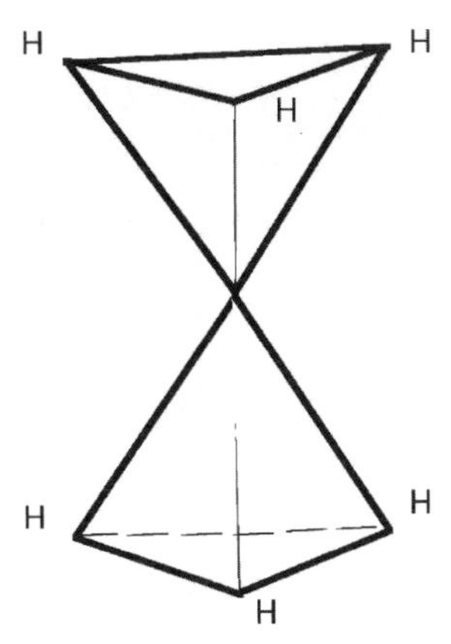

Figure 3: van't Hoff's symbol for ethane (ca. 1880).

positions of the linear carbons, again, are implied by the centers of the tetrahedra, and the two adjacent atom positions are signified by the two free corners of the joined tetrahedra.

As a general rule, early perspective projections, such as van't Hoff's, were later schematized, i.e., they were replaced by systems that conventionalized spatial representation. This was partly due to the difficulty of drawing perspective projections and partly due to the fact that in them some atoms obscure others. It was also because of the conceptual simplicity that could be achieved by conventionalizing spatial representation. The Fischer projection of a molecule, for example, represents the atom positions of its main chain vertically and its side groups horizontally using lines between nearest neighbors to show connectivity. The spatial conventions of this system include that the nearest neighbors of a carbon atom, when shown to the right or left of it, are positioned in space slightly closer to the observer than is the central carbon, and those shown vertically are positioned slightly farther away. This convention makes it possible for atom centers to be drawn in vertical columns and horizontal rows without obscuring atoms or bonds. In this system, as illustrated in the top of Figure 4, atom representations that are not close to each other on the page can, nevertheless, represent nearest neighbors if connected by a line. Perspective projection in these diagrams, then, was given up in exchange for ease of drawing, as well as for atom representations that do not obscure others. Figure 4 also suggests the conceptual simplicity that can be achieved by the fact that glucose in the straight chain form appears nearly identical with the cyclic form, to which it can be very easily converted.

A consequence of the conventionalization of the representation of spatial relations is that the geometries of the diagrams are sometimes no longer shared with two dimensions of the molecule and the method of coding the third dimension may depart from the way it is coded in perception. Therefore, unlike perspective projections, conventionalized representations require an explanation of the conventions used for showing spatial relations.

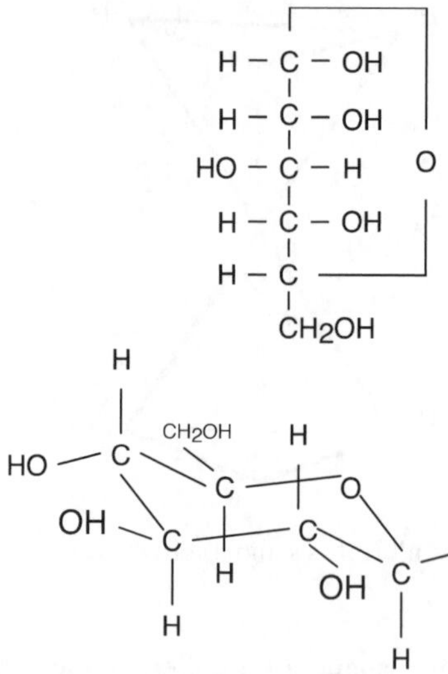

Figure 4: Fischer representation of D-(+)-Glucose, and an approximate perspective projection of the same molecule.

SUMMARY

Representation of the spatial relations between atoms by showing their geometries was introduced systematically into chemical representation by Dalton when he showed the spatial relations between atoms in a molecule by showing the spatial relations between the circles that represented them. The atoms of the molecules Dalton represented were co-planar, and thus the positions of circles in the plane of the paper showed the relative positions of the atoms they represented. In this system, the atoms and their representations share a geometry in two dimensions. When speculations about the three-dimensional shapes of molecules became commonplace, a method was needed to show the third dimension. Systems in which the diagrams continued to share a geometry in two dimensions with the atoms, and that also represented the third dimension, were introduced into chemical representation, such as van't Hoff's diagrams of the geometries of carbon centers using projections of solid tetrahedra.

In order to show all the atoms of a molecule clearly and in a way that can easily be drawn, alternatives, such as Fischer projections, were developed that conventionalized spatial relations. Because the spatial relations were conventionalized, reading them in these diagrams requires knowing the spatial conventions of the system.

The aspect of representation that was introduced by Dalton into chemical representation—showing the spatial relations between atoms by sharing those relations—this chapter suggests, made Dalton's molecular diagrams pictorial. What makes a representation of a three-dimensional object pictorial, this chapter also suggests, is that the object shares its geometry in two dimensions with its picture and the third dimension is represented, usually in the way it is in perception. On this account, representation by showing geometries in this way is what makes pictures in part natural. It is also what causes a picture and the object pictured, when seen by a perceiver like us, to be experienced as similar, and may be what distinguishes pictures from representations that are linguistic or conventional in other ways.

ACKNOWLEDGMENTS

Thanks to Peter Ramberg, Eric Francoeur, and Ursula Klein for many stimulating discussions of molecular representation. Thanks to Harold Kincaid and George Graham for encouragement and comments during the later stages of this work. Thanks to Nancy Hall for her advice and comments at all stages of this work. Thanks also to Ursula Klein and Max Planck Institute for the History of Science in Berlin for financial support during the early stages of this work and to the University of Alabama at Birmingham for support in the later ones.

REFERENCES

Bredekamp, H. 2000. *What's in a Line?* Conference at Max Planck Institute for the History of Science, Berlin.

Earle, J.H. 1989. *Graphics for Engineers*. Reading, Massachusetts: Addison-Wesley Publishing Company, 390–395.

Eco, U. 1976. *A Theory of Semiotics*. Bloomington, Indiana: Indiana University Press, 76–301.

Elkins, J. 1994. *The Poetics of Perspective*. Ithaca: Cornell University Press.

Goodman, N. 1976. *Languages of Art; An Approach to a Theory of Symbols*. Indianapolis, Indiana: Hackett Publishing Company.

Hopkins, R. 1998a. Depiction. In: Craig, E. (ed.), *Routledge Encyclopedia of Philosophy*. New York: Routledge, 892–896.

Hopkins, R. 1998b. *Picture, Image and Experience; A Philosophical Inquiry*. Cambridge: Cambridge University Press, 77–93.

Hopkins, R. 2003. What Makes Representational Painting Truly Visual? *Aristotelian Society* Supp 77: 151–167.

Hudson, J. 1992. *The History of Chemistry*. New York: Chapman and Hall.

Klein, U. 2001. Berzelian Formulas as Paper Tools in Early Nineteenth-Century Chemistry. *Foundations of Chemistry* 3: 7–32.

Klein, U., Lenoir, T., and Gumbrecht, H.U. 2003. *Experiments, Models, Paper Tools; Cultures of Organic Chemistry in the Nineteenth Century*. Stanford: Stanford University Press.

Kuhn, T. 1962. *The Structure of Scientific Revolutions*. Chicago: University of Chicago Press.

Lopes, D. 1996. *Understanding Pictures*. New York: Clarendon Press.

Peacocke, C. 1987. Depiction. *Philosophical Review* 96: 383–410.

Peirce, C.S. 1931–1958. In: Hartshorne, C. and Weiss, P. (eds.), *Collected Papers of Charles Sanders Peirce*. Cambridge: Harvard University Press.

Ramberg, P.J. and Somsen, G.J. 2001. The Young J.H. van't Hoff: The Background to the Publication of his 1874 Pamphlet on the Tetrahedral Carbon Atom, Together with a New English Translation. *Annals of Science* 58: 51–74.

Szabadváry, F. 1966. *History of Analytical Chemistry*, Translated by Svehla G. Oxford, New York: Pergamon Press.

Tyler, C. http://www.ski.org/CWTyler_lab/CWTyler/Art%20Investigations/
Vollmer, S. 2000. Two Kinds of Observation: Why van Fraassen was Right to Make a Distinction, But Made the Wrong One. *Philosophy of Science* 67: 355–365.
Vollmer, S. 2003. The Philosophy of Chemistry Reformulating Itself: Nalini Bhushan and Stuart Rosenfeld's of Minds and Molecules: New Philosophical Perspectives on Chemistry. *Philosophy of Science* 70: 383–390.

CHAPTER 16

VISUALIZING INSTRUMENTAL TECHNIQUES OF SURFACE CHEMISTRY

DANIEL ROTHBART

Department of Philosophy, George Mason University, Fairfax, Virginia

JOHN SCHREIFELS

Department of Chemistry and Biochemistry, George Mason University, Fairfax, Virginia

INTRODUCTION

What is chemical substance? According to the eminent 20th century chemist Friedrich Paneth, "chemical substance" is defined by the chemical elements of the periodic table, and elements are causally responsible for certain kinds of empirical attributes (Paneth 1962, 151). Based on this notion of substance, the instrumental techniques used by chemists to analyze substances are quasi-transparent, offering researchers efficient means for detecting the true properties of compounds. Such techniques are presumably separable, in principle, from the ontology of chemical substance. But based on the current state of analytical chemistry, the character of a chemical substance is inseparable from the methodological assumptions about instrumentation. Spectrometers are used for skillful manipulation of "pre-existing" material based on cunning interventions in the dynamic properties of a specimen. Laboratory technicians are agents who start a process, remove obstacles, and set free natural possibilities. Technologically induced manipulations are performed, various microscopic effects appear and then quickly disappear, and signals are produced. Photon streams function as manipulating probes in specimen's atomic or subatomic structure, as Ian Hacking (1983) reminds us. A specimen is poked, dissected, and disturbed through the use of instrumental techniques.

In this chapter, we argue that certain ontological commitments concerning the character of chemical substance are revealed in the design of instruments used in research. A clear historical precedent for this argument can be found in the microscopy of Robert Hooke, whose 17th century development of the compound microscope contributed profoundly to advances in modern science (Introduction). Hooke's explanation of his new optical devices includes a thought experiment that enabled researchers to visualize the causal mechanism underlying such techniques (Hooke's Design of the Compound Microscope). To underscore the centrality of thought experiments in contemporary instrumentation, we examine the developments preceding the design of

D. Baird et al. (eds.), Philosophy of Chemistry, 309–324.

scanning tunneling microscopes (Reading Between the Lines). We then explore the notion of a chemical specimen (The Design of Scanning Tunneling Microscopes), with concluding remarks about the centrality of instrumental technology to chemistry (What is a Specimen).

HOOKE'S DESIGN OF THE COMPOUND MICROSCOPE

For the natural philosophers of the 17th century, experimental devices were designed to extend the senses to otherwise inaccessible regions and to raise exactness of perceptions. Telescopes magnified the eye's natural powers, the ear trumpet improved hearing, and tactile senses are simulated in the barometer and thermometer. By the mid-1600s, instrument-makers understood how to enhance the magnifying power of a microscope by combining lenses. In one of the monumental advances of modern science, the optics of the compound microscope was revealed. Many scholars credit this discovery to a Dutch spectacle-maker named Jans Janssen who, along with his son Zacharias, stumbled upon the technique quite by accident, possibly from the chance act of holding two lenses together [The issue of historical originality is not settled (Bradbury 1967, 21–22)].

Before 1650, telescopes were made of wood, brass, lenses, and occasionally ivory and leather for decoration. The first use of lenses for telescopic vision was probably developed by Janssen's work in 1608, and was immediately put to military use (Charleston and Angus-Butterworth 1957, 230–231). Learning of this Dutch invention, Galileo converted it to study planets and stars, leading to great excitement with the discovery of Saturn's rings, Jupiter's satellites, spots on the Sun, and mountains on the Moon. Galileo is also known to have converted the telescope, with its concave lens, into a microscope (Brown 1985). Marcello Marpighi, one of the most original microscopists of the era, performed extensive examinations of the human lungs and the circulation of blood with this new device. The new philosophical instruments remedied the "infirmities" of the senses.

But how could the new optical instruments improve on God-given capacities of sensory perception? In their search for truths about the material world, natural philosophers of the 17th century were increasingly drawn to the shops of machinists who produced telescopes and microscopes. More than ever before, natural philosophers identified their professional status by the skills associated with manipulating and improving such devices, such as metal-working, machining, wood-working, glass-working, and tube-making.

Robert Hooke was the first natural philosopher to systematically record experimental discoveries using compound microscopes, stimulating tremendous enthusiasm throughout Europe. He exhibited considerable talent in the craft of lens-grinding, which led to his design of instruments for experimental inquiry. His diary records almost daily visits to manufacturers for the purpose of refining lenses (Price 1957, 630). He discovered a new kind of glass, which he used to construct new telescopes and microscopes. With their greater magnifying power, these new telescopes were able to disclose a visible world of heavens and stars, including the possible discovery of living creatures on the Moon or other planets.

But Hooke's true masterpiece came in microscopy. In one of the great works of science literature, *Micrographia: or some physiological description of minute bodies made by magnifying glasses with observations and enquiries thereon* (1961), Hooke provides detailed illustrations of microscopic discoveries. He displays the edge of a razor, the points of a needle, and moss growing on leather. He placed insects such as louse, gnat, and fly under the microscope. Indeed, *Micrographia* includes a 16-inch diagram of an insect (210). Hooke's rendering of the surface of a cork revealed small pores, or cells, and his description may be the first known use of the term "cell" in biology (55). So impressed were they by his demonstrations of the microscope's magnifying powers, The Royal Society solicited Hooke in 1663 to display at least one microscopic observation at every meeting (Bradbury 1967, 39).

In their widely heralded book, *Leviathan and the Air-Pump: Hobbes, Boyle, and the Experimental Life*, Shapin and Schaffer reduce Hooke's explanation of the compound microscope to empiricist standards for knowledge. They argue that, for Hooke, the power of new scientific instruments rests on their capacity to enlarge the dominion of the senses, revealing spectacular scenes to those skilled in instrumental techniques. According to these scholars, Hooke implores experimenters to use such devices to overcome the "infirmities" of the senses, and to enlarge the range of empirical attributes accessible to experimenters (Shapin and Schaffer 1985, 36–37).

Hooke's experimentalism, however, is not amenable to empiricist standards for knowledge. Conspicuously absent from the interpretation by Shapin and Schaffer is a reference to Hooke's mechanistic explanation of instrumental techniques, requiring commitment to unobservable mechanistic powers.

Several themes can be gleaned from Hooke's notion of experimental inquiry. First, Hooke views scientific instruments as an extension of causal mechanisms found in nature. One of the widely held fictions about modern science is that the new optical instruments of the 17th century were passive devices that provided a quasi-transparent window to an empirical landscape. Based on this view, unobservable attributes must remain inaccessible to experimenters even with the use of instruments. But Hooke explicitly endorsed the use of compound microscopes for revealing the workings of otherwise undetectable (unobservable) machines of nature. To Hooke, the natural world is a series of smaller machines embedded in larger ones. Motions of machines at one level are produced by machines at a deeper level, comprising the wheels, engines, and springs of the microworld. Each material body is endowed with powers that are causally responsible for detectable effects. Every instrument is an agent for change, that is, a source of movement that is responsible for certain occurrences. The power of a compound microscope to reveal truths rests on the designer's exploitation of causal mechanisms of nature.[1] In particular, the reliability of any microscope rests on the causal processes that are responsible for the images appearing at the eyepiece. The skilled use of a compound microscope can help answer the following analytical questions: what are the causal agents responsible for the fact that some fluids readily unite with, or dissolve within, others? How can we explain fermentation by yeast? How can the infection of one man lead to the destruction of thousands of others? Discovery of nature's causal mechanisms must be included in the mission of experimenters, even though such mechanisms are "secretly and far removed from detection by the senses" (Hooke 1961, 47).

Second, conjectures about any natural body can be drawn from analogical associations to the workings of artificial bodies. Natural philosophers should look to machinists for insight into the workings of the universe. Although mechanics is a practical enterprise, partly physical and partly mathematical, Hooke drew philosophical inspiration about the truths of nature from the discoveries of machinists. Even human diseases, he believed, could be diagnosed by discovering "what Instrument or Engine is out of order" (1961, 39). Natural philosophers should become acquainted with the machinists' techniques of hammering, pressing, pounding, grinding, cutting sawing, filing, soaking, dissolving, heating, burning, freezing, and melting (1961, 60). Of course, these techniques rely on a skilled manipulation of material bodies, and typically require physically connecting bodies through construction of joints and hooks. The image of a watch offers special insight into the workings of nature, with the balance beating, wheels running, and hammers striking. By analogy, "[it] may be possible to discover the motions of the Internal Parts of bodies, whether Animal, Vegetable, or Mineral, by the sound they make" (Hooke 1961, 39).

Third, for Hooke, any machine of the natural world is characterized by its agent capacities to produce movement, and its reagent properties to be moved by other bodies. Any machine is knowable through its physical powers to manipulate, agitate, or transform another machine, and its propensity to be altered itself. Such devices exhibit a power to move or to be moved by other machines. Also, such devices have the power to unite with or repel other bodies, as evident in the capacity of any fluid to unite with or repel other fluids (Hooke 1961, 46). Hooke explains the success of the microscope through the causal mechanisms that are responsible for observable events. Any causal mechanism is a system of entities with the capacity to generate movement in, or to be moved by, other bodies. The conditions of agitation are causal events and the resulting occurrences are effects.

Fourth, to apprehend the workings of a machine, experimenters must acquire a designer's skills of visualization. For a proper experimental philosophy, Hooke writes, "the intellect should, like a skilful Architect, understand what it designs to do..." (Waller 1705, 18). Seventeenth century designers resort to artists' skills for the purpose of registering the method and history of a particular scientific investigation. Designers are trained to visualize how research will be performed. The language of graphic design is central to such skills. Through schematic illustrations, designers offer a vividness of presentation that invites vicarious participation. Through visualizations, the fleeting events of the laboratory can be distinguished from the enduring "elements" that remain stable under technological interference.

Against the backdrop of four centuries, in hindsight *Micrographia* emerges as arguably one of the great works on the philosophy of experimental inquiry. In detailed descriptions of his experiments, Hooke directs experimenters in the skilled use of instruments. A robust notion of experimental inquiry emerges from his mechanistic orientation. Each experiment is itself a kind of mechanistic system that exploits the instrument's power, the specimen's capacities, and the experimenter's skills in operating machines, both artificial and natural. Of course, optical instruments do enhance sensory capacities. But such enhancements are explained by the techniques of the mechanical arts, revealing the power of "nature's machines" at the deepest regions. Underlying the optics of vision are the machinists' manipulations of material bodies. The fundamental prototype for experimental inquiry centers on a machinist's hand and eye in the skilled manipulation of machines, rather than an observer's sensory organs in the pure perception of empirical attributes.

As shown above, several themes can be extracted from Hooke's contributions to microscopy. First, Hooke explicitly endorsed the use of compound microscopes as dynamic tools for revealing the workings of otherwise unobservable machines. Second, suppositions about any natural body can be drawn from analogical associations to the workings of artificial machines. Third, any machine of the natural world is characterized by its capacities to produce movement, and by its reagent properties. Fourth, to apprehend a machine's capacities, experimenters must acquire a designer's skills of visualization.

READING BETWEEN THE LINES

The currency of Hooke's ideas to contemporary research is revealed in his work on instrumental design. Just as Hooke insisted upon the use of visual techniques for showing a machine's operations, designers today are expected to construct visual models of a device or process. In their graphic representations of machines, engineers frequently resort to a language of lines, shapes, and pictorial images. Schematic illustrations are commonly used to convey a visual model of a machine's function, showing how a device will perform under idealized conditions [The mind's eye requires a synthesis of past experiences and a projection of future contrivances (Ferguson 1992, 42.)] In their design plans for instruments, researchers visualize the functional relationships among a machine's components. These plans provide an idealized visualization of function conveyed in pictorial media. Such media are well suited to convert a mental vision into information. Of course, visualizations are ubiquitous in the theoretical sciences. For centuries, investigators have relied on conceptual models for visualizing real-world processes. Through modeling, scientists can simulate processes of the world in an abstract realm of ideas.[2] The categories of truth and falsity have no place in defining the success or failure of a theoretical model.

Visualization of a proposed technique is a common rhetorical device of persuasion. A designer's plan is often scrutinized by fellow researchers, prospective customers, and potential critics. Designers of instruments are charged with the task of explaining their work to manufacturers, justifying their research to funding agencies, or persuading reluctant experimenters to follow new techniques. We often read that thought experiments are confined to the realm of physics (Brown 1991, 31). But the design of an instrument can be tested through thought experiments without using metals, wires, and apparatus in a laboratory. Such experiments occur in a conceptual design space where certain processes are conceived and their effects anticipated. A design space is an abstract representational space used to replicate movement, where certain conceptual elements are imagined and the effects are anticipated.

Rather than providing a media for visual copies of actual machines, a design space defines a range of possible movements of "objects," based on principles of engineering and the physical sciences. Architects, artists, and map-makers work in such spaces.

Through these plans for instruments, designers encourage readers to participate vicariously in hypothetical experiments. Readers of design-reports typically become virtual witnesses as they engage in critical assessment of designs. The reader rehearses, at least privately, the kind of malfunctions of apparatus, mistakes of implementation, and interfering effects that plagued past experiments. Are such dangers relevant to the present experiment? If so, are they avoidable? The answers to these questions require a command of the subject matter that is usually limited to expert witnesses. A reader vicariously participates in an experiment for the purpose of evaluating the initial plan of action. A reader of schematic illustrations anticipates how the device would perform under experimental conditions. A thought experiment thus takes place, inviting critique of the experimental design. As a tool of persuasion, design plans provide readers with a cognitive vision of the laboratory events in ways that recommend endorsement. Readers are often persuaded that they *could* reproduce the same processes and would get the same correspondence of concepts to perceptions (Gooding 1990, 167). Even if the reader never performs the experiment described in the report, he or she typically recalls direct participation in experiments that are similar to those described. The reader is expected to follow a narrative that selects and idealizes certain steps of a procedure (205). This narrative transports the reader from the actual to the possible by vicariously re-enacting the significant features of the experiment, focusing on the instrument's design, material apparatus, and microscopic phenomena.

Design plans are often expressed through the arrangement of shapes. To read visual information about instruments from such drawings, a literate observer must understand the language of pictorial symbols. The vocabulary of design is defined by a grammar of points, lines, and shapes. Designers of instruments use this grammar to provide a graphic language for representing an experimental technique. This representation is often achieved with a nested series of drawings of increasing depth. As a reader unfolds layers of drawings, the information revealed becomes progressively more detailed. Certain pictorial symbols serve as interrogatives about one process and invite further inspection at a deeper level. In a design plan, lines of one schematic illustration are given specificity by other drawings representing deeper processes. The visual model of an instrument invites vicarious participation in the detection of a transition from microscopic to macroscopic events. Such a sequence replicates the lawful character of embeddedness (Rothbart 2003).

An engineer employs lines to depict features of the possible movements of machinery. To an untrained observer, schematic illustrations seem static, providing a still-life vision of a bulky device. But literate viewers can apprehend movement from such diagrams. An attentive observer anticipates how our perceptions would change from the possible movement of objects. When rendered in two dimensions, the lines of an engineer's design plan lie between exposed surfaces and hidden ones, from the reader's perspective.

THE DESIGN OF SCANNING TUNNELING MICROSCOPES

To explain how a machine works, designers refer to underlying causal mechanisms. In this context, a mechanism comprises a system of entities that is an agent for change, causally responsible for producing detectable effects. Consider the causal mechanisms underlying instrumental techniques of surface chemistry.

The properties of surfaces have provoked fascination and frustration among researchers. Through major advances in chemical instruments, the old dream of observing an arrangement of atoms at the surface of a compound has been realized. Armed with these new powers of detection, researchers have produced stunning results related to the surface of silicon, for example, the distinction coils of DNA, or the biochemical process associated with blood clots. Approximately, three thousand publications each year can be directly credited to such advances.

Gerg Binnig and Heinrich Röhrer, who discovered scanning tunneling microscopy, invoked the famous curse of Wolfgang Pauli: "The surface was invented by the devil!" (Binnig and Röhrer 1985, 50) In this section, we highlight three major advances in instrumental techniques of surface chemistry: scanning electron microscopy (SEM), field ion microscopy (FIM), and scanning tunneling microscopy (STM). In particular, we show how the design-plans for scanning tunneling microscopes function as mechanistic thought experiments. A mechanistic thought experiment relies on a conceptual explanation of an instrument's performance in terms of the underlying mechanisms for change. The mechanism underlying this technique includes a microprobe, in the form of a sharpened needle, traveling across a prepared surface. This probe, or needle, sweeps across the surface of a specimen, moving up and down, left and right, forward and back, in increments of a single angstrom, to reveal a specimen's properties.

Scanning electron microscopy

Familiar to researchers in chemistry, material science, geology, and biology, SEM provides detailed information about surfaces of solids on a submicrometer scale. The technique occurs as follows: After surface atoms of a metal are bombarded with high-energy electrons, the spatial distribution of electrons is measured. To obtain an image, the surface of a solid sample is swept in a raster pattern. A raster is a scanning pattern in which an electron beam (1) sweeps across a surface in a straight line, (2) returns to its starting position, and (3) shifts downward by a standard increment (Skoog and Leary 1992, 394).

This technique solved a major problem associated with image resolution. In performing an electron microprobe analysis, the spatial resolution of the image is limited by the fact that the probing particles must be smaller than the dimension of the atom structure being observed. As an analogy, imagine trying to detect scratches on a smooth surface by running a finger across the surface. Those scratches that are much smaller than the dimension of the fingernail would go undetected, while the scratches that are larger than the width of the fingernail would easily be felt. In SEM, the problem of spatial resolution arises from the fact that a typical wavelength of visible light is 600 nm wide, and the approximate diameter of an atom is 0.2 nm. Since the wavelength of visible light is several thousand times larger than the width of an individual atom, the placement of individual atoms is unobservable. To solve this problem, designers developed a technique in which the wavelength of the probing electrons is smaller than the atomic size of 0.2 nm (Bonnel 1993, 436).

Despite this important advance, designers quickly identified three significant deficiencies with SEM. First, the images of a surface can be produced only in a vacuum, which is difficult to produce; second, the observed images may not accurately represent the original surface structure; third, electrons with high kinetic energy penetrate relatively far into the surface and are then reflected back to the detector. This process of penetration and reflection distorts the image of the top one or two atomic layers. Because of such distortion, the best result that researchers can expect is an average of many atomic layers. To overcome these deficiencies, scientists explored alternative methods for detecting atomic surface properties.

Field ion microscopy

The first technique for detecting properties of individual atoms on metallic surfaces was developed by Erwin Müller in 1936 (Chen 1993, 412). (For a history of this invention, see Drechsler 1978). This instrument, known as a field ion microscope, has a simple design comprising a vacuum system, a needle tip, and a phosphorescent screen. Müller's diagrams and discussion enable readers to visualize the technique he developed (Müller and Tsong 1969, 99) (Figure 1).

During an experiment, high voltage is applied between the filament and screen. After positioning the needle tip on a specimen, helium or neon atoms are adsorbed directly on the atoms under investigation. A potential on the order of 10,000 volts

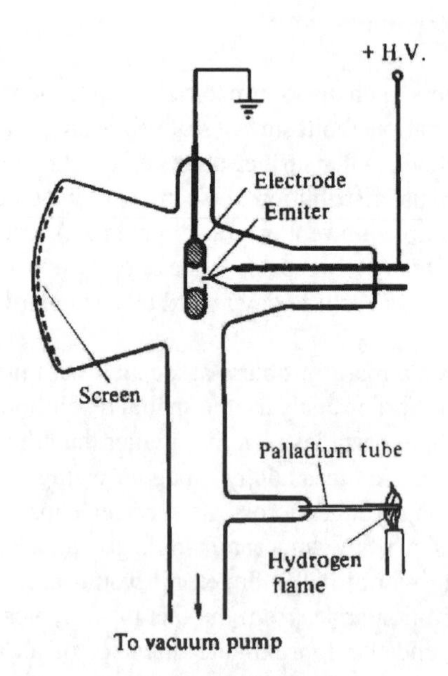

Figure 1: Original field ion microscope.

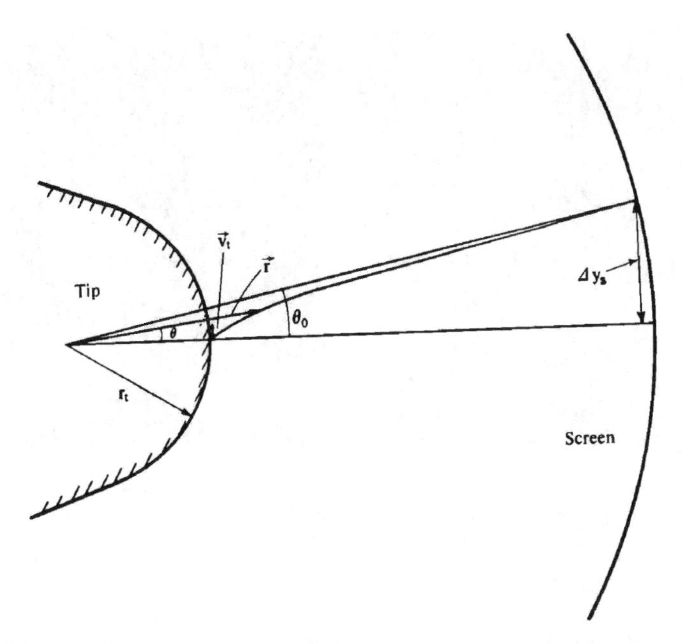

Figure 2: Representation of the path of an ion from the tip to the screen.

is then applied to the surface, causing helium or neon atoms to be released in the form of ions, and directed toward the fluorescent screen. On their way to the fluorescent screen, the distances between neighboring ions increase, causing a magnification of the distribution of atoms on the surface. This process is shown in Figure 2 (ibid.).

The atoms are shown to move farther away from one another after they hit the screen than when they lie dormant on a surface. As the helium or neon atoms hit the screen and produce an image on the surface structure, they retain their original relative positions.

To persuade the scientific community, designers of FIM constructed a physical model of a tungsten tip based on analogies to a hemispherical ball, as shown in Figure 3 (78).

By painting the kinks (protruding edges of individual planes) with a phosphorescent coating in the model, the pattern of light and darkness was observed to be similar to images obtained in FIM, as shown in the right diagram of Figure 3. The physical model replicates the pattern of images produced when xenon atoms are released from the kinks. Although this model helped to explain observations of such images, the model was impractical to construct due to the large number of atoms (balls) that were needed.

As with SEM, experimenters soon identified deficiencies in FIM. Like electron microscopy, FIM requires that images of a surface be produced in a vacuum. Additionally, the image area is limited to about 0.8 nm^2. Furthermore, low temperatures are required for operating the microscope; neon and helium will not adsorb on most surfaces except at either liquid nitrogen or liquid hydrogen temperatures.

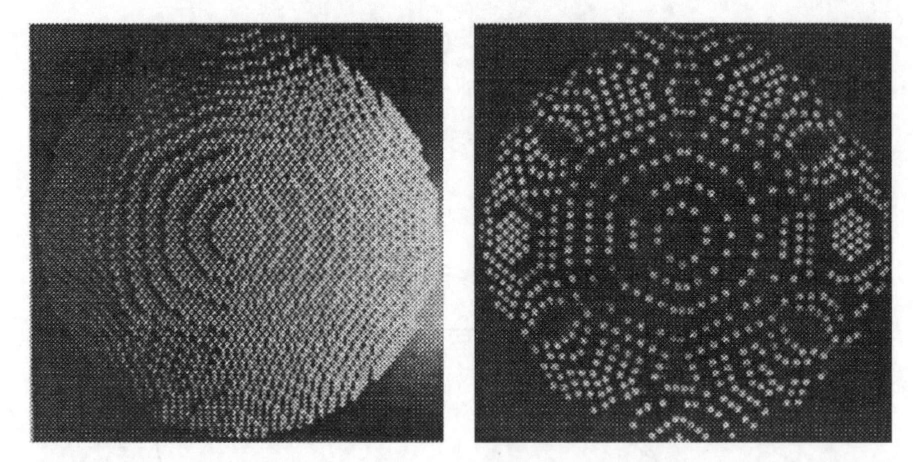

Figure 3: *Left*: Ball model for an ordered surface. *Right*: Ball model with kink site atoms painted with phosphorescent paint, then viewed in the dark.

Scanning tunneling microscopy

Binnig and Röhrer, both working for IBM in Zurich, discovered a means for producing high-resolution measurements that offered the following improvements over the topographic technique (its predecessor): (1) the distance between emitter tip and sample was decreased, (2) the required voltage between the two was lowered, (3) the emitter tip was sharpened, and (4) vibrational dampening was added (Binnig et al. 1982, 178–180). For their discovery, they won the 1986 Nobel Prize in chemistry.

In STM, as in the other techniques described, a needle rides across a prepared surface. By passing a current from the needle to the surface of the material under examination, one is able to "feel" the topography of the surface. Through a tunneling technique, the probing electrons in STM escape interaction with the surface atoms by traveling on a different route than they "normally" would, and thereby avoiding experimental interferences.

Under normal conditions, a relatively large amount of energy (a few eV) is required to remove an electron from an atom. An electron must normally acquire a certain amount of energy to free it from the influence of the surface atom. As shown in Figure 4, a minimum amount of energy is needed for the electron to surpass the so-called energy barrier.

The energy barrier is equal to the energy difference between the charge of an electron attached to the surface atom and the charge of an electron that is freed from the influence of a surface atom. If an energy barrier is surpassed, then the amount of energy that overcomes the barrier becomes attached to an atom on the tip. But this rarely occurs because the amount of energy needed to overcome the energy barrier is impractical to produce under laboratory conditions.

However, quantum mechanical processes allow an electron to travel from one atom (on the emitter tip) to another atom (on the surface) by a low-energy route. This route is made possible due to a small overlap of the wave functions of each electron (Atkins 1978, 402). From the perspective of energetics, the electron travels

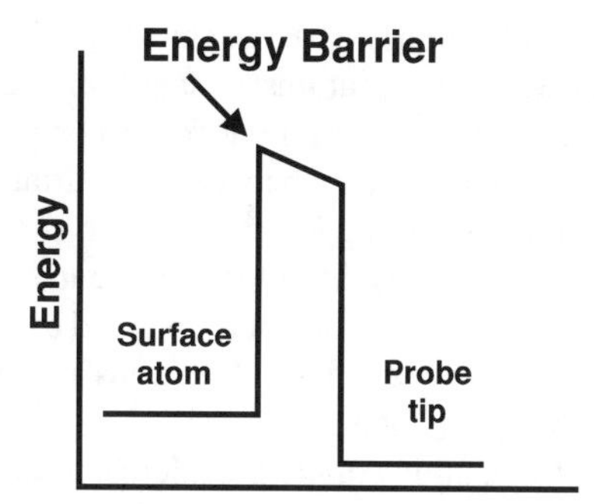

Figure 4: Energy barrier when an electron travels from a surface atom to the probe tip of an STM.

to a surface atom by tunneling *through, but not over*, the energy barrier. Tunneling occurs when electrons are freed from the potential barrier without the thermal energy that is normally required. In their 1982 patent, filed with the United States Patent Office, Binnig and Röhrer show how tunneling can occur between two metals (Binnig and Röhrer 1982, Sheet 1).

In Figure 5, the electrons of the first metal are bound in a first potential well, as indicated by (14) in the figure. A second metal is characterized by a second potential well (15), which is separated from the first potential well (14) by a potential barrier (16). For tunneling to occur, electrons must leave the first metal and enter into a vacuum when their energy (E) is raised to the value of the upper limit of the bound energy states, the so-called emission edge (17) (Binnig and Röhrer 1982, 6). Binnig and Röhrer discuss the dynamics of tunneling in the following passage:

> In an atomic system or in a solid body, if charged particles are subjected to an interaction composed of a long range repelling component and a short range attractive component, then the resulting force builds to a potential wall or a barrier. According to classical conceptions such a barrier can be crossed only by particles having energy greater than the barrier. There are nevertheless always a finite number of particles by a potential barrier which are capable of crossing the potential barrier even though they do not have sufficient energy. In a sense, they undercross the potential barrier via a tunnel. This so-called tunnel effect can be explained only by wave mechanics ... According to the tunnel effect there exists a calculable probability that a finite number of electrons bound by a potential can cross the tunnel barrier even at low voltage differences ... Some bound electrons are capable of tunneling through (barriers) (1–2).

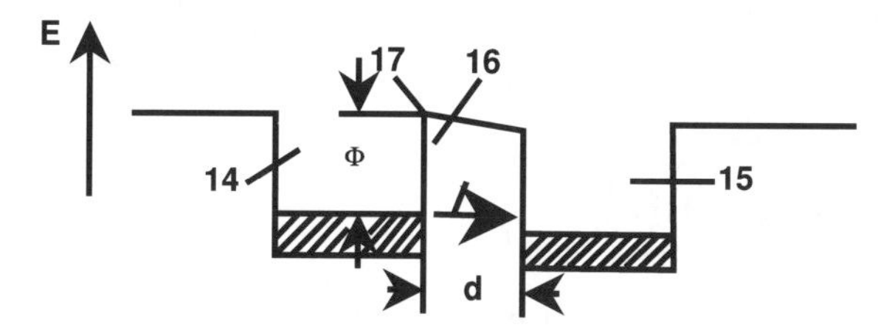

Figure 5: Electron tunneling between two metals.

Tunneling probability depends upon the barrier thickness, and occurs only when the distance between the surface and tip atoms is small and when a small voltage is applied between them. The tunneling current decreases exponentially by four orders of magnitude with respect to the distance between the surface and tip (gap) (Binnig et al. 1982, 178–180). These results suggest that the gap should be on the order of a few tenths of a nanometer, since very little tunneling current can occur with larger gaps. At larger distances, the tunneling current is obtained by producing a much larger potential, in the range of a few thousand volts. Electrons are stripped from the surface of the tip and flow to the sample (Chen 1993, 412).

Based on design-plans for scanning tunneling microscopes, tunneling is analogized to a process of boring through a material barrier with occluding edges. A material barrier is analogized to a physical impediment to electrons escaping the influence of a surface atom. Electrons are said to "undercross the potential barrier via a tunnel." The energy barrier has a surface that lies between an exposed volume and a hidden one. The thought experiment for tunneling replicates a process of moving through an occluding surface. Experimenters can visualize electrons moving through, and not over, an energy barrier as a graphic analogue to an experience of boring through a solid wall.

Figure 6 provides a schematic diagram of a scanning tunneling microscope.

Three piezoelectric ceramic pieces comprise the piezoscanner (Magonov and Whangbo 1996, 323). The voltage applied to each from the control electronics determines the exact position of the x, y, and z directions. Special circuitry is used to measure the current between the tip and the sample. Computers are deployed to manage the control signals to the electronics as well as the signal from the tunneling current.

In practical terms, a topographical experiment begins by positioning the tip properly in relation to a sample. The tip is then moved in the z-direction to adjust the sample-to-tip distance, which must not exceed a few tenths of a nanometer. A small bias voltage is then applied between the sample and the tip. As the tip approaches the surface, a tunneling current guides the final positioning in the z-direction. The

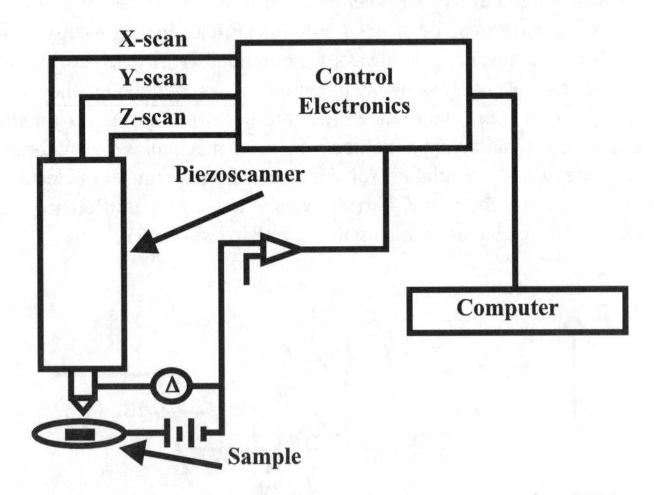

Figure 6: Block diagram of the major components of an STM.

X-scan voltage is then ramped, causing the tip to move in the x-direction but not in the y-direction.

WHAT IS A SPECIMEN?

Again, a specimen's detectable properties must be amenable to the kinds of instrumental techniques deployed during research. Analytical instruments do not operate on concealed objects directly, but rather, engage experimenters in various manipulations of a prepared substance. Even before such manipulations can be achieved, analytical chemists must determine whether a sample includes impurities. If it does, chemists attempt to remove substances that would interfere with signals from the (pure) substance analyzed. Purification comprises an array of laboratory techniques that ostensibly prepare the specimen for instrumental investigation. The deployment of these relatively mundane preparatory techniques can mask deeper commitments to fundamental categories of nature, and in so doing, can place significant constraints on the character of the substance. Purification techniques presuppose fundamental beliefs about the nature of properties and the mode of investigation.

In analytical chemistry, a sample is cleansed of extraneous elements through "sterilizing" methods. For example, a sample may be subjected to high temperatures, dissolved by hot mineral acids, or converted to an aqueous solution. In some experiments, a pure sample is one whose temperature remains constant in the course of further phase transitions. A material sample is regarded as pure if the sample exhibits properties that persist through technological interventions. To remove impurities from a sample presupposes a positive commitment to certain experimental techniques. In preparing a sample for instrumental analysis, an experimenter isolates and "puts on display" certain attributes of the compound. In contemporary spectroscopy, a pure specimen is knowable by its invariance throughout purification techniques, commonly associated with variations of thermodynamic conditions. In analytical chemistry, a specimen is dissected, manipulated physically, heated, cooled, separated, or synthesized. The invariance of a specimen throughout interventions allows experimenters to isolate chemical species as reaction products afterwards. The product of such procedures, in turn, can function as a component for further chemical investigation under new controlled conditions (Schummer 1998).

The experimenter is constantly comparing laboratory results to an idealized "picture" of the system's "true" properties. Certain elements of the experimental landscape are sought, others avoided, and still others ignored. Known similarities are made salient, dissimilarities are suppressed, and unknown relations are explored. In laboratory research, a specimen is never a finished product, never removed from the constraints associated with detection. Roughly speaking, the ontological categories that characterize a specimen are revealed in ideas associated with a new being. The notion of purity rests on an assumption that some substances are homogeneous with respect to reproducible properties.

For chemists, homogeneous substances are unchanging elements from which other more complex or compound substances are composed (Brock 1992, 174–175). A specimen is known by its capacity to produce, create, and generate detectable events under various interferences. Defined this way, the notion of a pure specimen does not refer to certain isolated, self-existing properties. A pure specimen, thus considered, is not passive; it is reactive. Purity is revealed in the reactions to interventions. The purified sample is defined by capacities that *permit* intrusion from external forces, and that *react* to such intrusion in ways that generate detectable effects. The notion of a reactive system becomes a prototype for a specimen's properties. The chemical and physical changes that occur rest on the conditions associated with a reactive system, such as the phase in which the reaction takes place, the nature of the solvent, the temperature, and the incidence of radiation (Suckling et al. 1978, 65–68).

In this context, the familiar metaphor of nature as machine retains its hold on current experimental research practices. The substance under investigation has a function, as if it were a tool in the service of the human goal of knowledge acquisition. A specimen functions as one of nature's machines, with capacities to generate movement when sufficiently agitated by mediating technologies: instruments.[3] In effect, the specimen itself becomes an experimental tool, mimicking artificial machinery: at times functioning as an agent for change, at other times as a reagent under the influence of other bodies. At all times, however, a specimen is grounded on its own capacities for change.

Nancy Cartwright's notion of a nomological machine offers guidance. For Cartwright, the physical world is a world of nomological machines. Underlying each machine are stable capacities that can be assembled and reassembled into different nomological machines (Cartwright 1999, 52). A machine's capacities give rise to the kind of regular behavior that we represent in our scientific laws (Cartwright 1999, 49). In classical mechanics, attraction, repulsion, resistance, pressure, and stress are capacities that are exercised when a machine is running properly. The machines of the 17th century were characterized by their powers to transmit, or modify, actions initiated from some other (mechanical) source. Mechanistic philosophers of the modern era found in Nature's machines capacities to transmit, or modify, action initiated from an external source. Robert Hooke's mechanistic philosophy was born from the stunning advances in the construction of lenses, magnifying glasses, and microscopes. Hooke declared that the Grand Watchmaker created the universe as a Cosmic Machine, in which small machines were embedded in large ones. In the 18th century, these mechanical, clock-like images of Nature were superseded by such technical advances as the steam engine and advances in electronics. Underlying the entities and activities of a biomolecular mechanism are productive properties that are causally responsible for the occurrence of an organism's state (Rothbart 1999). The ontological "ties that bind" causal events with effects are the capacities of particular entities that emanate from a generative mechanism (Harré and Madden 1977, 11).

CONCLUSION

Underlying STM are techniques of manipulation and production. As instrumental techniques improve, the objects under investigation become increasingly distant from experimenters: distant in space and in time, distant, in the end, because of their strangeness with respect to laws of the macroscopic world in which we live (Pomian 1998, 228). In an attempt to compensate for the widening "distance" between experimenter and specimen, experimenters presuppose that the portion of the world under investigation is endowed with the same kinds of capacities that designers attribute to their own creations. Experimenters get no closer to experimental phenomena than a sequence of analytical signals, and such a sequence should not be confused with physical states. Analytical signals are never completely cleansed of evaluations drawn from instrumental techniques. The decision to accept an experimental phenomenon as real follows a solution to an optimization problem, in efforts to maximize the signal-to-noise ratio, S/N. Such a solution requires judgment, evaluation, and criticism with respect to the goals of information retrieval (Rothbart 2000). But this assertion does not imply that such reactions are inherently contaminated, tainted, or distorted by the goal-oriented aspects of experimental design. Rather, the damaging effects of noise conflict with an experimenter's goals but are never completely absent from the scene (Coor 1968).

In their reliance on the physics of electron tunneling, Binnig and Röhrer illustrate an important aspect of instrumentation generally—the inseparability of technological design and ontological commitment. The technology is defined through models of real-world processes that in turn are knowable through technologically mediated inquiry. In this respect, a unity of minds and tools underlies the design plans for this, and other, instrumental techniques.

NOTES

1. In *The Posthumous Works of Robert Hooke*, he writes: "The Business of Philosophy is to find out a perfect Knowledge of the Nature and Properties of bodies, and of the Causes of natural productions. . . ." (Waller 1705, 3).
2. The most compelling argument on the centrality of modeling in science appears in Rom Harré's *Principles of Scientific Thinking* (1970). Morrison and Morgan (1999) have recently contributed to this subject in their "Models as Mediating Instruments," appearing in their co-edited volume *Models as Mediators: Perspectives in Natural and Social Science*.
3. A causal conception of substance advanced in the important work by Eva Zielonacka-Lis (1999), and I thank her for valuable and inspiring discussions on this topic.

REFERENCES

Atkins, P.W. 1978. *Physical Chemistry*. San Francisco: W.H. Freeman.

Binnig, G. and Röhrer, H. 1982. *United States Patent: Scanning Tunneling Microscope. August 10, 1982*. Assignee: International Business Machines Corporation, Armonk, NY. Patent Number: 4,343,993.

Binnig, G. and Röhrer, H. 1985. The Scanning Tunneling Microscope. *Scientific American* (August): 50–56.

Binnig, G., et al. 1982. Tunneling through a controllable vacuum gap. *Applied Physics Letters* 40(2): 178–180.

Bonnel, D. (ed.). 1993. *Scanning Tunneling Microscopy and Spectroscopy: Theory, Techniques and Applications*. New York: VCH Publishers, Inc.

Bradbury, S. 1967. *The Evolution of the Microscope*. Oxford: Pergamon Press.

Brock, W. 1992. *Chemistry*. New York: W.W. Norton and Company.

Brown, H. 1985. Galileo on the Telescope and the Eye. *Journal of the History of Ideas*. 46: 487–501.

Brown, J.R. 1991. *The Laboratory of the Mind*. New York: Routledge.

Cartwright, N. 1999. *The Dappled World: A Study of the Boundaries of Science*. Cambridge: Cambridge University Press.

Charleston, R.J. and Angus-Butterworth, L.M. 1957. Glass. In: Singer, C., Holmyard, E.J., Hall, A.R., and Williams, T.I. (eds.), *A History of Technology: Volume III From the Renaissance to the Industrial Revolution c 1500–c1750*. New York: Oxford University Press, 206–244.

Chen, C.J. 1993. *Introduction to Scanning Tunneling Microscopy. Oxford Series in Optical and Imaging Sciences*. New York: Oxford University Press.

Coor, T. 1968. Signal to Noise Optimization in Chemistry—Part One. *Journal of Chemical Education* 45: A533–A542.

Drechsler, M. 1978. Erwin Müller and the early development of field emission microscopy. *Surface Science* 70: 1–18.

Ferguson, E. 1992. *Engineering and the Mind's Eye*. Cambridge, Massachusetts: The MIT Press.

Gooding D. 1990. *Experiments and the Making of Meaning*. Dordrecht: Kluwer Academic Publishers.

Hacking, I. 1983. *Representing and Intervening*. Cambridge: Cambridge University Press.

Harré, R. 1970. *Principles of Scientific Thinking*. Chicago: University of Chicago Press.

Harré, R. and Madden, E.H. 1977. *Causal Powers*. Oxford: Blackwell.

Hooke, R. 1961. *Micrographia or Some Physiological Descriptions of Minute Bodies Made by Magnifying Glasses with Observations and Inquiries thereupon*. New York: Dover Publications.

Magonov, S.N. and Whangbo, M. 1996. *Surface Analysis with STM and AFM: Experimental and Theoretical Aspects of Image Analysis*. New York: VCH.

Morrison, M. and Morgan, M. (eds.). 1999. *Models as Mediators: Perspectives in Natural and Social Science*. Cambridge: Cambridge University Press.

Müller, E.W. and Tsong, T.T. 1969. *Field Evaporation: Field Ion Microscopy*. New York: American Elsevier Publishing.

Paneth, F.A. 1962. The Epistemological Status of the Chemical Conception of Element. *The British Journal for the Philosophy of Science* XIII(May): 1–14,144–160.

Price, D.J. 1957. The Manufacture of Scientific Instruments from c1500 to c1700. In: Singer, C., Holmyard, E.J., Hall, A.R., and Williams, T.I. (eds.), *A History of Technology: Volume III: From the Renaissance to the Industrial Revolution, c1500–c1750*. Oxford: Oxford University Press, 620–647.

Pomian, K. 1998. Vision and Cognition. In: Jones, C.A. and Galison, P. (eds.), *Picturing Science/Producing Art*. New York: Routledge.

Rothbart, D. 1999. On the Relationship Between Instrument and Specimen in Chemical Research. *Foundations of Chemistry* 1: 257–270.

Rothbart, D. 2000. Substance and Function in Chemical Research. In: Bhushan, N. and Rosenfeld, S. (eds.), *Of Minds and Molecules: New Philosophical Perspectives on Chemistry*. Oxford: Oxford University Press, 75–89.

Rothbart, D. 2003. The Design of Instruments and the Design of Nature. In: Radder, H. (ed.), *The Philosophy of Scientific Experimentation*. Pittsburgh, Pennsylvania: University of Pittsburgh Press, 236–254.

Schummer, J. 1998. The Chemical Core of Chemistry I: A Conceptual Approach. *Hyle: An International Journal for the Philosophy of Chemistry* 4: 129–162.

Shapin, S. and Schaffer, S. 1985. *Leviathan and the Air-Pump: Hobbes, Boyle, and the Experimental Life*. Princeton: Princeton University Press.

Skoog, D.A. and Leary, J.J. 1992. *Principles of Instrumental Analysis*. Fort Worth: Harcourt Brace Jovanovich College Publisher.

Suckling, C.J., Suckling, K.E., and Suckling, C.W. 1978. *Chemistry through Models: Concepts and Applications of Modeling in Chemical Science*. New York: Cambridge University Press.

Waller, R. (ed.). 1705. *The Posthumous Works of Robert Hooke*. London: Frank Cass & Co. Ltd.

Zielonacka-Lis, E. 1999. Research Practice of Modern Bioorganic Chemistry and the Erotetic Conception of Explanation. In: N. Psarros and K. Gavroglu (eds.), *Ars Mutandi. Issues in Philosophy and History of Chemistry*. Leipziger Universitätsverlag, 153–157.

SECTION VII

CHEMISTRY AND ONTOLOGY

CHAPTER 17

ARE CHEMICAL KINDS NATURAL KINDS?

NALINI BHUSHAN
Department of Philosophy, Smith College, Northampton, Massachusetts 01063, USA

INTRODUCTION

The concept of a natural kind is often invoked to settle debates in the philosophy of science and in metaphysics, debates which raise skepticism about a variety of our practices such as the justifiability of inductive reasoning, the projectibility of predicates, or the correct meaning of some of our terms. Thus, for example, the predicates that are projectible are just the ones that are true of natural kind terms,[1] these being the ones that might figure in natural laws. The natural kinds literature in philosophy has typically identified natural kinds with chemical kinds. Are chemical kinds natural kinds? The answer may appear obvious—of course, they are! In fact, one could perfectly reasonably believe that if there were any kinds at all that would be the ideal candidate for being natural kinds, they would be chemical kinds such as gold and H_2O. This is not surprising. For if there are any properties of a natural object that one may regard as essential to its makeup, chemical properties would seem to be the most likely, in being the least controversial candidate.[2] Although philosophers have used zoological kinds (tigers, *Homo sapiens*), botanical kinds (elms, beeches), and psychological kinds even (colors like red and yellow) as examples of natural kinds, the chemical examples predominate. And while one might expect debate as to whether a mind is a natural kind or whether milk or tigers or species are natural kinds, the domain of chemical kinds seems by and large not subject to serious challenge.

In this paper, I briefly go over some of the contemporary history of the role of chemical kinds in the argument for natural kinds; give some reasons for skepticism by using as my case study the area of chemical synthesis, the designing of molecules; and finally, sketch the outlines of an alternative approach to kinds in chemistry. In this connection, I contrast the new theory of reference made respectable by Kripke and Putnam with the conventionality of reference made popular by Quine, and argue that Quine was right in his skepticism about natural kinds, although he was wrong in his characterization of chemical kinds. However, I part company with him in choosing to go for a particularism,[3] rather than a conventionalism with respect to kinds. Thus, what we would have on my proposal is a kind for every occasion: kinds are real but particular to the occasion of the individuals who choose to work with them, their choices of models and strategies, and what happens to function best in the environment

D. Baird et al. (eds.), Philosophy of Chemistry, 327–336.

in question. In chemistry, what this means very generally is that a kind might be a substance, or a set of molar properties, or a micro-structural feature. To give a more specific example, a ligand–receptor complex might be considered by chemists to be a chemical (or biochemical) kind because it fits better into a particular site, is functional, or is "right" in some sense that does not reduce to thinking of it as natural. While the grouping is not natural in any unproblematic sense of the term, it is nonetheless real rather than (merely) conventional.

The position of particularism challenges a traditional dichotomy that, crude and tired as it may seem, continues to exercise a powerful influence on the psyche. This is the dichotomy between what is discovered (or essential or real or natural) and what is constructed (or conventional or created).[4] There is no doubt an interesting sociological and psychological lesson here that I am not entirely equipped to provide: certainly it has been useful (and is perhaps practically indispensable) to have twin pegs such as these upon which to hand two different constellations of ideas that resonate together. Thus, to give up the "discovery picture" (LaPorte 1996) is not just to give up a crude folk theory but to give up a deeply permeating sensibility—a way of thinking about entities as they relate to their environment; a way of approaching solutions to problems—of method, of modeling, of categorization; a way of making sense of one's own relationship to the world and one's hand in altering it.

To consider seriously the notion of making in the successful synthesis of a range of molecules from the completely novel to the already abundantly existent in nature is to give up the discovery/creation dichotomy, not because of a logical argument to the effect that everything is at bottom created or constructed (as some have argued), but rather because the very practical lesson of the case of chemical synthesis is that there is literally no difference between what is discovered and what is created!

NATURAL KINDS, REALISM, AND ESSENTIALISM

What makes something a *kind*? And what makes a kind *natural*? Although the first question is an interesting one, for the purposes of this paper, I will assume that we can give a relatively unproblematic answer in terms of groupings of objects based on some salient property or set thereof. My concern here is with the second question. The philosophical literature has traditionally recognized two sorts of distinction: (a) natural versus arbitrary groups of objects and (b) natural versus artifactual groups of objects. What justifies these distinctions? In the first case, the idea is that there is some theoretically important property that members of the "natural" group share; in the second, natural is understood as "naturally occurring" (found in nature).[5] A bit of reflection on the sense of "natural" understood in (a), however, shows that it will not suffice to give us natural kinds. If the salient feature is some theoretically important property not otherwise specified, one could conceive of an arbitrary group of objects sharing a theoretically important property under some concocted theory. Thus, all things could potentially be part of a group, so long as one could come up with "respects" in which they are related. It does not help to require that the property be projectible, for this requirement could be satisfied under the concocted theory.

I contend that it is the distinction of type (b) that has *de facto* been relied upon to answer the skeptic, "natural" understood as the occurrence of the grouping (that shares a theoretically important property) *in nature*. In other words, (b) entails (a) in that if the grouping occurs in nature, it follows that there must be some theoretically important property that its members share. But the reverse does not hold, i.e., it does not follow that the sharing of a theoretically important property implies a natural grouping. As Guttenplan explains: "To understand the concept of a natural kind, one must focus on the difference between kinds as represented by the set of typewriters and those as represented by the set of tigers" (Guttenplan 1994, 449). His choice of examples is telling of the kind of distinction that is really doing the work in articulating the concept of a natural kind. Presumably, typewriters share some theoretically important property that explains their operation and can be used to predict their behavior, but they do not occur in nature. Since they are artifacts, they are not natural kind material.

Natural kinds and realism

Not surprisingly, but underscoring the point made above about the sense of natural that is really at work in the notion of a natural kind, a belief in natural kinds tends to go hand in hand with a belief in a robust form of realism and against most forms of conventionalism. Here again is Guttenplan: "What is crucial to the notion [of a natural kind] is that the shared properties have an *independence* from any particular human way of conceiving of the members of the kind" (1994, 450; italics added). This tight connection between natural kinds and realism has been relatively uncontroversial. For if the grouping of objects is "natural", then it is reasonable to infer that there are underlying properties that actually exist and are shared, even if we do not yet know what they are. This form of realism is robust in that it is stronger than a realist position that for example insists upon contingencies of various stripes—historical, social, theoretical, individual—in making sense of any kind of existence claim. Such a realism also tends to be universalist in that it subscribes to the existence of objects with a determinate set of properties regardless of the particularities of circumstance. Such universal properties can however be contingent in one specific respect—contingent upon the way the world actually is.

Natural kinds and essentialism

The view that things have essences has a very long history in the philosophical literature, going all the way back to Aristotle. On Aristotle's view, there are species of things that contain inherently the power to undergo different kinds of "change" (depending upon different kinds of cause) in a law-like way. These changes (and the causes which ground them) are different aspects of the same underlying generative principle that defines the species of thing (the substance)—its identity conditions—and that constitutes its persistence conditions. Here, we get the earliest connection between substances [what some have taken to be synonymous with real or natural kinds (Millikan 2000)] and essentialism.

Belief in essences (and in the metaphysical enterprise more generally) received a severe setback with the logical positivist attack on the intelligibility of entities that were not open to verification. What were these essences anyway, if understood as being something in addition to all observable properties? Interestingly enough, the concept experienced a revival in the 1970s and the 1980s with Putnam (1975) and Kripke (1980) as the idea of essence was systematically detached from earlier conceptual apparatus and connected specifically to the microscopic properties sanctioned by science, possessed by objects and grouped by such properties into natural kinds. Here is Sober, articulating a view that is today relatively uncontroversial among philosophers: "*Essentialism* is a standard philosophical view about natural kinds. It holds that each natural kind can be defined in terms of properties that are possessed by all and only members of that kind" (Sober 1993, 145). A natural kind thus requires a property that is both necessary and sufficient for membership. We may have properties or sets thereof that justify grouping together individuals as automobiles, students, or games, i.e., as kinds, but nobody (after Wittgenstein, at any rate) would wish to argue that these properties were *essential* to these kinds, for one could always imagine that there might come into being a new model of car or a newly invented game that would lack that property or set thereof (although we might not be able to actually come up with automobiles or students or games that lack said properties). Also, the necessary and sufficient condition on the said property cannot be simply stipulated, i.e., agreed upon by convention. It might be contingent that the object (e.g., water) exists in the first place, but given that it exists it necessarily has that property (its H_2O-ness) (Kripke 1980). That is, the property must in some sense actually inhere in the object. So the concept of a natural kind is distinguished from the concept of a mere kind by its link to the metaphysical thesis of essentialism.

There is thus a constellation of concepts that are traditionally intertwined: essentialism, realism, and necessity Sidelle (1989), for instance, uses the concepts of essence and necessity almost interchangeably;[6] the overarching concept that justifies these links is that of a natural kind. What justifies the deployment of this crucial category?

THOUGHT EXPERIMENTAL EVIDENCE FOR NATURAL KINDS

The natural kinds literature in philosophy has been dominated by thought experiments created by Putnam and Kripke that have linked notion of natural kinds with the possession of essential properties. For instance, Putnam's famous twin-earth thought experiment uses water (or the H_2O-ness of water) as a critical component in his argument for factoring "deep structure" properties of a natural object into the meaning of the term that serves to pick it out. Thus, the meaning of the term "water" is not exhausted by the description "the stuff that is colorless, odorless, flows in rivers, quencher of thirst," etc. in addition, and, in light of the famous thought experiment, crucially, the H_2O-ness of water, in being essential to its makeup, is essential to its meaning as well.[7]

Here, we have a rationale for the dichotomy of properties—the contingent ones—surface or molar properties captured in our descriptions of what we see, touch, feel, hear, or smell, and the essential properties inherent in the nature of things themselves, independent of our perceptions. Essential properties are real, although not vice versa. For real properties may be essential or contingent—these latter may be transitory or fleeting or relational in some respect but are real for all that.

This link between natural kinds and essentialism has been criticized by many in the philosophy of science but curiously enough an exception is often made when the kind of natural kind on the table for discussion is a chemical kind. And so it is that many philosophers do not balk when Sober (1993) argues in a section provocatively entitled "The death of essentialism," that essentialism is *not* a plausible view of biological species because the notion of a natural kind upon which it rests is not one to which biologists are committed, and states, at the same time, and almost in passing: "Essentialism is arguably a plausible doctrine about the *chemical* elements" (146; italics added).

That said, there exists a body of literature that criticizes the thought experimental evidence for the existence of *chemical* kinds as natural kinds. The criticisms take various forms. I will mention a few of them primarily to provide a contrast with the line of criticism taken in this paper. One has to do with the methodology of thought experiments: "Who knows what we ought to say in such a fantastic situation?" (Dupre 1981, 71) Another challenges the existence of pure entities such as gold or water, arguing that these terms in fact cannot refer to a single micro-structural composition (the impurities objection) (Aune 1994). A third and more recent criticism is that the meanings of our chemical kind terms are often indeterminate (LaPorte 1996). According to LaPorte, "Paradigmatic instances of a kind share many properties, macro- and micro-; which of these is definitive of the kinds will always remain indeterminate to some extent" (LaPorte 1996, 115). Groupings into kinds take place on the basis of a likeness relation that is chosen out of a range of possible likeness or similarity relations and could happen at either the molar or the molecular level. If so, the groupings are not into "natural" kinds in any sense that is clear and unproblematic.[8] (This focus on similarity judgments as the basis for groupings is very much in the spirit of Quine, a point to which I will return, in agreement, later.)

A goal of this essay is to contribute to this body of critical literature on chemical kinds, but from a different direction. It considers the impact of chemical synthesis on the natural/non-natural divide. The point is simply that the properties of synthesized compounds can satisfy criteria just as easily as naturally occurring compounds, and thus that the intuitive connection between essences (or necessities or rigidities) and natural kinds is not in fact warranted. For example, take naturally synthesized proteins versus artificially synthesized proteins that are in every other way identical to the former. Is one a natural kind, the other not? What is the basis for categorization?

THE PRACTICE OF SYNTHESIS AND NATURAL KINDS

Might a closer look at the implications of the possibility of chemical synthesis bring fresh insights into the discussion regarding natural kinds?[9] I think so. Natural kinds research has historically proceeded independently of a consideration of the implications of the possibility of synthesis. Ruminating on what it means to create entities at the microscopic level as chemists do, entities that behave in every way as do naturally occurring counterparts, forces us to re-evaluate the weight traditionally given by philosophers to natural (or naturally occurring) kinds. The goal of this section is to raise questions about the philosophical practice of automatically identifying natural kinds with chemical kinds. If the reasoning here is plausible, it should incline the reader to rethink the naturalness of being a certain kind of realist in the field of chemistry as well.

Consider the following three cases of chemical synthesis:

(1) Synthesized compounds that have an analog in nature. For example, take naturally synthesized proteins versus artificially synthesized proteins that are in every other way identical to the former.
(2) Synthesized compounds that have no natural analog. This group might include molecules for which the analog is a complex macroscopic object (brakes, switches, gears),[10] molecules for which the analog is a conceptual or mathematical object (Platonic solids), molecules for which the analog is a macroscopic object with aesthetic appeal, molecules for which there is no analog and where the molecular structure itself has aesthetic appeal. For example, pentacyclo[4.2.0.$0^{2,5}$.$0^{3,8}$.$0^{4,7}$]octane, also known as cubane, has eight carbons located with respect to each other as the vertices of a cube. First synthesized in 1964 (Eaton and Cole), cubane does not occur in nature nor was any use known or anticipated for it at the time of its synthesis.
(3) A compound that is synthesized first and later discovered to occur naturally. Here, there are two possibilities: (a) the compound was already present in the environment or (b) is newly introduced into the environment, but by "natural" means. For example, around 1985 the compound now called buckminsterfullerene was prepared synthetically. This is a 60-carbon compound in which the carbons are at the vertices of five- and six-membered rings positioned lie those inscribed on the surface of a soccer ball. Subsequently, buckminsterfullerene was found to be a component of soot, including ancient samples of soot.

In all of these cases, what is the relationship between the synthesized molecules and their naturally occurring counterparts? This is a discussion yet to take place in the philosophical literature, presumably because the answer has been taken to be obvious: they are all of a kind, synthesized or no, since the synthesized ones are indistinguishable from the "natural" ones. But it seems far from obvious that this is the case, once one thinks through, on the one hand, the notion of natural kind that is at work, and on the other, the extent of the "making" that is involved in synthesis and the level at which this occurs. So what is the basis for categorization?

It seems that one can adopt one of the following positions:

(a) Bite the bullet and continue to stipulate natural kind to mean naturally occurring, but this now seems *ad hoc*, in light of the indistinguishability of the natural and synthetic compounds.

(b) Broaden the notion of natural kind so that it now comes to mean naturally occurring things and those identical to them. This move has some possibilities. But, how about a molecule *first* synthesized by humans (so non-naturally occurring) and then created by a plant or animal [3(b) above]? Does its category change at that moment and become a natural kind where it was not before? One might respond by conceding that some chemical compounds are not natural kinds—those synthesized ones that do not occur in nature perhaps—but hold to the basic intuition that the paradigmatic examples of natural kinds are still chemical kinds. But this strategy is not persuasive. For, there is an additional feature of chemical kinds that distinguishes them from their zoological and botanical counterparts: Any chemical compound potentially could be made by us. This is not the case with the other examples of natural kinds. So there is a sense in which, far from being paradigmatic, chemical kinds are really more suspect as natural kinds than their zoological, botanical, and psychological cousins.

(c) Conclude that the category of natural kind is suspect to begin with.

In cases (1)–(3) considered above, every attempt at categorization seems problematic. We are tempted to go the route of (c), i.e., to give up the category of natural kind as explanatorily, epistemologically or ontologically basic. Indeed, philosophers have adopted critiques of type (c)—that one cannot get at a suitable criterion for separating off the natural from non-natural kinds—but on grounds independent of the implications of chemical synthesis. We have shown, by going a different route, i.e., by exploring the nature of synthesis and its products, what is wrong with a metaphysic of categorization that takes as central the concept of a natural kind.

I have argued above that the possibility of "making" at the chemical level serves to blur the natural/artifact divide. This has three consequences:

(1) Chemical kinds are not the unproblematic, paradigmatic instances of natural kinds they have been taken to be in the philosophical literature.
(2) We are left without a clear and uncomplicated example of a "natural" kind.
(3) Brands of realism, which are tightly tied to natural kinds, are no longer automatically the status quo position to take vis-a-vis chemistry.

FROM QUINE TO CARTWRIGHT: CONVENTIONALISM TO PARTICULARISM

In contrast to the concept of natural kind sketched above that is suggested by Kripke and Putnam's writings, Quine's use of phrase "natural kinds" in his paper with the same title (Quine 1969) is to be understood as "kinds that come naturally i.e., instinctively or innately to us." On this use of the phrase, chemical kinds would be a subset of natural kinds in being the most theoretically sophisticated of kinds to be found in the sciences; paradoxically, therefore, the least natural in the sense that they come the least naturally to us. Also, Quine's use of kinds has no link to essentialism either. For him, a kind is the consequence of a similarity judgment that we are innately predisposed to make rather than a consequence of groupings that are in some sense an essential aspect of the natural world. Quine, in his paper "Natural Kinds" suggests however that when it comes to the science of chemistry, questions about categorization do not arise since the "structural conditions" of the mechanisms involved are completely known and uncontroversial. In a previous paper (Rosenfeld and Bhushan 2000), we have argued against Quine that categorization is a live issue in chemistry and we use the area of synthesis to show in what ways this is so. Here, we register this fact in order to bring chemical kinds back under the fold of the kinds that come innately to us. That is, the rationale for groupings in chemistry is as much a matter of what seem to go together as it is in other domains. Quine was wrong to make an exception of chemistry.

The concept of similarity in organizing combinatorial libraries of compounds in chemistry has largely pragmatic roots (Wilson 1998). In fact, the sense in which two compounds are considered to be similar depends on context. For example, one may wish to find compounds that are similar to a given compound in terms of the ability to bind to a particular receptor (because this might be an indicator of certain pharmacological properties). Here, the positioning of various functional groups, the binding sites, becomes the crucial characteristic for comparison and therefore this is a purely structural similarity where the important aspects of structure are things like the distance between two functional groups. A different sort of descriptor might be a set of properties like whether the structure includes an aromatic ring, a chlorine atom, etc. The selected properties could even include molar properties such as solubility. In a sense, and to extend the library metaphor a bit further, one might think of the set of all descriptors for a particular compound as being akin to keywords used in a (computer) search of a library of books.

Similarity is a robust concept that is very much in use by chemists, but in many cases it is intuitive and not the mathematical version that Quine talks about. It is true that similarity could be very well defined in a narrow context and satisfy the criteria for rigor required by Quine. The notion of reaction distance is a case in point. Here, talk of reaction distance in mathematical terms replaces talk of similarity, exactly as Quine wants. However, this quantitative expression of similarity of structure uses as a vehicle processes (called unit reactions) that interconvert structures, and clearly, this is not the only way to frame structural similarity. To see how the means of comparison used here is embedded in the specific concerns of chemists, compare the following macroscopic analog to this situation. One might compare a pair of pants to a bag and see the two as similar in virtue of the fact that simply cutting off the legs and sewing up the pant holes would do the interconversion trick. However, this seems an arbitrary basis for similarity (between pants and bags) even though the "interconversion" is straightforward. The fact that chemists do use interconversion as a mark of similarity for its utility in synthesis, therefore, does not mean that there are not other, perhaps better, measures of chemical similarity. Thus, there is a real need for discussion and evaluation of the "respects" in which two molecules can be said to be similar.

Although we have seen that a good case can be made for the ongoing use of a range of similarity judgments in chemistry, the kind of conventionalism espoused by Quine, however, does not comfortably fit with the kind of work that kinds do in chemistry. Although the choice of kind in a spectrum of possible groupings depends upon context, they figure in causal and explanatory narratives and therefore are real, although particular to interests and histories of the individuals involved and to the particular models and strategies they choose to solve their problems. My hunch is that the particularism proposed by Cartwright in recent work (Cartwright 1999) may be a good way to make sense of kinds in chemistry.

CONCLUSION

The argument in this essay against taking chemical kinds to be natural kinds has two parts:

(1) It registers what is empirically the case—that chemists use similarity judgments in grouping items together as do practitioners from other disciplines. Quine is wrong to make chemistry an exception. Kinds in *any* discipline are not naturally occurring. Furthermore, the similarity judgments used by chemists vary from occasion to occasion.

(2) It underscores the philosophical significance of the very possibility of chemical synthesis. Organic synthesis results in the creation of compounds identical to ones found in nature. This undermines the discovery model in an interesting way. The point is not that in fact no entities are ever really discovered (a good point and one that has been made a number of times in the literature). It is rather that given that entities occurring naturally can be matched in the makings and designings of the synthetic organic chemist, there is no longer any justification for the extra explanatory or ontological weight given to natural (or naturally occurring) kinds. If so, in the phrase "natural kinds" the emphasis must shift instead to kinds and a search for varieties thereof independent of the natural/conventional divide.

My proposal is that we look to a particularism in kind term reference as an alternative.[11]

NOTES

1. Thus, "green" rather than "grue," for example, where "grue" means "observed before 2005 and green, or blue" (Goodman 1955).
2. In general, this is perhaps why there has been so little philosophy of chemistry: all the fundamental questions in chemistry seem settled or at least without deep controversy.
3. For a detailed explication and defense of particularism, albeit in a different context, see Cartwright (1999).
4. To see the pull of the idea of a natural kind, consider the question whether the mind is a natural kind. The question is an invitation to consider whether the mind is the way that it is beyond what society/culture have built into it.
5. I do not distinguish here between the view that natural kinds are substances that have properties, over which a variety of predicates are projectible, and the view that natural kinds are the set of properties themselves captured by predicate concepts such as being gold, or being green. See Millikan (2000) for

a very nice defense of this distinction and for her argument in favor of bringing back into the discussion about kinds the traditional Aristotelian notion of substance.

6. This raises a very interesting issue in a debate between realism and conventionalism, but we will not pursue it here. See Sidelle (1989).
7. I note in passing that although one might consider it an open question whether or not the chemical kind is to be identified at the molar (water) or molecular (H_2O) level, that inhere no matter what, this is not so for Putnam who would take molar level properties to describe superficial characteristics that are contingent of the kind.
8. In a recent paper, LaPorte argues that kind terms rigidly designate abstract kinds (LaPorte 2000). On this view, chemical kind terms refer, and refer rigidly, but not in virtue of picking out natural kinds. This is an interesting position, but I do not have the space to discuss it here.
9. This section on chemical synthesis appears as part of a longer essay entitled "Chemical Synthesis: Complexity, Similarity, Natural Kinds, and the Evolution of a 'Logic' " (Rosenfeld and Bhushan 2000).
10. For the very interesting and provocative example of a "molecular ratchet," see Carpenter (2000).
11. Particularists are pluralists, holding that there is more than one kind-relevant property. Thus properties have variable relevance. A property that makes an entity belong to one kind in a certain scientific context can be irrelevant in another. See my 'What is a Chemical Property?' (forthcoming) for a development of this proposal regarding particularism.

REFERENCES

Aune, B. 1994. Determinate Meaning and Analytic Truth. In: Debrock G. and Hulswit M. (eds.), *Living Doubt*. Netherlands: Kluwer Academic Publishers.

Carpenter, B. 2000. Models and Explanations: Understanding Chemical Reaction Mechanisms. In: Bhushan and Rosenfeld (eds.), *Of Minds and Molecules*. New York: Oxford University Press.

Cartwright, N. 1999. *The Dappled World*. Cambridge: Cambridge University Press.

Dupre, J. 1981. Natural Kinds and Biological Taxa. *The Philosophical Review* 90: 66–90.

Goodman, N. 1955. *Fact, Fiction and Forecast*. Cambridge: Harvard University Press.

Guttenplan, S. 1994. Natural Kind. In: Guttenplan, S. (ed.), *A Companion to the Philosophy of the Mind*. Oxford: Blackwell.

Kripke, S. 1980. *Naming and Necessity*. Cambridge: Harvard University Press.

LaPorte, J. 2000. Rigidity and Kind. *Philosophical Studies* 97(3): 293–316.

LaPorte, J. 1996. Chemical Kind Term Reference and the Discovery of Essence. *Nous* 30(1): 112–132.

Millikan, R. 2000. *On Clear and Confused Ideas*. Cambridge: Cambridge University Press.

Putnam, H. 1975. *Philosophical Papers Vol. 2: Mind, Language, and Reality*. Cambridge: Cambridge University Press.

Quine, W.V.O. 1969. *Ontological Relativity and Other Essays*. New York: Columbia University Press.

Rosenfeld, S. and Bhushan, N. 2000. Chemical Synthesis: Complexity, Similarity, Natural Kinds, and the Evolution of a 'Logic'. In: Bhushan N. and Rosenfeld S. (eds.), *Of Minds and Molecules*. New York: Oxford University Press.

Sidelle, A. 1989. *Necessity, Essence, and Individuation*. Ithaca: Cornell University Press.

Sober, E. 1993. *The Philosophy of Biology*. Boulder, CO: Westview Press.

Wilson, E.K. 1998. Computers Customize Combinatorial Libraries. *Chemistry and Engineering News* (April 27): 31–37.

CHAPTER 18

WATER IS *NOT* H_2O

MICHAEL WEISBERG
University of Pennsylvania

INTRODUCTION

In defending semantic externalism, philosophers of language have often assumed that there is a straightforward connection between scientific kinds and the natural kinds recognized by ordinary language users.[1] For example, the claim that water is H_2O assumes that the ordinary language kind water corresponds to a chemical kind, which contains all the molecules with molecular formula H_2O as its members. This assumption about the coordination between ordinary language kinds and scientific kinds is important for the externalist program, because it is what allows us to discover empirically the extensions of ordinary language kind terms.

While I am sympathetic to the semantic externalist project, I think that the discussion of chemical kinds by philosophers of language has been rather badly oversimplified, hiding difficulties that arise when we try to coordinate scientific kinds with the natural kinds recognized by ordinary language users.[2] In this paper, I will examine these difficulties by looking more closely at the chemist's notion of water.

To help with this examination, I will begin by making explicit a principle on which I believe semantic externalists rely. The *coordination principle* is the thesis that scientific kinds and the natural kinds recognized by natural language users line up or can be mapped onto one another one-to-one. A brief examination of an externalist picture of kind reference will show how the coordination principle is relied on.

At some point in human history, when in causal contact with water, someone baptized water with the word "water." This person need not have had any correct beliefs about the kind of thing water was, she merely needed to attach the kind term "water" to the token of water that was in front of her. Whether subsequent samples of liquid are actually water or not depends on whether they are of the same natural kind and hence share the same micro-structure or essence as the original sample of water. The reference of the term "water" is fixed for the community of language users in virtue of their causal connections to the baptizer. By being causally connected to the baptizer, the community of language users is connected to the original baptism event. In virtue of this, they need not have any true beliefs about water to refer to water. All that is required is that they are part of a linguistic community that has the right kind of causal links through time.

D. Baird et al. (eds.), Philosophy of Chemistry, 337–345.

While the ordinary language user and even the baptizer need not have true beliefs about water in order to refer to it, someone in the community eventually will need to have these beliefs if the meaning of "water" is to be made explicit. The obvious people to ask about the nature of water are chemists, and semantic externalists have often assumed that this is where we must turn. Putnam and Kripke are not very explicit about the details of the role chemists play in semantics, but this is where I believe the coordination principle is implicitly relied upon. Chemists discover the natural kinds of the material world, which I will call "chemical kinds." The coordination principle presumes that the very same kinds that chemists discover are the ones relevant to ordinary language. If this is the case, then when you describe a chemical kind in detail, you will have nailed down the semantics of the ordinary kind term associated with that chemical kind. Putnam and Kripke seem to believe that in the case of water, chemists describe a chemical kind whose members include all and only the molecules with molecular formula H_2O. Appealing to this fact and to the coordination principle, they conclude that water is H_2O.

Closer examination of what water really is, I believe, shows that for chemists, water is not just the set of all molecules with molecular formula H_2O. There are multiple chemical kinds that might reasonably be coordinated with the ordinary language kind water. Because chemistry provides us with many different types of natural kinds, and because it does not provide us with rules favoring one set of kinds over another, deference to the findings of chemistry will not unambiguously allow us to discover the extensions of natural kind terms in ordinary language. A more nuanced version of the coordination principle, which has specific rules for picking out the appropriate chemical kind in particular circumstances, will be needed to carry forward the semantic externalist project.

INDIVIDUATION CRITERION

Linus Pauling famously told us that chemistry is "the science of substances—their structure, their properties, and the reactions that change them into other substances" (Pauling, 1947). Using slightly more contemporary language, we can say that chemistry studies the *structure* and *reactivity* of substances. Structure is studied at three compositional levels. Molar structure consists of the macroscopic or bulk properties of a substance. Molecular structure is the spatial configuration of atoms connected by chemical bonds. Atomic structure, from the chemical point of view, includes both the kinds of atoms from which a substance is composed and their quantum mechanical state. "Reactivity" is a general term referring to the transformations of substances. This involves both the intrinsic properties of a substance to transform itself over time and the ways in which it is transformed when brought in contact with other substances.

This description of chemistry's subject matter leads us to a criterion for the individuation of chemical kinds. In deciding whether two samples of a substance are of the same kind, a chemist examines their structure and reactivity at all three compositional levels. For example, the two alcohols methanol and ethanol are distinct chemical kinds in virtue of their different molecular structures. Drawing on Pauling's description, we can give a simple criterion for chemical individuation:

> Chemical kinds are to be individuated with respect to structure and reactivity at the molar, molecular, and atomic levels.

In this chapter, I will appeal to this individuation criterion in order to determine which groupings are legitimate chemical kinds and which are not. For example, all the solvents in a laboratory manufactured by the Aldrich chemical company do not constitute a chemical kind. They might be a legitimate grouping if we wanted to determine how much money was owed to Aldrich; however, given that they were not individuated by similarities in structure or reactivity, they do not constitute a chemical kind.

When we apply the individuation criteria to water, it is clear that we ought to investigate all three structural levels as well as water's reactivity in order to have a complete picture of how water should be individuated. In this paper, I will confine myself to discussing atomic level structure as well the related properties of reactivity. A full treatment of what exactly water is, however, would require far more detail about the higher structural levels than I will give in this chapter.[3]

ISOTOPIC ISOMERS

Both hydrogen and oxygen are found in a variety of isotopes in nature, giving rise to the phenomenon of isotopic isomerism. After describing how this phenomenon applies to water, I will discuss the implications of isotopic isomerism for the coordination principle. Although the presentation of the scientific and philosophical issues here is my own, I am indebted to the insights found in Mellor (1974), Stroll (1998), and Needham's (2000) discussions of these issues.

Isotopes are sets of atoms with the same numbers of protons and electrons, but different numbers of neutrons. In most cases a hydrogen atom has one proton and no neutrons. Chemists generally just call this "hydrogen" and symbolize it as "H". When it is necessary to distinguish between isotopes chemists call it "hydrogen-1" or symbolize it as "^1H". A second isotope of hydrogen called "deuterium" was discovered by Harold Urey in 1931. Deuterium, which is symbolized as "D", has one neutron, one proton, and one electron. Subsequent research also discovered a third isotope of hydrogen called "tritium" (symbolized as "T") which has two neutrons, one proton and one electron. Naturally occurring samples of hydrogen contain a mixture of hydrogen-1 and deuterium. Tritium does not occur naturally but can be found in samples of hydrogen for many different reasons, such as being generated as decay products of other isotopes, by cosmic rays, or even by nuclear fallout. Oxygen also has three isotopes—^{16}O, ^{17}O, and ^{18}O. All three of these stable isotopes have 8 protons and 8 electrons. They have 8, 9, and 10 neutrons respectively.

Since isotopes have very similar chemical and thermodynamic behavior, the naturally-occurring chemical reactions that produce molecules like water cannot, for the most part, distinguish between the isotopes.[4] Because of this fact and because all samples of hydrogen in nature contain a mixture of the isotopes, an elemental analysis of natural water would reveal a certain fraction of deuterium mixed with hydrogen-1. The ratio would most likely mirror the natural or background ratio of hydrogen-1 to deuterium, 99.985:0.015. Similarly, many samples of terrestrial water will contain trace amounts of tritium because of the trace amounts of tritium found on Earth. If we look at enough samples of enough water, we will find $H_2{}^{17}O$, $H_2{}^{18}O$, $HD^{16}O$, $D_2{}^{17}O$, $T_2{}^{18}O$, etc., in addition to $H_2{}^{16}O$. In fact, natural samples of water almost always contain a mixture of these other isomers. In figuring out how to individuate the kind water, then, we need to ask several questions: Is pure $H_2{}^{16}O$ a chemical kind? How about pure $D_2{}^{16}O$? In normal, terrestrial samples that are mostly $H_2{}^{16}O$, how much tolerance of isotopic variation is allowed? If the substances described in all these other questions are chemical kinds, how do we decide which one corresponds to the ordinary language kind water?

Pure $H_2{}^{16}O$ is a chemical kind because it can be individuated with respect to structure at both the molecular and the atomic levels. Similarly, pure $H_2{}^{17}O$ is a chemical kind and so is pure $D_2{}^{16}O$. This is a problem because the coordination principle requires that chemistry generate a single kind that can be associated with the ordinary language kind. Although this looks like it might merely be a manifestation of the *qua problem*,[5] I will argue that it is actually a symptom of a deeper problem. Before turning to this, let us consider what would happen if we just picked one of these kinds to coordinate with our ordinary language kind. One principled way to do this is to pick the major component of samples of naturally occurring water. Following this rule, we would pick the kind that consisted of all and only the molecules with molecular formula $H_2{}^{16}O$.

This solution, however, is problematic because there are *always* isomers present in natural samples. One of these isomers ($HD^{16}O$) constitutes about 0.03% of Earth's natural water. This may seem insignificantly small, but it means that in one sip of water (about 18 mL) there are about 1.8×10^{22} molecules of $HD^{16}O$. That is ten million billion molecules of $HD^{16}O$ per sip. For this reason, it seems wrong to ignore outright the contributions of other isotopes.

Perhaps one way to save the claim that water should be identified with all and only molecules with molecular formula $H_2{}^{16}O$ is to insist that the other isomers are impurities. This suggestion seems initially plausible, for we have no trouble regarding sea water, dirty water, acid rain and the like as water, at least in the every day sense, although they are not homogenous at the molecular level.

Attractive as this possibility may seem, I believe that we cannot regard isomers as impurities. To regard some substance token as an impurity or containing impurities, we have to begin with a conception of a pure substance type. If isomers like $D_2{}^{16}O$ are impurities, then this suggests that a sample completely made up of $H_2{}^{16}O$ molecules is the pure substance. Are there good reasons to conclude this?

I believe that there are not. In fact, if purity means something like "without changes or additions," then there is a sense in which isotopically *homogenous* samples are impure. Standard isotopic ratios have been measured for all stable isotopes and built into the elemental masses reported on the periodic table. These values reflect the outcome of a set of geological, biological, chemical and nuclear fractionation processes that have taken place through the history of our planet. The reason these values allow us to make accurate calculations is because all the natural samples that we have measured have undergone the same fractionation processes throughout their histories. Deviations from this background distribution of abundances must be explained as the intervention of some *further* geological, biological, chemical, or nuclear fractionation process.

If the notion of a pure substance is to be useful, it cannot merely be a measure of homogeneity. Rather, our notion of a pure substance must take natural isotopic variation into account. This is important if we want to discover the extension of the ordinary language natural kind term "water." A homogeneous sample of $H_2{}^{16}O$ molecules is a chemical kind, but when we are trying to find a chemical kind that is close to the kind recognized by ordinary speakers, there is no justification that I can see for simply ignoring the standard isotopic rations.

It is clear that we cannot just ignore isomers in individuating chemical kinds. We also have learned that we cannot just claim that isomers are impurities. A more plausible possibility is to claim that "H_2O" is actually a higher-order term, a genus which includes as its constituent species all of the isomers of H_2O. We then could preserve the claim that water is H_2O, but with a slightly different understanding of what "H_2O" means. Something seems right about this suggestion, for we are starting to see how our kind terms in ordinary language may not map neatly onto chemical kinds. The genus/species solution, however, will not work well in this case.

The main problem with the H_2O-as-genus solution is that it makes no mention of the relative abundances of the isomeric species and such differences can have large impacts on the properties of substances. If we treat "H_2O" as a genus term and preserve the claim that "water is H_2O," then we have to conclude that a sample consisting solely of $D_2{}^{16}O$ is water. No doubt pure $D_2{}^{16}O$ is a chemical kind. In fact, it is usually called "heavy water" and is used in some nuclear reactors. Pure $D_2{}^{16}O$, however, is saliently different from ordinary water. Although $D_2{}^{16}O$ can undergo similar reactions to $H_2{}^{16}O$, the reaction rate is different enough to make ingestion of $D_2{}^{16}O$ lethal. The D from $D_2{}^{16}O$ exchanges with hydrogen-1 atoms in our body, disrupting critical metabolic processes. In addition, some of the molar structural properties, such as freezing point and viscosity, are different among samples with different mixtures of isotopic isomers. The freezing point of pure $D_2{}^{16}O$, for example, is about 2°C as opposed to 0°C for a sample of ordinary terrestrial water. This is also a salient difference according to our individuation criteria. Our proposal to treat "H_2O" as a higher-order term, picking out all of its isomers, however, treats pure $D_2{}^{16}O$ and pure $H_2{}^{16}O$ as the same kind of thing. And while they are species of a common genus, they are clearly distinguished both in chemistry and in everyday contexts.

Neither treating isomers as impurities nor treating "H_2O" as a higher order term picking out all its isomers seems to be an acceptable way to find the single chemical kind demanded by the coordination principle. It is clear that the system of kinds recognized within chemistry is very complex and multi-faceted, which is at odds with the coordination principle's demand for a single chemical kind to be associated with the ordinary language kind water. Perhaps a closer look at how chemists themselves deal with the complex system of kinds will help to resolve the tension.

Chemists deal with the multiplicity of chemical kinds in two different ways. Most often, they deal with it by using context-sensitive kind terms. These terms pick out different chemical kinds in different explanatory and conversational contexts. "Water," as uttered by a chemist, will sometime refer to the isomers of H_2O in their natural abundances, sometimes to any isomer of H_2O, and sometimes, perhaps, to a homogenous sample of $H_2{}^{16}O$ depending on the circumstances of the utterance.

When forced to be explicit, chemists use a set of very specific kind terms corresponding to very specific chemical kinds. The results of chemists being forced to be explicit can be found in the extremely precise and often very complex nomenclature established by the International Union of Pure and Applied Chemistry. In this nomenclature, even the different phases of water, a concept which I have not discussed, are given their own designations like water (I), water (VI), etc. The vast majority of chemists and the chemical literature, however, refers to substances in a more straightforward manner—using terms like "water," or "ethanol," or "tetrahydrafuran," which are clear in context.

REVISING THE COORDINATION PRINCIPLE

The discussion of isotopic isomerism is just once source of chemical complexity that poses problems for the coordination principle. Many other fascinating and complicating factors arise when we consider the macroscopic properties of water.[6] These are especially important as chemists usually think of water as a macroscopic substance with macroscopic properties, not merely as a collection of water molecules. While it would be fascinating to explore these complexities further, I believe we already have enough information to reexamine the coordination principle.

Our very brief examination of the nature of water has revealed that there is no single kind for water that is useful in all chemical contexts. In particular, we have seen that the set of substances with molecular formula H_2O is often not a very useful chemical kind. It fails to make distinctions among substances that both chemists and ordinary language users would want to make. Even if the coordination principle is acceptable without revision, we should choose a kind with more carefully determined membership conditions. Perhaps, we could choose a kind that takes into account standard isotopic distributions. I am skeptical that this will be an adequate solution, however, because the problems with the coordination principle run deeper then this solution addresses.

Two results of our investigation put pressure on the coordination principle in its current form. The first result is that chemistry cannot just hand us a single kind with which we can associate the ordinary language kind water, because in chemistry there is a more complex system of kinds. The second result is that chemists often deal with these issues by using kind terms in context sensitive ways. I believe that each of these holds a key to refining the coordination principle.

The first result suggests that the one-to-one match between kind terms in natural language and kind terms in the natural sciences required by the coordination principle may not be possible. Unlike some critics, I have not been arguing that ordinary language and the natural sciences use terms with different extensions. An externalist can evade this kind of criticism by defending an attitude of deference to experts, ignoring our folk conception of the extensions of natural kind terms. My argument is that even if we defer to experts about cases like water, they cannot give us a *single* natural kind to associate with the kind used in everyday contexts by ordinary language users. The system of kinds recognized within chemistry is much more complex than the system of kinds recognized by users of natural language. There are many more types of kinds in chemistry, useful for different theoretical purposes.

The second result suggests a way that an externalist philosopher of language might refine the coordination principle to take into account the first result. Chemists' ordinary use of natural kind terms are highly context sensitive. When they need to be more precise so as to make fine distinctions between subtly different phenomena, they use a more robust set of kind terms corresponding more closely to the multiplicity of kinds recognized within chemistry. In many situations, however, context sensitive terms are sufficient. Say a chemist needs to use a warm water bath to keep a reaction at a particular temperature. Her request for more warm water from an associate will be taken to mean that she wants a substance that is composed primarily of H_2O molecules with an isotopic abundance somewhere in the normal background range. On the other hand, if she is doing a very isotopically sensitive kinetic study, her request for water would be interpreted in a different way, perhaps requiring additional distinctions.

These considerations suggest that the coordination principle should include a mechanism for picking out the right chemical kind to be coordinated with ordinary language kinds in different contexts. Like chemistry itself, and perhaps more so, ordinary language admits of a lot of tolerance for things like isotopic distributions in many contexts. Watering the lawn, filling a swimming pool, and even bathing don't put very strenuous requirements on such things as isotopic ratios. However in some cases, like when we want to prepare pharmaceuticals, protect an ecosystem, or even just have a drink, our tolerance for variation is limited. In contexts where extreme variation is acceptable, the coordination principle can be fairly relaxed, associating the kind term in an utterance with a set of chemical kinds or a higher level genus kind. In contexts where such variation is not acceptable, the coordination principle must have a mechanism for picking out the most appropriate chemical kind for that context.

While an externalist could simply pick some chemical kind to coordinate with the ordinary language kind in all contexts, ignoring the complexity of the system of chemical kinds, I believe that this strategy is short-sighted. Since the externalist strategy I have been discussing contains an attitude of deference to the natural sciences, I believe that our closer look at the types of kinds actually recognized in chemistry is relevant. The complexity of material substances demands that chemists, the people who interact with them at the greatest level of detail, use a multi-faceted and often context sensitive set of kind terms. We would do well to mirror this practice in our discussions of ordinary language.

CONCLUSIONS

In this essay I have discussed an assumption of semantic externalist theories which I called the coordination principle. This is the idea that natural language kinds and scientific kinds line up or can be mapped onto one another one-to-one. A closer look at water shows that there is not this type of simple one-to-one match between chemical and ordinary language kinds. In fact, the use of kind terms in chemistry is often context sensitive and in cases where chemists want to ensure no ambiguity, they use a very complex and nuanced set of kind terms, none of which could be reasonably associated with the ordinary language kind term "water" alone. Since we cannot just turn to chemistry to find a *single* chemical kind that can be used to determine the extension of "water," there is not any strict sense in which water is H_2O, because exactly what water is depends on the context in which "water" is uttered.

ACKNOWLEDGMENTS

Many thanks to Peter Godfrey-Smith, Michael Strevens, Anthony Everett, and Deena Skolnick who helped me to clarify the main ideas of this paper tremendously. I would also like to thank Philip Kitcher, Sandra Mitchell, Paul Churchland, Roald Hoffmann, John Brauman, Paul Needham, Tania Lombrozo, and Daniel Corbett for helpful discussions of earlier drafts. The work in this paper was partially supported by an NSF Graduate Research Fellowship.

NOTES

1. There is, of course, a good deal of variation among semantic externalists. In this paper, I will primarily be discussing the views of Putnam (1975) and Kripke (1980).
2. One philosopher who has appreciated the problems with this view for biological kinds is John Dupré (1993).
3. Many discussions of natural kinds in the philosophical literature treat the macroscopic as the domain of common sense and the microscopic as the domain of science. Johnson (1997), for example, makes this claim in his distinction between chemical kinds and manifest kinds. Although I will not be saying much about chemical treatments of the macroscopic properties of water, these are very important. While there may be such a thing as a manifest kind, it is important to see that chemical kinds can also be individuated at the macroscopic level in virtue of the ensemble structures of substances.
4. A more formal way of making this point is to say that the chemical behavior of the two isotopic isomers is the same to a first approximation. Roughly speaking this is because the chemical behavior of the different isotopes is proportional to $\sqrt{m}$, where m is the reduced mass of the molecule. This is exemplified in molecular velocity, the vibrational frequencies of IR spectra, and other chemically important properties of molecules. Isotopic differences typically only change the reduced mass slightly. The different isotopes of hydrogen, however, are significantly different in mass, enough to give rise to significant chemical differences between isomers.
5. The *qua problem* has to do with the multiplicity of kinds associated with any particular object. Sterelny imagines that on a mission to Mars, he spots a catlike animal and calls it a "schmat." He writes: "... the schmat will be a member of many kinds. A non-exhaustive list would include: physical object, animate object, animate object with certain structural properites, ... " (Sterelny, 1983, p. 120) The problem is

how to determine what kind was baptized as a schmat on Sterelny's visit to Mars. Semantic externalists writing today are sensitive to this issue and take seriously the need to address this problem. See, for example, Devitt and Sterelny. (1990, p. 90).

6. See Needham (2000) for a thorough discussion of these issues.

REFERENCES

Devitt, M. and Sterelny, K. 1999. *Language and Reality.* Cambrdige, MA: MIT Press.

Dupré, J. 1993. *The Disorder of Things.* Cambridge, MA: Harvard University Press.

Johnson, M. 1997. Manifest kinds. *Journal of Philosophy* 94: 564–583.

Kripke, S.A. 1980. *Naming and Necessity.* Cambrdige, MA: Harvard University Press.

Mellor, D. N. 1974. Natural kinds. *British Journal for the Philosophy of Science* 28: 299–312.

Needham, P. 2000. What is water? *Analysis* 60: 13–21.

Pauling, L. 1947. *General Chemistry.* San Francisco: W. H. Freeman.

Putnam, H. 1975. The meaning of 'meaning'. In *Mind, Language, and Reality. Collected Papers. Vol. 1.* Cambridge: Cambridge University Press.

Sterelny, K. 1983. Natural kind terms. *Pacific Philosophical Quarterly*, 64: 110–125.

Stroll, A. 1998. *Sketches of Landscapes: Philosophy by Example.* Cambridge, MA: MIT Press.

CHAPTER 19

FROM METAPHYSICS TO METACHEMISTRY[1]

ALFRED NORDMANN
Department of Philosophy, Technische Universität Darmstadt

THE PROMISE OF METACHEMISTRY

In 1940 appeared *La Philosophie du Non* by Gaston Bachelard. The American edition of 1968 translates the title obviously enough as *The Philosophy of No* and 10 years later, the German followed with *Die Philosophie des Nein*. And yet, *Die Philosophie des Nicht* would have been more appropriate and in English—impossible though it sounds—*Philosophy of Non*. After all, taking his cue from non-Euclidean geometry, Bachelard revels in the "non" of non-Aristotelian logic, non-Cartesian epistemology, non-Baconian science, non-Kantian ontology, non-Newtonian mechanics, and non-Lavoisian chemistry. In all these cases, the "non" does not signal a negation or antithesis but marks Euclidean geometry as a special case of a differentiated non-Euclidean geometry, Lavoisian chemistry as a limited set of practices which is dialectically reflected in non-Lavoisian chemistry, etc. (Bachelard 1968, 55, 115)

According to Bachelard, new experimental procedures and practices of the sciences introduce new ways of identifying, positioning, inferring, or stabilizing events. The sciences thus add over time new layers of conceptualization for properties to "take root" (Bachelard 1968, 45), new spectro-lines to the "epistemological profile" of notions like "mass," "energy," or "substance." Bachelard, therefore, introduces his philosophy of the "non" not as a general theory of science but as an attempt to capture and articulate the significance of an emerging new science that creates in its wake also a new philosophy (cf. Bachelard 1984, 3): Bachelard's "non" gives "some pre-sentiment of a profound revolution in chemical philosophy." Signaling this imminent revolution, Bachelard continues, "metachemistry would already seem to be a possibility."[2] And: "Metachemistry would be to metaphysics in the same relation as chemistry to physics" (Bachelard 1968, 45). This essay explores Bachelard's promise of metachemistry. Along the way, it assembles a series of clues that suggest that in the meantime, metachemistry has been more fully articulated or realized in the work of Bruno Latour.[3] Though he does not use that term, Latour's *Pandora's Hope* (Latour 1999), for example, is a metachemical treatise. However, while Bachelard tries to determine for a new scientific age the relation between metaphysics and metachemistry, Latour offers metachemistry as a way to dissolve

D. Baird et al. (eds.), Philosophy of Chemistry, 347–362.

metaphysical pseudo-problems for science in general. This difference calls for an exploration of intellectual contexts. Beginning with the challenge issued by chemist-turned-philosopher Émile Meyerson, this exploration might continue with the response to that challenge by some-time-chemist Gaston Bachelard,[4] and then perhaps conclude with Bruno Latour's inheritance of Meyerson's and Bachelard's problematics even as he rejects their rationalism.[5] However, instead of reconstructing contexts and trajectories of influence, the following remarks primarily attempt to get past the idiosyncracies of Bachelard's style—the excess of neologisms, in particular—and clarify his contrast of metaphysics and metachemistry:

> Metaphysics could have only one possible notion of substance because the elementary conception of physical phenomena was content to study a geometrical solid characterized by general properties. Metachemistry will benefit by the chemical knowledge of various substantial activities. It will also benefit by the fact that true chemical substances are the products of technique rather than bodies found in reality. This is as much as to show that the real in chemistry is a realization. (1968, 45)

Bachelard's suggestion can be unpacked by highlighting the various stages of this movement from different conceptions of "substance" to the physical, social, as well as conscious "realization of the real." Implicitly and explicity, the metaphysics and metachemistry of science will be juxtaposed throughout.

THE SUBSTANCE OF "SUBSTANCE"

Metaphysics, Bachelard suggests, operates with an impoverished, insubstantial notion of substance which it inherited—as did classical physics—from the Greek conception of science and its interest in that which persists through change.[6] "Metaphysics could have only one possible notion of substance because the elementary conception of physical phenomena was content to study a geometrical solid characterized by general properties" (Bachelard 1968, 45).

The general properties of elements are the properties of matter, whether considered as extension and impenetrability or in terms of force or energy. From the spatio-temporal arrangements and re-arrangements of these elements, everything is thought to be composed. This notion of substance is entirely undifferentiated; it does not distinguish anything in particular but characterizes everything material. At the same time, it is generously hypothesized as a pervasive substrate of reality. According to physics and metaphysics, for everything that happens and for far more that could happen, there are latent, immutably lawful general properties waiting to be activated and to manifest themselves. Nature has thus become overpopulated with innumerable dormant powers that are semantically significant yet physically inconsequential.[7] The varied critiques of metaphysics, therefore, targeted the hypothetical character of substance (though rarely its multiplication beyond necessity), but kept maintaining that all that could be meant by the term "substance" is a persistent constituent of reality.[8] So, while some critics now claimed that reason or subjectivity is the substance of the world, and while others took the category "substance" in a Kantian manner as a conceptual pre-condition for the possibility of scientific knowledge, the term kept referring to something self-sufficient and undifferentiated: It is the stuff of the world that persists through change; it is an immutable carrier of accidental properties that is never directly or perceptually cognizable. Since substance is always, perhaps necessarily *hypothesized*, it is attended by the metaphysical question concerning its existence all the way down to the contemporary contest between the "cantankerous twins" of realism (or objectivism) and relativism (Feyerabend 1991, 515).[9]

This *physical* conception of substance was questioned by scientific philosophers like F.W. Schelling, Charles Sanders Peirce, Émile Meyerson, and Alfred North Whitehead.[10] With Bachelard, one might say that their critical questions introduce a *chemical* conception of substance into philosophy. For the chemist, the term "substance" designates first and foremost the particular elementary or compound stuff that stands at the beginning and at the end of a chemical process (cf. Bachelard 1968, 45, 49, 60, 70). As such, chemical substance is no hypothetical substrate but presents itself in chemical practice. Questions regarding its reality concern not its existence but how it makes itself known. Since chemical substance presents itself at different levels of laboratory experience, Bachelard posits a "laminated reality" for chemical substance—"substance does not have, at all levels, the same coherence" (Bachelard 1968, 46):

> In the early days of organic chemistry people used to like to believe that synthesis merely served to verify the exactitude of a piece of analysis. Practically the reverse is true now. Chemical substances only get to be truly defined at the moment of their reconstruction. (1968, 47)

As long as synthesis was strictly an analog to analysis, chemistry retained a limited focus on the elementary substrate of particular compounds. At this level, substances are still individualized such that each chemical element might have its own substance (Stengers 1994; Bensaude-Vincent 1994). The coherence of substance increased, however, when synthesis came into its own. The multiple techniques of realization established new chemical relations, suggested functional groupings, allowed for combinations produced in the laboratory to illuminate the combinations found in nature.[11] To the extent that this chemical notion of substance became generalized only through the development of converging chemical techniques, the success of the chemist is at odds with the conceptualizations of the (metaphysical) philosopher:

> In the face of a reality which has been so surely constructed, let philosophers equate substance, if they will, with that which evades cognition in the process of construction, let them continue, if they will, to define reality as a mass of irrationality.[12] For a chemist who has just realized a synthesis, chemical substance must, on the contrary, be equated with what one knows about it . . . (1968, 47)

This opposition is also present, for example, in Whitehead's critique of the metaphysical notion of substance, a critique that employs chemical metaphors and ultimately refers to chemistry. According to Whitehead, when we posit that our sense perceptions are merely attributes of a substance, i.e., merely the effects in our minds of an underlying reality that is already given, "a distinction has been imported into nature which is in truth no distinction at all":

> what is a mere procedure of the mind in the translation of sense-awareness into discursive knowledge has been transmuted into a fundamental character of nature. In this way matter has emerged as being the metaphysical substratum of its properties... (1920, 16)[13]

On Whitehead's account, the "substance" or "substratum" of classical metaphysics results from a process of translation and transmutation, a procedure of the mind that ought to be recognized as such. "If we are to look for substance anywhere," he concludes, "I should find it in events which are in some sense the ultimate substance of nature" (Whitehead 1920, 19). An example from chemistry helps define what an "event" is: It is "a nexus of actual occasions... For example, a molecule is a historic route of actual occasions; and such a route is an 'event'" (Whitehead 1978, 80).[14]

Through a somewhat circuitous historic route of their own, Whitehead's remarks would leave their mark on Bachelard's *Philosophy of Non*.[15] Bruno Latour refers to them more explicitly. He, too, shifts from "substance" with all its metaphysical baggage to "event" or "institution." While this shift applies to all science, it originates in Latour's discussion of the work of a chemist:[16]

> What Pasteur made clear... is that we slowly move from a series of attributes to a substance. The ferment began as attributes and *ended up being a substance*, a thing with clear limits, with a name, with obduracy, which was more than the sum of its parts. The word 'substance' does not designate what 'remains beneath,' impervious to history, but what gathers together a multiplicity of agents into a stable and coherent whole. A substance is more like the thread that holds the pearls of a necklace together than the rock bed that remains the same no matter what is built on it... substance is a name that designates the *stability* of an assemblage. (1999, 151, *cf.* 167 and 170)[17]

Whitehead thus anticipated and Latour echoes Bachelard's remark, quoted above, "that true chemical substances are the products of technique rather than bodies found in reality" (Bachelard 1968, 45). Now, if events, synthetic reconstructions, and technical realizations are in some sense the ultimate substance of nature, what happened to the original observation that chemical substance is first and foremost the particular, elementary or compound stuff that stands at the beginning and at the end of a chemical process? According to Bachelard, science adds to the "naively" realistic identification of substance (where "substance" is simply predicated) another, "rationalized" layer of meaning (which takes "substance" to be a category of the understanding). While these two layers of meaning exist together in each individual mind and while for each layer substance is what one knows about it, the two layers conjoined do not yield a single coherent notion of "substance." This is how Bachelard's conception of laminated or layered reality finally arrives at the third layer of "non-substantialism": Just as a molecule traverses a historic route of actual occasions, so does the notion of 'substance' itself. And since that route itself is the event and since the event is in some sense the ultimate substance of nature, Bachelard finds that the ultimate substance of 'substance' has dissolved into its own history of rationalizations and conceptualizations (1968, 44, 72f., 76).[18]

LAVOISIAN SCIENCE

Bachelard's construction extrapolates not only from the history of philosophy but also, more significantly, from that of chemistry. According to Bachelard and Meyerson before him, the development of chemistry itself offers a philosophical history of "substance." Like Meyerson, Bachelard believed that only with Lavoisier "(t)he scientific mind has... completely supplanted the pre-scientific mind" (Bachelard 1968, 47f.). According to Meyerson, "science in its entirety" takes place in the interval between on the one hand perceived, sensible every-day reality which science destroys, and on the other hand the eventual "disappearance of matter (or the dissolution of substance) into the ether" which concludes the project of science (Meyerson 1991, 407). While "the claims of phlogiston theorists were also based on observation" (Meyerson 1991, 207), they did not employ the principle of conservation consistently, did not use that principle to destroy sensible reality, did not "impoverish reality to create legalistic science" (Meyerson 1991, 407), and they, therefore, spoke of chemical change in qualitative terms as if properties passed from one body to another (Meyerson 1991, 206f.). Lavoisier employed a principle of conservation to institute a legalistic science which then creates "theoretical science by stripping reality as much as possible of any qualitative elements" (Meyerson 1991, 407), substituting the motion of invisible entities for the passage of sensible properties (Meyerson 1991, 206f., cf. 62ff.).[19] "In a certain sense," Meyerson adds, the quantitative beings of theoretical science are "even more substantial" than "the things we believe we perceive": "They are assumed to be actual substances, and science, by taking away their qualitative aspect... added to their perdurability" (Meyerson 1991, 407). Thus, the essential form of our science appears to us to be shaped above all by the concern to explain that which changes by that which persists (Meyerson 1991, 130, cf. 119).

On Meyerson's interpretation, then, Lavoisier took an important step toward the fulfillment or realization of the Greek program in philosophy and science: He dissolves time into space and shows that qualitative change is not real while the underlying invisible elements are all the more real as they persist through mere displacements or changes of location.[20] Meyerson thus takes the Lavoisian shift from one conception of substance to another as instituting just that idealized conception of science which Meyerson embraces. While Meyerson is explicitly committed to a metaphysical conception of science, his historical account yields an implicitly metachemical claim. Indifferent to the problem of existence and the ultimate foundation of reality, metachemistry concerns the processes by which reality is transformed. According to Meyerson, Lavoisian science is guided by a physical ideal and thereby transforms reality by dissolving matter into ether, time into space, that which is perceived into that which is inferred and, of course, chemical substance into physical substance. Meyerson's implicit metachemistry came into its own when the further development of chemistry revealed that the "Lavoisian" ideal is just that—a particular metachemical stance among others, compelling but limited. According to Bachelard, these limitations became apparent and an explicit metachemistry became a possibility, once certain tensions within the Lavoisian conception of science and substance became productive and served to differentiate Lavoisier's substantialism or—in Bachelard's terminology—once they "dialectized" it.[21]

As with all dialectical movements, the development of alternative or differentiated metachemical stances and of non-Lavoisian science accentuates ambiguities that were already present in Meyerson's conception of science as well as in Lavoisier's conception of chemical substance.[22] On the one hand, as described by Meyerson, Lavoisier physicalizes chemical substance by ruling that "in all the operations of art and nature, nothing is created." On the other hand, as Bachelard points out, Lavoisier's scientific practice and his definition of "element" establish that as the products of technique, substances are realized.[23]

> Realizations have to be multiplied. One has more chance of knowing sugar by making sugars than by analyzing a particular sugar. In this plan of realizations, one is not looking for a generalization anyway, one is looking for a *systematization*, a *plan*. The scientific mind has *then* completely supplanted the pre-scientific mind. To *our* way of thinking, then (as opposed to Meyerson's), *this* is reverse realism . . . *It* is the foundation of chemical rationalism. (1968, 47f.; emphases added)

Bachelard's remark articulates tensions within Lavoisian science: The creative work of scientists in Lavoisier's laboratory is supposed to establish that nothing is created and everything merely discovered. The nature of "sugar" arises from the making of "sugars," and the substance in the singular appears not as an immediate likeness among the plurality of particulars but as a co-ordination of practices. The presumed nature informs research only as a category of possible experience for the representation of "sugar." Instead of generalizing from the experience of an unchanging reality, the systematic course of inquiry and the plans of science carve out an unchanging reality. This is "reverse realism" in that reality appears not as the cause of perception but as the product of inquiry.[24] And while Meyerson celebrates the intended product of Lavoisier's science, Bachelard emphasizes its procedure. According to Bachelard, it is this tension between intended representation and the making of it which gives rise (within the rationalized substantialism of Lavoisian science) to the non-substantialism of a non-Lavoisian science that reflects this tension in its practice.

NON-LAVOISIAN SCIENCE

Bachelard only hints at this non-Lavoisian practice, and these hints can be patched together in a tentative manner at best.[25] The most prominent among them strikes right at the heart of Lavoisian science with its reliance on principles of conservation:

> As there are geometries which do not obey the displacement group, which are organized around other invariants, it is to be foreseen that there are chemistries which do not obey the conservation of matter[26] and which could, therefore, be organized around some invariant other than that of mass. (1968, 54)

Bachelard says little about what these invariants might be, whether one of them will replace "mass" or whether non-Lavoisian science varies the invariants in order to gain a multi-perspectival, properly dispersed access to the layers of reality. The latter possibility may be implicit in his suggestion of an alternative to Meyerson's Lavoisian ideal:

> it was believed that structural conditions decided everything, the idea being, no doubt, that one masters time when one is well organized in space, with the result that all temporal aspects of chemical phenomena came to be neglected. There was no appreciation of the fact that time was itself structured; no pains were taken to study rates, unfoldings, operations, transformations—along these lines, therefore, there is new knowledge to be gained. (1968, 72)

Depending on the level of organization or experimental intervention at which these transformations are studied, the new knowledge to be gained will differentiate the notion of substance. Bachelard suggests that in metachemical substance converge three separate notions, one of which is the traditional metaphysical conception of "substance." It is complemented by "sur-stance" and "ex-stance" (Bachelard 1968, 66). "Sub-stance" refers to what stands behind, beneath, or before the observed phenomena; "sur-stance" refers to what emerges in the process of realization, namely what Latour calls an institution which co-ordinates human and non-human practices.[27] "Ex-stance" finally refers to the excess of meaning that is not absorbed within a single coherent notion of substance and that tends to be overlooked by Latour.[28]

Metachemistry would thus "*disperse* substantialism," where the metaphor of dispersion is borrowed from spectrographic analysis (Bachelard 1968, 45). Extending this spectroscopic analogy, Bachelard represents the differentiation of "substance" by means of a chemical and not at all Freudian "psycho-analysis." The spectrographic dispersion of substantialism produces spectro-lines of "substance" where the ensemble of the lines provides the psycho-analysis of the mental construction of that concept. What Bachelard calls the epistemological profile of "substance" is at the same time a representation of its laminated reality, namely the successive layers of naive realism with its predicative use of "substance," of rationalism or Kantianism in which substance is a category, of its "dynamization" in terms of sub-stance, sur-stance, ex-stance. The resultant profile would be similar to the one he produced for his personal notion of "mass" (Figure 1). Since "(p)hotochemistry, with the spectroscope, seems

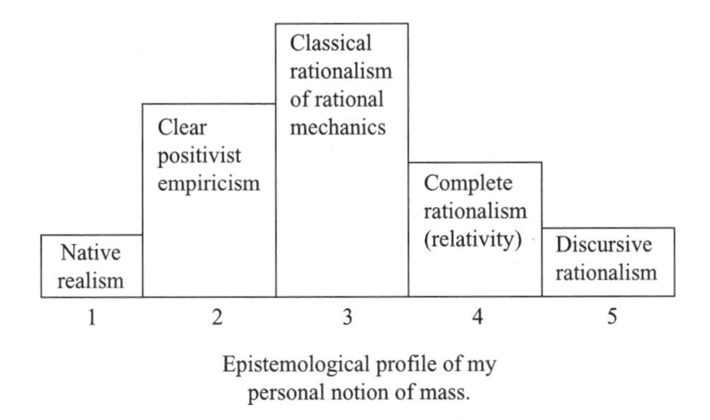

Figure 1: Bachelard's spectroscopic "psycho-analysis" of mass (Bachelard 1968, 36).

to be a non-Lavoisian chemistry" (Bachelard 1968, 58), Bachelard here draws in a self-exemplifying manner on a non-Lavoisian technique to advance his argument for a non-Lavoisian metachemistry.

THE CHEMICAL TRAJECTORY OF PHYSICAL REALIZATION

As opposed to metaphysics, metachemistry does not attempt to decide between naive realism and the various shades of rationalism; instead, it produces dispersion analyses of notions like substance or mass, fire or air, at various stages on their route toward the realization of the real. It is this route and the development of propositions over time that finally needs to be elucidated.

"(W)here shall we find the facts that foreshadow, as we have come to believe, the non-Lavoisian aspect of generalized chemistry?" Bachelard asks and immediately provides an answer: "They are to be found in the *dynamization* of chemical substance" (Bachelard 1968, 55). He then begins to elaborate this dynamic and temporal, if not historical character of chemical substance:

> (T)he reaction must henceforward be represented as describing a course, as forming a chain of various substantial states, or a movie film of substances... Immediately a *becoming* defines itself underneath being. Now this becoming is neither unitary nor continuous. It presents itself as a sort of dialogue between matter and energy... Energy is as *real* as substance and substance it *not more real* than energy. Through the intermediary of energy time puts its mark on substance. The former conception of a substance by definition, outside of time, cannot be maintained. (1968, 56f.)

The former conception of substance rendered it a hypothetical entity precisely because it was posited as that which persists through time. Even among those who agreed that the world is intelligible to the human mind only if the persistence of substantial characteristics is assumed, the gulf between realists and anti-realists opened up: With Whitehead, Locke, and most Kantians, anti-realists suspect that realists are transmuting a posit of the mind into a fundamental character of nature (see the discussion of Whitehead in the section on "The Substance of 'Substance'" earlier). However, as soon as one lets "time put its mark on substance," the hypothetical character of persistence drops out, and the old debate of realists versus anti-realists becomes obsolete. It is replaced by the question of how substance is instituted and how its reality becomes physically, socially, consciously realized over the course of time.

Whitehead referred the institution of "substance" to the nexus of actual occasions or operations, i.e., to an intrinsically historical event. Bachelard goes on to pursue "the dynamization of substance" and considers its history as a *series* of events by adopting Paul Renaud's metaphor of a "chemical trajectory" (Bachelard 1968, 61) that can be represented as a continuous line or curve:

> It is quite natural to say... that the substance being purified *passes through* successive states, and it is no far cry from here to the supposition that purification is *continuous*. If one hesitates to postulate this continuity, at least it is not difficult to accept... that the purification can be *represented* by a continuous line. (1968, 61)

This continuous line represents the "incorporation, within the definition of substances, of the conditions needed to detect them" (Bachelard 1968, 59). In other words, "when one of the variables included in the representation is *time* and the other variable corresponds to some characteristic of substance," a chemical trajectory becomes visible (Bachelard 1968, 64).

What Bachelard is suggesting here is that one might graph the definition or institution of substance. This graph would represent a route or passage of purification. The choice of variables avoids the metaphysical pre-supposition that there is a stable "it" that is being purified. The apparent constancy of this "it" emerges only from the actually observed persistence of the characteristics over time, i.e., only from the tentative continuity of the line. In this sense, the "representation" provided by the graph is not a representation *of* reality but expresses the "*supremacy of representation over reality*" (Bachelard 1968, 62):

> The representation of the purifying passage vouchsafes that there is a continuous "it" there.[29] Substances thus emerge from the acquisition of more and more characteristics: They become more articulate and better articulated as they incorporate "more and more of the conditions needed to detect them." (1968, 59)

In other words, substances become increasingly reliable or stable actors in experimental and technological interactions, i.e., as the situations are defined and become defined in which they will assert themselves in certain ways.[30] The trajectory is, therefore, graphed in reference to two variables: The time that passes as the work of science goes on, and a scale that registers the increasing specificity of the characteristics with which the substance becomes identified.

In the course of his own metachemical investigations, Bruno Latour produced Bachelard's graph. It represents not the discovery of Tasmania but its "construction" through the collaboration of navigators, explorers, ships, currents, coasts, mapmakers, etc. (Figure 2).

Over time and through the collective work of scientists, some vague "it out there" acquires more and more characteristics, more and more associations, becomes

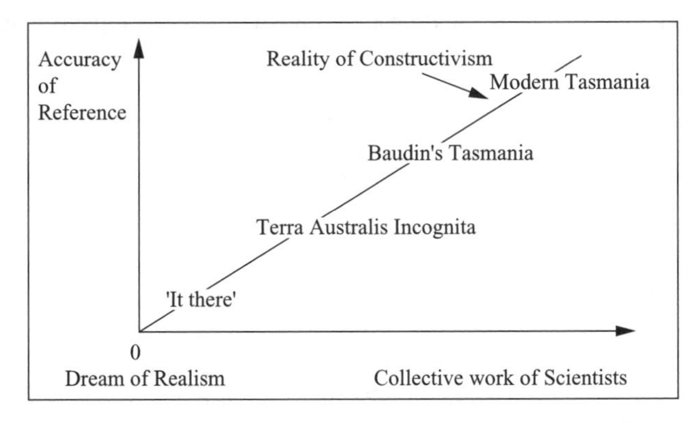

Figure 2: The realization ("discovery", exploration, mapping, etc.) of Tasmania (Latour 1990, 68).

institutionalized at the nexus of numerous occasions until it becomes "modern tasmania" to which we now refer with great accuracy.[31] Latour's graph attempts to capture also how the gulf between realism and anti-realism is bridged by this representation of substance as vouchsafed by a "historic route of actual occasions" (Whitehead 1978, 80).

The "dream of realism" views the trajectory prospectively as the unfolding of a given substantial reality. Metaphysical realists would claim that Tasmania always was what it is, that it was merely discovered and has not substantially changed from the time when it was completely unknown and void of any specified characteristics, to today when it is articulated in great detail. Latour and the metachemistry of science consider a peculiar obsession this attempt to insist that "modern tasmania" is identical to the eternal timeless substance that, on the dream of realism, must have been there all along.[32] One can understand this obsession if one understands that for the metaphysician the timelessness of substances serves as the foundation of the real and that, therefore, the denial of this identity would cast us into the abyss of relativism and deprive us of all reality (cf. Latour 1999, 3–9, 296).

The "reality of constructivism," on the other hand, views the graph's point of origin as a fictitious common referent, the vague "it out there" to which characteristics gradually accrue and are imputed retroactively. The trajectory itself, i.e., the history of "Tasmania" links the dream of realism (world) and the reality of constructivism (language). Instead of pitting realism against constructivism, Latour argues that the dream of a stable reality is realized as stability is forged, and one of the characters that "Tasmania" acquires in the course of its interactions with navigators, map-makers, inhabitants is the character of obduracy or persistence.[33] If things, objects, substances, and facts have histories, they may well have been different in the past and did not yet have certain defined traits before humans encountered them. However, Tasmania proved itself to be so steady and reliable that we readily extrapolate its existence into the future and the past, claiming by a kind of inductive argument that it possessed its most stable characteristics even before they acquired the character of stability-in-interactions.[34] The metachemist Latour thus agrees with Whitehead, Bachelard, and—philosophically more radical and sophisticated than either of them—Charles Sanders Peirce. They all view reality as standing at the end of inquiry or appearing in its course.

METAPHYSICAL PHILOSOPHY, METACHEMICAL HISTORY OF SCIENCE

If metachemistry stands to metaphysics as chemistry to physics, what then is their relation? The answer to this question refers us to the thorny issue of whether physics is somehow fundamental or prior to chemistry—a loaded question in that the very notions of the "fundamental" and the "prior" have "metaphysics" and not metachemistry inscribed in them. Here is a tentative sketch of the relationship: Bachelard and Meyerson agree (but Latour tends to overlook[35]) that metaphysics posits the scientific picture of the world, i.e., conditions of intelligibility. Metaphysics is, in that sense, conceptually prior and ulterior; it formulates (as Meyerson emphasized most explicitly) an idealized anticipation of what science intends. In the language of Peirce, one should say that metaphysics introduces the hypothesis of reality. How must we conceive world and nature if we want to arrive at or realize a stable and representable reality? And in Kantian terms, one might say that metaphysics specifies the conditions for the possibility of objective knowledge. The world it posits is all that is the case, i.e., the static world of Wittgenstein's *Tractatus* which is composed of discrete states of affairs and which is amenable to representation. The modern science of chemistry also intends such a world, and (meta)physics has, therefore, come into chemistry by way of conservation principles, by way of the periodic table and its interpretation in terms of substantial atoms or in terms of molecules and an insubstantial plurality of elements. It also comes into chemistry by way of quantum mechanics and its attendant tools of analysis.[36]

In contrast, metachemistry informs and traces the practice of science. Moving alongside science, it characterizes the stuff of science, namely the operations in the heads and laboratories of scientists: What kinds of transformation produce a durable representation of the world that can be fixed in thought as a world of representation? Any theory of inquiry, therefore, must treat objects and substances, instruments and propositions, models and theories, nature and culture, mind and matter metachemically on their historical trajectories. In the language of Peirce, one should say that metachemistry develops the hypothesis of reality: How do humans and nature interact in the fixation of belief and reality?[37] If there were Kantian terms available here, they would concern the conditions for the synthesis of apperception. The later Wittgenstein, at any rate, considered metachemistry by trying to relate sentences like "all is flux" to the static world of the *Tractatus*. To the extent that chemistry is still a science of becoming (Earley 1981; Müller 1994) and to the extent that all sciences are engaged in the realization of the real, metachemistry extends from the practice of chemistry into the laboratories of physics.

The priority of (meta)physics, therefore, consists in the scientifically intended image of an objectively knowable world, the final and formal causes of scientific inquiry. The priority of (meta)chemistry, on the other hand, consists in the synthetic making of this stable world of representation, the material, and efficient causes of scientific inquiry.

NOTES

1. Small papers with grand titles are bound to be programmatic. This one is no exception and, along with Nordmann (2000), marks only the beginning of what should become an extended course of inquiry.
2. I am here following the French original: "Dès maintenant, une métachimie nous paraît possible . . . La métachimie serait à la métaphysique dans le même rapport que chimie à la physique" (Bachelard 1981, 53).
3. See Latour (1999, 142–144) on the relation between articulation and realization (a relation emphasized already by Charles Sanders Peirce). A metachemical stance has also been suggested in the later work of Feyerabend (1991, 507–521).
4. He taught physics and chemistry from 1919 to 1930 at the Collège Bar-sur-Aube; one of his two dissertations of 1927 concerned thermal propagation in solids; and in 1932 he published *Le Pluralisme Coherent de la Chimie Moderne*. On Bachelard's philosophy of science see Tiles 1984.

5. While Bachelard rejects Meyerson's conservative or Lavoisian ontology of timeless identities rather than historicized substances, he shares with Meyerson the conviction that science rationalizes nature. Latour completes the critique of Meyerson when he rejects Bachelard's rationalism and criticizes his view that epistemological breaks objectify science by displacing intuitive and mythologically "realistic" views (Latour 1995, 81 and 124ff.).
6. Aside from some of the following quotes, see Ted Benfey's opening remarks at the 1999 International Conference on the Philosophy of Chemistry.
7. Bachelard describes how "each simple substance actually received a substructure. And the characteristic thing was that this substructure revealed itself as having a totally different essence from the essence of the phenomenon being studied. In explaining the chemical nature of an element by an organization of electric particles [...] [a] sort of non-chemistry constituted itself to sustain chemistry" (Bachelard 1968, 52).
8. This can be seen, for example, in two historical dictionaries of philosophy that represent the state of philosophical discussion at the time when Whitehead enters the scene with Bachelard following soon thereafter. See the entries on "Substanz" in Michaëlis (1907, 607–610) and in Eisler (1904, vol. 2, 450–464).
9. These are "cantankerous twins" because they arose together along with the metaphysical conception of substance. According to Bachelard, for the metaphysical realist "*existence is a one-toned function*," that is, everything is real or unreal in the same way, in that it either exists or does not exist—"the electron, the nucleus, the atom, the molecule, the colloidal particle, the mineral, the planet, the star, the nebula" (Bachelard 1968, 46).
10. Indeed, the physical concept of substance was questioned implicitly also by the development of the sciences throughout the long 19th century (1780–1920), by the development of modern chemistry, statistical mechanics, Darwinism, electrodynamics, relativity and quantum theory. According to Wilhelm Ostwald, therefore, the notion of "substance" merely sets the task for scientists to determine what does and what does not possess the property of conservation or persistence (cf. Eisler 1904, vol. 2, 464). It would be a worthwhile project to see just how far back into the 19th century can be traced the notion that the chemical notion of substance is so deeply at odds with the physical one as to require a re-orientation of philosophy. As early as 1863, for example, Ernst Mach criticizes the mechanical conception of substance by suggesting its inapplicability to chemistry. One should not "imagine the chemical elements lying side by side in a space of three dimensions" or else "a crowd of the relations of the elements can escape us" (Mach 1911, 53, see 43, 54, and 86ff.). Lange quotes Mach in his *History of Materialism* (Lange 1925, 388). According to Bertrand Russell's 1925-preface to the English edition, Lange's own work contributed to the critique of substance, a concept that "persisted in the practice of physics" until the arrival of relativity theory. At the present time, according to Russell, physics can finally agree that "[n]othing is permanent, nothing endures; the prejudice that the real is the persistent must be abandoned" (Lange 1925, xii).
11. Bachelard quotes a particular example concerning the chains of groups of CH_2 from Mathieu (1936, vol. 1, p. 9).
12. Bachelard's strong claim that metaphysians "define reality as a mass of irrationality" is echoed and perhaps clarified by Bruno Latour. According to the latter, the metaphysical conception of substance recommends itself on first sight because it gives "a clear meaning to the truth-value of a statement": Scientific statements can be said to refer if and only if substantial states of affairs correspond to it (for the tight connection between physical substantialism and the representability of states of affairs, see Wittgenstein's *Tractatus* 2.0211). With the notion of reference, however, arises the problem of how to bridge the gap between language and world: "In spite of the thousands of books, philosophers of language have thrown into the abyss separating language and world, the gap shows no sign of being filled [...] except that now we have an incredibly sophisticated version of what happens at one pole—language, mind, brain, and now even society—and a totally impoverished version of what happens at the other, that is, *nothing*" (Latour 1999, 148).
13. "The history of the doctrine of matter has yet to be written. It is the history of Greek philosophy of science. [...] The entity has been separated from the factor that is the terminus of sense awareness. It has become the substratum for that factor, and the factor has been degraded into an attribute of

the entity. In this way, a distinction has been imported into nature that is in truth no distinction at all" (Whitehead 1920, 16). Elsewhere, Whitehead characterizes this transmutation as the fallacy of misplaced concreteness. According to Bachelard, Kantianism moved from naive realism and traditional metaphysics to a critical rationalism that considers "substance" a category of the understanding (see Kant's *Critique of Pure Reason*, note on B xxi). Whitehead's critical question places him between Kantianism and Bachelard's third stage of chemical philosophy (Bachelard 1968, 45, 50ff.).

14. Joseph Earley (1981) amplifies this chemical dimension of the "event." Whitehead qualifies the "nexus of actual occasions" in a manner that threatens to sneak the abandoned metaphysics back in: The nexus is "inter-related in some determinate fashion in some extensive quantum" (1980, 80). Quite in the spirit of Bachelard's "philosophy of 'non'," however, Whitehead's non-metaphysical and metachemical conception of substance appears to *explain* metaphysics in metachemical terms, namely as a transmutation (as, for example, through misplaced concreteness).
15. Bachelard (1968, 66) refers for inspiration to a remark "Sur Whitehead" by Jean Wahl: "La réflexion sur Whitehead me fournit aujourd'hui un nouvel élément: à l'idée de substance pourrait-on substituer l'idée de sur-stance? (idée qu'un ensemble organique est quelque chose de supérieur à ses éléments). On retrouverait l'entéléchie aristotélicienne, la 'vérité' hégélienne, l'émergence d'Alexander" (Wahl 1938, 931). A philosophical history of metachemistry would, therefore, have to consider wider contexts such as French existentialism.
16. Latour may have been introduced to Whitehead through Isabelle Stengers, cf. Latour (1996, 1999) and Prigogine and Stengers (1984).
17. Latour's mixing of metaphors indicates just how difficult it is even for him to get away from appeals to prior substances. For Latour's "search for a figure of speech," see Latour (1999, 133–144).
18. Latour details this process in a chapter on "The Historicity of Things: Where Were Microbes Before Pasteur?" (Latour 1999, especially 164–172; see also Latour 2000).
19. Compare Bernadette Bensaude-Vincent's characterization of Lavoisier's achievement: "In the act of weighing, Lavoisier sought to create an experimental space that was entirely under the experimenter's control. Once balanced with weights on Lavoisier's scale, substances were transformed from objects of nature to objects of science. The balance divested substances of their natural history. Their geographical and geological origins, their circumstances of production made little difference. They were transformed into samples of matter made commensurable by a system of standardized weights. [...] In translating the conservation of elements' qualities into quantitative and ponderal terms, the principle dodges the question how—in what form—the elements are conserved. How do they subsist in compounds and how do they move from one compound to another during a reaction?" (Bensaude-Vincent 1992, 222ff.)
20. According to Meyerson, Carnot's principle concerning the irreversibility of time is factually true but irrational in that rational science follows the principle of causality that pre-supposes reversibility and the identity of antecedent and consequent; science thus "tends to the elimination of time." Carnot's principle asserts reality as it resists, from without, our scientific attempts at rationalizing it (cf. Meyerson 1930, 278, 286, 317). According to Meyerson, Carnot's principle signifies a limit of science.
21. However, cf. Bachelard's remark that "a metachemistry came into being with the Mendeleef table" which appears to be at odds with the sustained emphasis on metachemistry as part of a new scientific spirit (Bachelard 1968, 49).
22. "It is to be understood—we cannot stress this often enough—that a non-Lavoisian chemistry, like all the activities of the philosophy of non, does not deny the utility of classical chemistry, either for the past or for the future. It tends merely toward the organization of a more general chemistry, a pan-chemistry, just as pan-geometry tends to give the plan for the possibilities of geometrical organization" (Bachelard 1968, 55).
23. Latour refers to Bachelard's "un fait est fait" when he embarks upon his own project to show how it can be that facts are at the same time "fabricated" and "real" (Latour 1999, 127), how the statement "the ferment has been fabricated in my laboratory" can be understood as synonymous with "the ferment is autonomous from my fabrication" (Latour 1999, 135).
24. According to Peirce, this "reverse" or Kantian realism is just realism plain and simple as opposed to nominalism that posits reality as prior to inquiry. Peirce articulates this in some of his earliest papers in Peirce (1992). Peirce reflects the tension between Meyerson's explicitly metaphysical stance with its

"nominalist" conception of a substantial and persistent reality and his implicitly metachemical account with its "realist" conception of science as effecting transformations and creatively producing reality in the metaphysical or nominalist image.

25. A more complete account might be reconstructible if one considered Bachelard (1932), the work of Georges Matisse who appears to have coined the term "non-Lavoisian" (Matisse 1938), perhaps Paul Renaud's *Structure de la Pensée et Définitions Expérimentales* (1934) or the influence on Bachelard of Leo Brunschvicg (1937).
26. The English editon accidentally prints "water" instead of "matter," but see Bachelard (1981, 64).
27. See note 15 above for the "emergence" of "sur-stance" (in the sense of an Aristotelian or Hegelian entelechy) and on how it was inspired by Wahl and Whitehead.
28. See notes 31 and 35 below.
29. To Bachelard's interest in "representation" (Bachelard 1968, 62–64) corresponds Latour's appropriation of Whitehead's "proposition" (Latour 1999, 141, 148): Both terms elude the misleadingly dichotomized spheres of noumenon and phenomenon (Bachelard), of subject and object (Latour).
30. The preceding sentences have begun to conflate Bachelard's and Latour's vocabularies.
31. While the construction and maintenance of a fact requires an unbroken trajectory, why should the trajectory also be straight? Jeff Ramsey raises this point (cf. Ramsey 1992): Does Latour's anti-mentalism commit him to the view of an ineluctable conspiracy of associations that tends to agreement, stability, accrual of properties? If ideas or "epistemological breaks" play any rôle at all, can they do anything but introduce instability, open black boxes, unravel an accomplished reality? Substituting railway tracks for trajectories, an alternative approach is suggested by Max Weber: "Interests (material and ideal), not: ideas, immediately govern the actions of people. However, the worldviews that have been created by these ideas have often determined, like switchmen, the tracks along which action is propelled by the dynamics of interest" (Weber 1920, 252). The notion of ideas providing *direction* and altering the course of events is consistent with Latour's (and Peirce's) emphasis on continuity and the *unbroken* chain of operations: "To scientific facts pertains as to frozen fish that the chain of coldness which keeps them fresh may not be interrupted, not even for a moment" (Latour 1995, 159). Thus, while the trajectory has to be unbroken, there is no need for it to be straight, and indeed it ambulates in Latour's more recent graphs (Latour 2000, 256).
32. Latour's critique of this obsession is discussed in Nordmann (2000). His critique exemplifies another dimension of his literally non-Lavoisian commitments, namely his view that experimentation should not be considered a zero-sum-game in which nothing is gained or lost.
33. Compare Bachelard (1968, 13): "Chemical substance will come to be represented as a part—a mere part—of a process of differentiating; the real will come to be represented as a moment of realization well carried out."
34. "A little history spawns relativism," writes Latour, "a great deal engenders realism" (Latour 1996, 91). If we historicize our ways of knowing only, we remain bound to metaphysics and open an abyss between eternal truth and constructed knowledge; if, however, we historicize the production as well as the objects of knowledge, a Peircean realism becomes possible. Peirce and Latour detail this possibility, both attempting to explain how the objects become known or determined in such a way that their acquired specifications appear timeless. What the object acquires over time is substance; in other words, it takes on a nature, and thus a substance can *become* something that it has *always been*. These proposals obviously require greater critical scrutiny; the brief account provided here is far too sketchy. Cf. note 24 above, Latour (1999, 145–173).
35. Perhaps, I should say "as Latour must overlook" since he sets to out to provide without reference to mental entities (ideas, beliefs) a symmetrical "anthropology" that can *explain* the world as it is today; that is, it can explain what is and *through what activity it has become differentiated*. In contrast, the metaphysical stance *reflects* on our *representations* of the world. Once one considers thinking a powerful, consequential, and continuous human activity that occasionally produces representations, one can appreciate the relation between Latour's dynamically continuous metachemistry and a statically reflective metaphysics.
36. Bernadette Bensaude-Vincent has detailed the struggles and trade-offs that came with this fashioning of chemistry after a metaphysical image of science.

37. Physics and metaphysics articulate a timeless, perhaps quantitative ontological framework, chemistry and metachemistry study genuine qualitative change, that is, processes that cannot be represented as displacements of material points. The picture I am invoking here of the relation between (meta)physics and (meta)chemistry has all the hallmarks of late 19th-century scientific philosophy. It is questionable, of course, whether this picture has survived the 20th century.

REFERENCES

Bachelard, G. 1932. *Le Pluralisme Coherent de la Chimie Moderne.* Paris: Vrin.

Bachelard, G. 1968. *The Philosophy of No: A Philosophy of the Scientific Mind.* New York: Orion Press.

Bachelard, G. 1981. *La Philosophie du non: Essai d'une Philosophie du Nouvel Esprit Scientifique.* Paris: Presse Universitaire de France.

Bachelard, G. 1984. *The New Scientific Spirit.* Boston: Beacon Press.

Bensaude-Vincent, B. 1992. The Balance: Between Chemistry and Politics. *The Eighteenth Century* 33(2): 217–237.

Bensaude-Vincent, B. 1994. Mendeleev: Die Geschichte einer Entdeckung. In: Serres, M. (ed.), *Elemente einer Geschichte der Wissenschaften.* Frankfurt: Suhrkamp, 791–827.

Brunschvicg, L. 1937. *Le Rôle du Pythagorisme dans l'évolution des Ideés.* Paris: Hermann.

Earley, J. 1981. Self-Organization and Agency: In Chemistry and in Process Philosophy. *Process Studies* 11(4): 242–258.

Eisler, R. 1904. *Wörterbuch der Philosophischen Begriffe*, 2nd ed., 2 vols. Berlin: Mittler.

Feyerabend, P. 1991. Concluding Unphilosophical Conversation. In: *Beyond Reason: Essays on the Philosophy of Paul Feyerabend.* Dordrecht: Kluwer, 487–527.

Lange, F.A. 1925. *The History of Materialism and Criticism of Its Present Importance.* New York: Harcourt.

Latour, B. 1990. The Force and the Reason of Experiment. In: LeGrand, H. (ed.), *Experimental Inquiries.* Dordrecht: Kluwer, 49–80.

Latour, B. 1996. Do Scientific Objects Have a History? Pasteur and Whitehead in a Bath of Lactid Acid. *Common Knowledge* 5(1): 76–91.

Latour, B. 1995. *Wir Sind nie Modern Gewesen: Versuch Einer Symmetrischen Anthropologie* Berlin: Akademie Verlag.

Latour, B. 1999. *Pandora's Hope: Essays on the Reality of Science Studies.* Cambridge: Harvard University Press.

Latour, B. 2000. On the Partial Existence of Existing *and* Nonexisting Objects. In: Daston, L. (ed.), *Biographies of Scientific Objects.* Chicago: University of Chicago Press, 247–269.

Mach, E. 1911. *History and Root of the Principle of Conservation of Energy.* Chicago: Open Court.

Mathieu, M. 1936. *Les Réactions Topochimique*, 3 vols. Paris: Hermann.

Matisse, G. 1938. *Le Primat du Phénomène dans la Connaissance.* Vol. 2 *of La Philosophie de la Nature.* Paris: Presses Universitaires de France.

Meyerson, É. 1930. *Identity and Reality.* London: George Allen & Unwin.

Meyerson, É. 1991. *Explanation in the Sciences.* Dordrecht: Kluwer.

Michaëlis, C. 1907. *Kirchner's Wörterbuch der Philosophischen Grundbegriffe*, 5th revised edition by Carl Michaëlis. Leipzig: Dürr'sche Buchhandlung.

Müller, A. 1994. Supramolecular Inorganic Species: An Expedition into a Fascinating, Rather Unknown Land Mesocopia with Interdisciplinary Expectations and Discoveries. *Journal of Molecular Structure* 325: 13–35.

Nordmann, A. 2000. Blinded to History? Science and the Constancy of Nature. In: Carrier, M., Massey, G., and Ruetsche, L. (eds.), *Science at Century's End: Philosophical Questions on the Progress and Limits of Science.* Pittsburgh: University of Pittsburgh Press, 150–178.

Peirce, C.S. 1992. In: Houser, N. and Kloesel, L. (eds.), *The Essential Peirce.* Bloomington: Indiana University Press.

Prigogine, I. and Stengers, I. 1984. *Order out of Chaos: Man's New Dialogue with Nature*. Toronto: Bantam.
Ramsey, J. 1992. On Refusing to be an Epistemologically Black Box: Instruments in Chemical Kinetics during the 1920s and 30s. *Studies in History and Philosophy of Science* 23(2): 283–304.
Renaud, P. 1934. *Structure de la Pensée et Définitions Expérimentales*. Paris: Hermann.
Stengers, I. 1994. Die doppelsinnige Affinität: Der newtonsche Traum der Chemie im achtzehnten Jahrhundert. In: Serres, M. (ed.), *Elemente Einer Geschichte der Wissenschaften*. Frankfurt: Suhrkamp, 527–567.
Tiles, M. 1984. *Bachelard: Science and Objectivity.* Cambridge: Cambridge University Press.
Wahl, J. 1938. Satire. *La Nouvelle Revue Francaise* 50: 927–934.
Weber, M. 1920. *Gesammelte Aufsätze zur Religionssoziologie*, Vol. 1. Tübingen: Klostermann.
Whitehead, A.N. 1920. *The Concept of Nature.* Cambridge: Cambridge University Press.
Whitehead, A.N. 1978. *Process and Reality: An Essay in Cosmology.* New York: The Free Press.

Boston Studies in the Philosophy of Science

Editor: Robert S. Cohen, *Boston University*

1. M.W. Wartofsky (ed.): *Proceedings of the Boston Colloquium for the Philosophy of Science, 1961/1962.* [Synthese Library 6] 1963 ISBN 90-277-0021-4
2. R.S. Cohen and M.W. Wartofsky (eds.): *Proceedings of the Boston Colloquium for the Philosophy of Science, 1962/1964.* In Honor of P. Frank. [Synthese Library 10] 1965 ISBN 90-277-9004-0
3. R.S. Cohen and M.W. Wartofsky (eds.): *Proceedings of the Boston Colloquium for the Philosophy of Science, 1964/1966.* In Memory of Norwood Russell Hanson. [Synthese Library 14] 1967 ISBN 90-277-0013-3
4. R.S. Cohen and M.W. Wartofsky (eds.): *Proceedings of the Boston Colloquium for the Philosophy of Science, 1966/1968.* [Synthese Library 18] 1969 ISBN 90-277-0014-1
5. R.S. Cohen and M.W. Wartofsky (eds.): *Proceedings of the Boston Colloquium for the Philosophy of Science, 1966/1968.* [Synthese Library 19] 1969 ISBN 90-277-0015-X
6. R.S. Cohen and R.J. Seeger (eds.): *Ernst Mach, Physicist and Philosopher.* [Synthese Library 27] 1970 ISBN 90-277-0016-8
7. M. Čapek: *Bergson and Modern Physics.* A Reinterpretation and Re-evaluation. [Synthese Library 37] 1971 ISBN 90-277-0186-5
8. R.C. Buck and R.S. Cohen (eds.): *PSA 1970.* Proceedings of the 2nd Biennial Meeting of the Philosophy and Science Association (Boston, Fall 1970). In Memory of Rudolf Carnap. [Synthese Library 39] 1971 ISBN 90-277-0187-3; Pb 90-277-0309-4
9. A.A. Zinov'ev: *Foundations of the Logical Theory of Scientific Knowledge (Complex Logic).* Translated from Russian. Revised and enlarged English Edition, with an Appendix by G.A. Smirnov, E.A. Sidorenko, A.M. Fedina and L.A. Bobrova. [Synthese Library 46] 1973 ISBN 90-277-0193-8; Pb 90-277-0324-8
10. L. Tondl: *Scientific Procedures.* A Contribution Concerning the Methodological Problems of Scientific Concepts and Scientific Explanation.Translated from Czech. [Synthese Library 47] 1973 ISBN 90-277-0147-4; Pb 90-277-0323-X
11. R.J. Seeger and R.S. Cohen (eds.): *Philosophical Foundations of Science.* Proceedings of Section L, 1969, American Association for the Advancement of Science. [Synthese Library 58] 1974 ISBN 90-277-0390-6; Pb 90-277-0376-0
12. A. Grünbaum: *Philosophical Problems of Space and Times.* 2nd enlarged ed. [Synthese Library 55] 1973 ISBN 90-277-0357-4; Pb 90-277-0358-2
13. R.S. Cohen and M.W. Wartofsky (eds.): *Logical and Epistemological Studies in Contemporary Physics.* Proceedings of the Boston Colloquium for the Philosophy of Science, 1969/72, Part I. [Synthese Library 59] 1974 ISBN 90-277-0391-4; Pb 90-277-0377-9
14. R.S. Cohen and M.W. Wartofsky (eds.): *Methodological and Historical Essays in the Natural and Social Sciences.* Proceedings of the Boston Colloquium for the Philosophy of Science, 1969/72, Part II. [Synthese Library 60] 1974 ISBN 90-277-0392-2; Pb 90-277-0378-7
15. R.S. Cohen, J.J. Stachel and M.W. Wartofsky (eds.): *For Dirk Struik.* Scientific, Historical and Political Essays in Honor of Dirk J. Struik. [Synthese Library 61] 1974 ISBN 90-277-0393-0; Pb 90-277-0379-5
16. N. Geschwind: *Selected Papers on Language and the Brains.* [Synthese Library 68] 1974 ISBN 90-277-0262-4; Pb 90-277-0263-2
17. B.G. Kuznetsov: *Reason and Being.* Translated from Russian. Edited by C.R. Fawcett and R.S. Cohen. 1987 ISBN 90-277-2181-5

Boston Studies in the Philosophy of Science

18. P. Mittelstaedt: *Philosophical Problems of Modern Physics.* Translated from the revised 4th German edition by W. Riemer and edited by R.S. Cohen. [Synthese Library 95] 1976
ISBN 90-277-0285-3; Pb 90-277-0506-2
19. H. Mehlberg: *Time, Causality, and the Quantum Theory.* Studies in the Philosophy of Science. Vol. I: *Essay on the Causal Theory of Time.* Vol. II: *Time in a Quantized Universe.* Translated from French. Edited by R.S. Cohen. 1980 Vol. I: ISBN 90-277-0721-9; Pb 90-277-1074-0
Vol. II: ISBN 90-277-1075-9; Pb 90-277-1076-7
20. K.F. Schaffner and R.S. Cohen (eds.): *PSA 1972.* Proceedings of the 3rd Biennial Meeting of the Philosophy of Science Association (Lansing, Michigan, Fall 1972). [Synthese Library 64] 1974 ISBN 90-277-0408-2; Pb 90-277-0409-0
21. R.S. Cohen and J.J. Stachel (eds.): *Selected Papers of Léon Rosenfeld.* [Synthese Library 100] 1979 ISBN 90-277-0651-4; Pb 90-277-0652-2
22. M. Čapek (ed.): *The Concepts of Space and Time.* Their Structure and Their Development. [Synthese Library 74] 1976 ISBN 90-277-0355-8; Pb 90-277-0375-2
23. M. Grene: *The Understanding of Nature.* Essays in the Philosophy of Biology. [Synthese Library 66] 1974 ISBN 90-277-0462-7; Pb 90-277-0463-5
24. D. Ihde: *Technics and Praxis.* A Philosophy of Technology. [Synthese Library 130] 1979
ISBN 90-277-0953-X; Pb 90-277-0954-8
25. J. Hintikka and U. Remes: *The Method of Analysis.* Its Geometrical Origin and Its General Significance. [Synthese Library 75] 1974 ISBN 90-277-0532-1; Pb 90-277-0543-7
26. J.E. Murdoch and E.D. Sylla (eds.): *The Cultural Context of Medieval Learning.* Proceedings of the First International Colloquium on Philosophy, Science, and Theology in the Middle Ages, 1973. [Synthese Library 76] 1975 ISBN 90-277-0560-7; Pb 90-277-0587-9
27. M. Grene and E. Mendelsohn (eds.): *Topics in the Philosophy of Biology.* [Synthese Library 84] 1976 ISBN 90-277-0595-X; Pb 90-277-0596-8
28. J. Agassi: *Science in Flux.* [Synthese Library 80] 1975
ISBN 90-277-0584-4; Pb 90-277-0612-3
29. J.J. Wiatr (ed.): *Polish Essays in the Methodology of the Social Sciences.* [Synthese Library 131] 1979 ISBN 90-277-0723-5; Pb 90-277-0956-4
30. P. Janich: *Protophysics of Time.* Constructive Foundation and History of Time Measurement. Translated from German. 1985 ISBN 90-277-0724-3
31. R.S. Cohen and M.W. Wartofsky (eds.): *Language, Logic, and Method.* 1983
ISBN 90-277-0725-1
32. R.S. Cohen, C.A. Hooker, A.C. Michalos and J.W. van Evra (eds.): *PSA 1974.* Proceedings of the 4th Biennial Meeting of the Philosophy of Science Association. [Synthese Library 101] 1976 ISBN 90-277-0647-6; Pb 90-277-0648-4
33. G. Holton and W.A. Blanpied (eds.): *Science and Its Public.* The Changing Relationship. [Synthese Library 96] 1976 ISBN 90-277-0657-3; Pb 90-277-0658-1
34. M.D. Grmek, R.S. Cohen and G. Cimino (eds.): *On Scientific Discovery.* The 1977 Erice Lectures. 1981 ISBN 90-277-1122-4; Pb 90-277-1123-2
35. S. Amsterdamski: *Between Experience and Metaphysics.* Philosophical Problems of the Evolution of Science. Translated from Polish. [Synthese Library 77] 1975
ISBN 90-277-0568-2; Pb 90-277-0580-1
36. M. Marković and G. Petrović (eds.): *Praxis.* Yugoslav Essays in the Philosophy and Methodology of the Social Sciences. [Synthese Library 134] 1979
ISBN 90-277-0727-8; Pb 90-277-0968-8

Boston Studies in the Philosophy of Science

37. H. von Helmholtz: *Epistemological Writings.* The Paul Hertz / Moritz Schlick Centenary Edition of 1921. Translated from German by M.F. Lowe. Edited with an Introduction and Bibliography by R.S. Cohen and Y. Elkana. [Synthese Library 79] 1977 ISBN 90-277-0290-X; Pb 90-277-0582-8
38. R.M. Martin: *Pragmatics, Truth and Language.* 1979 ISBN 90-277-0992-0; Pb 90-277-0993-9
39. R.S. Cohen, P.K. Feyerabend and M.W. Wartofsky (eds.): *Essays in Memory of Imre Lakatos.* [Synthese Library 99] 1976 ISBN 90-277-0654-9; Pb 90-277-0655-7
40. Not published.
41. Not published.
42. H.R. Maturana and F.J. Varela: *Autopoiesis and Cognition.* The Realization of the Living. With a Preface to "Autopoiesis' by S. Beer. 1980 ISBN 90-277-1015-5; Pb 90-277-1016-3
43. A. Kasher (ed.): *Language in Focus: Foundations, Methods and Systems.* Essays in Memory of Yehoshua Bar-Hillel. [Synthese Library 89] 1976 ISBN 90-277-0644-1; Pb 90-277-0645-X
44. T.D. Thao: *Investigations into the Origin of Language and Consciousness.* 1984 ISBN 90-277-0827-4
45. F.G.-I. Nagasaka (ed.): *Japanese Studies in the Philosophy of Science.* 1997 ISBN 0-7923-4781-1
46. P.L. Kapitza: *Experiment, Theory, Practice.* Articles and Addresses. Edited by R.S. Cohen. 1980 ISBN 90-277-1061-9; Pb 90-277-1062-7
47. M.L. Dalla Chiara (ed.): *Italian Studies in the Philosophy of Science.* 1981 ISBN 90-277-0735-9; Pb 90-277-1073-2
48. M.W. Wartofsky: *Models.* Representation and the Scientific Understanding. [Synthese Library 129] 1979 ISBN 90-277-0736-7; Pb 90-277-0947-5
49. T.D. Thao: *Phenomenology and Dialectical Materialism.* Edited by R.S. Cohen. 1986 ISBN 90-277-0737-5
50. Y. Fried and J. Agassi: *Paranoia.* A Study in Diagnosis. [Synthese Library 102] 1976 ISBN 90-277-0704-9; Pb 90-277-0705-7
51. K.H. Wolff: *Surrender and Cath.* Experience and Inquiry Today. [Synthese Library 105] 1976 ISBN 90-277-0758-8; Pb 90-277-0765-0
52. K. Kosík: *Dialectics of the Concrete.* A Study on Problems of Man and World. 1976 ISBN 90-277-0761-8; Pb 90-277-0764-2
53. N. Goodman: *The Structure of Appearance.* [Synthese Library 107] 1977 ISBN 90-277-0773-1; Pb 90-277-0774-X
54. H.A. Simon: *Models of Discovery* and Other Topics in the Methods of Science. [Synthese Library 114] 1977 ISBN 90-277-0812-6; Pb 90-277-0858-4
55. M. Lazerowitz: *The Language of Philosophy.* Freud and Wittgenstein. [Synthese Library 117] 1977 ISBN 90-277-0826-6; Pb 90-277-0862-2
56. T. Nickles (ed.): *Scientific Discovery, Logic, and Rationality.* 1980 ISBN 90-277-1069-4; Pb 90-277-1070-8
57. J. Margolis: *Persons and Mind.* The Prospects of Nonreductive Materialism. [Synthese Library 121] 1978 ISBN 90-277-0854-1; Pb 90-277-0863-0
58. G. Radnitzky and G. Andersson (eds.): *Progress and Rationality in Science.* [Synthese Library 125] 1978 ISBN 90-277-0921-1; Pb 90-277-0922-X
59. G. Radnitzky and G. Andersson (eds.): *The Structure and Development of Science.* [Synthese Library 136] 1979 ISBN 90-277-0994-7; Pb 90-277-0995-5

Boston Studies in the Philosophy of Science

60. T. Nickles (ed.): *Scientific Discovery.* Case Studies. 1980
ISBN 90-277-1092-9; Pb 90-277-1093-7
61. M.A. Finocchiaro: *Galileo and the Art of Reasoning.* Rhetorical Foundation of Logic and Scientific Method. 1980 ISBN 90-277-1094-5; Pb 90-277-1095-3
62. W.A. Wallace: *Prelude to Galileo.* Essays on Medieval and 16th-Century Sources of Galileo's Thought. 1981 ISBN 90-277-1215-8; Pb 90-277-1216-6
63. F. Rapp: *Analytical Philosophy of Technology.* Translated from German. 1981
ISBN 90-277-1221-2; Pb 90-277-1222-0
64. R.S. Cohen and M.W. Wartofsky (eds.): *Hegel and the Sciences.* 1984 ISBN 90-277-0726-X
65. J. Agassi: *Science and Society.* Studies in the Sociology of Science. 1981
ISBN 90-277-1244-1; Pb 90-277-1245-X
66. L. Tondl: *Problems of Semantics.* A Contribution to the Analysis of the Language of Science. Translated from Czech. 1981 ISBN 90-277-0148-2; Pb 90-277-0316-7
67. J. Agassi and R.S. Cohen (eds.): *Scientific Philosophy Today.* Essays in Honor of Mario Bunge. 1982 ISBN 90-277-1262-X; Pb 90-277-1263-8
68. W. Krajewski (ed.): *Polish Essays in the Philosophy of the Natural Sciences.* Translated from Polish and edited by R.S. Cohen and C.R. Fawcett. 1982
ISBN 90-277-1286-7; Pb 90-277-1287-5
69. J.H. Fetzer: *Scientific Knowledge.* Causation, Explanation and Corroboration. 1981
ISBN 90-277-1335-9; Pb 90-277-1336-7
70. S. Grossberg: *Studies of Mind and Brain.* Neural Principles of Learning, Perception, Development, Cognition, and Motor Control. 1982 ISBN 90-277-1359-6; Pb 90-277-1360-X
71. R.S. Cohen and M.W. Wartofsky (eds.): *Epistemology, Methodology, and the Social Sciences.* 1983. ISBN 90-277-1454-1
72. K. Berka: *Measurement.* Its Concepts, Theories and Problems. Translated from Czech. 1983
ISBN 90-277-1416-9
73. G.L. Pandit: *The Structure and Growth of Scientific Knowledge.* A Study in the Methodology of Epistemic Appraisal. 1983 ISBN 90-277-1434-7
74. A.A. Zinov'ev: *Logical Physics.* Translated from Russian. Edited by R.S. Cohen. 1983
[*see also* Volume 9] ISBN 90-277-0734-0
75. G-G. Granger: *Formal Thought and the Sciences of Man.* Translated from French. With and Introduction by A. Rosenberg. 1983 ISBN 90-277-1524-6
76. R.S. Cohen and L. Laudan (eds.): *Physics, Philosophy and Psychoanalysis.* Essays in Honor of Adolf Grünbaum. 1983 ISBN 90-277-1533-5
77. G. Böhme, W. van den Daele, R. Hohlfeld, W. Krohn and W. Schäfer: *Finalization in Science.* The Social Orientation of Scientific Progress. Translated from German. Edited by W. Schäfer. 1983 ISBN 90-277-1549-1
78. D. Shapere: *Reason and the Search for Knowledge.* Investigations in the Philosophy of Science. 1984 ISBN 90-277-1551-3; Pb 90-277-1641-2
79. G. Andersson (ed.): *Rationality in Science and Politics.* Translated from German. 1984
ISBN 90-277-1575-0; Pb 90-277-1953-5
80. P.T. Durbin and F. Rapp (eds.): *Philosophy and Technology.* [*Also* Philosophy and Technology Series, Vol. 1] 1983 ISBN 90-277-1576-9
81. M. Marković: *Dialectical Theory of Meaning.* Translated from Serbo-Croat. 1984
ISBN 90-277-1596-3
82. R.S. Cohen and M.W. Wartofsky (eds.): *Physical Sciences and History of Physics.* 1984.
ISBN 90-277-1615-3

Boston Studies in the Philosophy of Science

83. É. Meyerson: *The Relativistic Deduction.* Epistemological Implications of the Theory of Relativity. Translated from French. With a Review by Albert Einstein and an Introduction by Milić Čapek. 1985 ISBN 90-277-1699-4
84. R.S. Cohen and M.W. Wartofsky (eds.): *Methodology, Metaphysics and the History of Science.* In Memory of Benjamin Nelson. 1984 ISBN 90-277-1711-7
85. G. Tamás: *The Logic of Categories.* Translated from Hungarian. Edited by R.S. Cohen. 1986 ISBN 90-277-1742-7
86. S.L. de C. Fernandes: *Foundations of Objective Knowledge.* The Relations of Popper's Theory of Knowledge to That of Kant. 1985 ISBN 90-277-1809-1
87. R.S. Cohen and T. Schnelle (eds.): *Cognition and Fact.* Materials on Ludwik Fleck. 1986 ISBN 90-277-1902-0
88. G. Freudenthal: *Atom and Individual in the Age of Newton.* On the Genesis of the Mechanistic World View. Translated from German. 1986 ISBN 90-277-1905-5
89. A. Donagan, A.N. Perovich Jr and M.V. Wedin (eds.): *Human Nature and Natural Knowledge.* Essays presented to Marjorie Grene on the Occasion of Her 75th Birthday. 1986 ISBN 90-277-1974-8
90. C. Mitcham and A. Hunning (eds.): *Philosophy and Technology II.* Information Technology and Computers in Theory and Practice. [*Also* Philosophy and Technology Series, Vol. 2] 1986 ISBN 90-277-1975-6
91. M. Grene and D. Nails (eds.): *Spinoza and the Sciences.* 1986 ISBN 90-277-1976-4
92. S.P. Turner: *The Search for a Methodology of Social Science.* Durkheim, Weber, and the 19th-Century Problem of Cause, Probability, and Action. 1986. ISBN 90-277-2067-3
93. I.C. Jarvie: *Thinking about Society.* Theory and Practice. 1986 ISBN 90-277-2068-1
94. E. Ullmann-Margalit (ed.): *The Kaleidoscope of Science.* The Israel Colloquium: Studies in History, Philosophy, and Sociology of Science, Vol. 1. 1986 ISBN 90-277-2158-0; Pb 90-277-2159-9
95. E. Ullmann-Margalit (ed.): *The Prism of Science.* The Israel Colloquium: Studies in History, Philosophy, and Sociology of Science, Vol. 2. 1986 ISBN 90-277-2160-2; Pb 90-277-2161-0
96. G. Márkus: *Language and Production.* A Critique of the Paradigms. Translated from French. 1986 ISBN 90-277-2169-6
97. F. Amrine, F.J. Zucker and H. Wheeler (eds.): *Goethe and the Sciences: A Reappraisal.* 1987 ISBN 90-277-2265-X; Pb 90-277-2400-8
98. J.C. Pitt and M. Pera (eds.): *Rational Changes in Science.* Essays on Scientific Reasoning. Translated from Italian. 1987 ISBN 90-277-2417-2
99. O. Costa de Beauregard: *Time, the Physical Magnitude.* 1987 ISBN 90-277-2444-X
100. A. Shimony and D. Nails (eds.): *Naturalistic Epistemology.* A Symposium of Two Decades. 1987 ISBN 90-277-2337-0
101. N. Rotenstreich: *Time and Meaning in History.* 1987 ISBN 90-277-2467-9
102. D.B. Zilberman: *The Birth of Meaning in Hindu Thought.* Edited by R.S. Cohen. 1988 ISBN 90-277-2497-0
103. T.F. Glick (ed.): *The Comparative Reception of Relativity.* 1987 ISBN 90-277-2498-9
104. Z. Harris, M. Gottfried, T. Ryckman, P. Mattick Jr, A. Daladier, T.N. Harris and S. Harris: *The Form of Information in Science.* Analysis of an Immunology Sublanguage. With a Preface by Hilary Putnam. 1989 ISBN 90-277-2516-0
105. F. Burwick (ed.): *Approaches to Organic Form.* Permutations in Science and Culture. 1987 ISBN 90-277-2541-1

Boston Studies in the Philosophy of Science

106. M. Almási: *The Philosophy of Appearances.* Translated from Hungarian. 1989
ISBN 90-277-2150-5
107. S. Hook, W.L. O'Neill and R. O'Toole (eds.): *Philosophy, History and Social Action.* Essays in Honor of Lewis Feuer. With an Autobiographical Essay by L. Feuer. 1988
ISBN 90-277-2644-2
108. I. Hronszky, M. Fehér and B. Dajka: *Scientific Knowledge Socialized.* Selected Proceedings of the 5th Joint International Conference on the History and Philosophy of Science organized by the IUHPS (Veszprém, Hungary, 1984). 1988 ISBN 90-277-2284-6
109. P. Tillers and E.D. Green (eds.): *Probability and Inference in the Law of Evidence.* The Uses and Limits of Bayesianism. 1988 ISBN 90-277-2689-2
110. E. Ullmann-Margalit (ed.): *Science in Reflection.* The Israel Colloquium: Studies in History, Philosophy, and Sociology of Science, Vol. 3. 1988
ISBN 90-277-2712-0; Pb 90-277-2713-9
111. K. Gavroglu, Y. Goudaroulis and P. Nicolacopoulos (eds.): *Imre Lakatos and Theories of Scientific Change.* 1989 ISBN 90-277-2766-X
112. B. Glassner and J.D. Moreno (eds.): *The Qualitative-Quantitative Distinction in the Social Sciences.* 1989 ISBN 90-277-2829-1
113. K. Arens: *Structures of Knowing.* Psychologies of the 19th Century. 1989
ISBN 0-7923-0009-2
114. A. Janik: *Style, Politics and the Future of Philosophy.* 1989 ISBN 0-7923-0056-4
115. F. Amrine (ed.): *Literature and Science as Modes of Expression.* With an Introduction by S. Weininger. 1989 ISBN 0-7923-0133-1
116. J.R. Brown and J. Mittelstrass (eds.): *An Intimate Relation.* Studies in the History and Philosophy of Science. Presented to Robert E. Butts on His 60th Birthday. 1989
ISBN 0-7923-0169-2
117. F. D'Agostino and I.C. Jarvie (eds.): *Freedom and Rationality.* Essays in Honor of John Watkins. 1989 ISBN 0-7923-0264-8
118. D. Zolo: *Reflexive Epistemology.* The Philosophical Legacy of Otto Neurath. 1989
ISBN 0-7923-0320-2
119. M. Kearn, B.S. Philips and R.S. Cohen (eds.): *Georg Simmel and Contemporary Sociology.* 1989 ISBN 0-7923-0407-1
120. T.H. Levere and W.R. Shea (eds.): *Nature, Experiment and the Science.* Essays on Galileo and the Nature of Science. In Honour of Stillman Drake. 1989 ISBN 0-7923-0420-9
121. P. Nicolacopoulos (ed.): *Greek Studies in the Philosophy and History of Science.* 1990
ISBN 0-7923-0717-8
122. R. Cooke and D. Costantini (eds.): *Statistics in Science.* The Foundations of Statistical Methods in Biology, Physics and Economics. 1990 ISBN 0-7923-0797-6
123. P. Duhem: *The Origins of Statics.* Translated from French by G.F. Leneaux, V.N. Vagliente and G.H. Wagner. With an Introduction by S.L. Jaki. 1991 ISBN 0-7923-0898-0
124. H. Kamerlingh Onnes: *Through Measurement to Knowledge.* The Selected Papers, 1853-1926. Edited and with an Introduction by K. Gavroglu and Y. Goudaroulis. 1991
ISBN 0-7923-0825-5
125. M. Čapek: *The New Aspects of Time: Its Continuity and Novelties.* Selected Papers in the Philosophy of Science. 1991 ISBN 0-7923-0911-1
126. S. Unguru (ed.): *Physics, Cosmology and Astronomy, 1300–1700.* Tension and Accommodation. 1991 ISBN 0-7923-1022-5

Boston Studies in the Philosophy of Science

127. Z. Bechler: *Newton's Physics on the Conceptual Structure of the Scientific Revolution.* 1991 ISBN 0-7923-1054-3
128. É. Meyerson: *Explanation in the Sciences.* Translated from French by M-A. Siple and D.A. Siple. 1991 ISBN 0-7923-1129-9
129. A.I. Tauber (ed.): *Organism and the Origins of Self.* 1991 ISBN 0-7923-1185-X
130. F.J. Varela and J-P. Dupuy (eds.): *Understanding Origins.* Contemporary Views on the Origin of Life, Mind and Society. 1992 ISBN 0-7923-1251-1
131. G.L. Pandit: *Methodological Variance.* Essays in Epistemological Ontology and the Methodology of Science. 1991 ISBN 0-7923-1263-5
132. G. Munévar (ed.): *Beyond Reason.* Essays on the Philosophy of Paul Feyerabend. 1991 ISBN 0-7923-1272-4
133. T.E. Uebel (ed.): *Rediscovering the Forgotten Vienna Circle.* Austrian Studies on Otto Neurath and the Vienna Circle. Partly translated from German. 1991 ISBN 0-7923-1276-7
134. W.R. Woodward and R.S. Cohen (eds.): *World Views and Scientific Discipline Formation.* Science Studies in the [former] German Democratic Republic. Partly translated from German by W.R. Woodward. 1991 ISBN 0-7923-1286-4
135. P. Zambelli: *The Speculum Astronomiae and Its Enigma.* Astrology, Theology and Science in Albertus Magnus and His Contemporaries. 1992 ISBN 0-7923-1380-1
136. P. Petitjean, C. Jami and A.M. Moulin (eds.): *Science and Empires.* Historical Studies about Scientific Development and European Expansion. ISBN 0-7923-1518-9
137. W.A. Wallace: *Galileo's Logic of Discovery and Proof.* The Background, Content, and Use of His Appropriated Treatises on Aristotle's *Posterior Analytics.* 1992 ISBN 0-7923-1577-4
138. W.A. Wallace: *Galileo's Logical Treatises.* A Translation, with Notes and Commentary, of His Appropriated Latin Questions on Aristotle's *Posterior Analytics.* 1992 ISBN 0-7923-1578-2
     Set (137 + 138) ISBN 0-7923-1579-0
139. M.J. Nye, J.L. Richards and R.H. Stuewer (eds.): *The Invention of Physical Science.* Intersections of Mathematics, Theology and Natural Philosophy since the Seventeenth Century. Essays in Honor of Erwin N. Hiebert. 1992 ISBN 0-7923-1753-X
140. G. Corsi, M.L. dalla Chiara and G.C. Ghirardi (eds.): *Bridging the Gap: Philosophy, Mathematics and Physics.* Lectures on the Foundations of Science. 1992 ISBN 0-7923-1761-0
141. C.-H. Lin and D. Fu (eds.): *Philosophy and Conceptual History of Science in Taiwan.* 1992 ISBN 0-7923-1766-1
142. S. Sarkar (ed.): *The Founders of Evolutionary Genetics.* A Centenary Reappraisal. 1992 ISBN 0-7923-1777-7
143. J. Blackmore (ed.): *Ernst Mach – A Deeper Look.* Documents and New Perspectives. 1992 ISBN 0-7923-1853-6
144. P. Kroes and M. Bakker (eds.): *Technological Development and Science in the Industrial Age.* New Perspectives on the Science–Technology Relationship. 1992 ISBN 0-7923-1898-6
145. S. Amsterdamski: *Between History and Method.* Disputes about the Rationality of Science. 1992 ISBN 0-7923-1941-9
146. E. Ullmann-Margalit (ed.): *The Scientific Enterprise.* The Bar-Hillel Colloquium: Studies in History, Philosophy, and Sociology of Science, Volume 4. 1992 ISBN 0-7923-1992-3
147. L. Embree (ed.): *Metaarchaeology.* Reflections by Archaeologists and Philosophers. 1992 ISBN 0-7923-2023-9
148. S. French and H. Kamminga (eds.): *Correspondence, Invariance and Heuristics.* Essays in Honour of Heinz Post. 1993 ISBN 0-7923-2085-9
149. M. Bunzl: *The Context of Explanation.* 1993 ISBN 0-7923-2153-7

Boston Studies in the Philosophy of Science

150. I.B. Cohen (ed.): *The Natural Sciences and the Social Sciences.* Some Critical and Historical Perspectives. 1994 ISBN 0-7923-2223-1
151. K. Gavroglu, Y. Christianidis and E. Nicolaidis (eds.): *Trends in the Historiography of Science.* 1994 ISBN 0-7923-2255-X
152. S. Poggi and M. Bossi (eds.): *Romanticism in Science.* Science in Europe, 1790–1840. 1994 ISBN 0-7923-2336-X
153. J. Faye and H.J. Folse (eds.): *Niels Bohr and Contemporary Philosophy.* 1994 ISBN 0-7923-2378-5
154. C.C. Gould and R.S. Cohen (eds.): *Artifacts, Representations, and Social Practice.* Essays for Marx W. Wartofsky. 1994 ISBN 0-7923-2481-1
155. R.E. Butts: *Historical Pragmatics.* Philosophical Essays. 1993 ISBN 0-7923-2498-6
156. R. Rashed: *The Development of Arabic Mathematics: Between Arithmetic and Algebra.* Translated from French by A.F.W. Armstrong. 1994 ISBN 0-7923-2565-6
157. I. Szumilewicz-Lachman (ed.): *Zygmunt Zawirski: His Life and Work.* With Selected Writings on Time, Logic and the Methodology of Science. Translations by Feliks Lachman. Ed. by R.S. Cohen, with the assistance of B. Bergo. 1994 ISBN 0-7923-2566-4
158. S.N. Haq: *Names, Natures and Things.* The Alchemist Jābir ibn Ḥayyān and His *Kitāb al-Aḥjār* (Book of Stones). 1994 ISBN 0-7923-2587-7
159. P. Plaass: *Kant's Theory of Natural Science.* Translation, Analytic Introduction and Commentary by Alfred E. and Maria G. Miller. 1994 ISBN 0-7923-2750-0
160. J. Misiek (ed.): *The Problem of Rationality in Science and its Philosophy.* On Popper vs. Polanyi. The Polish Conferences 1988–89. 1995 ISBN 0-7923-2925-2
161. I.C. Jarvie and N. Laor (eds.): *Critical Rationalism, Metaphysics and Science.* Essays for Joseph Agassi, Volume I. 1995 ISBN 0-7923-2960-0
162. I.C. Jarvie and N. Laor (eds.): *Critical Rationalism, the Social Sciences and the Humanities.* Essays for Joseph Agassi, Volume II. 1995 ISBN 0-7923-2961-9
Set (161–162) ISBN 0-7923-2962-7
163. K. Gavroglu, J. Stachel and M.W. Wartofsky (eds.): *Physics, Philosophy, and the Scientific Community.* Essays in the Philosophy and History of the Natural Sciences and Mathematics. In Honor of Robert S. Cohen. 1995 ISBN 0-7923-2988-0
164. K. Gavroglu, J. Stachel and M.W. Wartofsky (eds.): *Science, Politics and Social Practice.* Essays on Marxism and Science, Philosophy of Culture and the Social Sciences. In Honor of Robert S. Cohen. 1995 ISBN 0-7923-2989-9
165. K. Gavroglu, J. Stachel and M.W. Wartofsky (eds.): *Science, Mind and Art.* Essays on Science and the Humanistic Understanding in Art, Epistemology, Religion and Ethics. Essays in Honor of Robert S. Cohen. 1995 ISBN 0-7923-2990-2
Set (163–165) ISBN 0-7923-2991-0
166. K.H. Wolff: *Transformation in the Writing.* A Case of Surrender-and-Catch. 1995 ISBN 0-7923-3178-8
167. A.J. Kox and D.M. Siegel (eds.): *No Truth Except in the Details.* Essays in Honor of Martin J. Klein. 1995 ISBN 0-7923-3195-8
168. J. Blackmore: *Ludwig Boltzmann, His Later Life and Philosophy, 1900–1906.* Book One: A Documentary History. 1995 ISBN 0-7923-3231-8
169. R.S. Cohen, R. Hilpinen and R. Qiu (eds.): *Realism and Anti-Realism in the Philosophy of Science.* Beijing International Conference, 1992. 1996 ISBN 0-7923-3233-4
170. I. Kuçuradi and R.S. Cohen (eds.): *The Concept of Knowledge.* The Ankara Seminar. 1995 ISBN 0-7923-3241-5

Boston Studies in the Philosophy of Science

171. M.A. Grodin (ed.): *Meta Medical Ethics*: The Philosophical Foundations of Bioethics. 1995 ISBN 0-7923-3344-6
172. S. Ramirez and R.S. Cohen (eds.): *Mexican Studies in the History and Philosophy of Science.* 1995 ISBN 0-7923-3462-0
173. C. Dilworth: *The Metaphysics of Science.* An Account of Modern Science in Terms of Principles, Laws and Theories. 1995 ISBN 0-7923-3693-3
174. J. Blackmore: *Ludwig Boltzmann, His Later Life and Philosophy, 1900–1906* Book Two: The Philosopher. 1995 ISBN 0-7923-3464-7
175. P. Damerow: *Abstraction and Representation.* Essays on the Cultural Evolution of Thinking. 1996 ISBN 0-7923-3816-2
176. M.S. Macrakis: *Scarcity's Ways: The Origins of Capital.* A Critical Essay on Thermodynamics, Statistical Mechanics and Economics. 1997 ISBN 0-7923-4760-9
177. M. Marion and R.S. Cohen (eds.): *Québec Studies in the Philosophy of Science.* Part I: Logic, Mathematics, Physics and History of Science. Essays in Honor of Hugues Leblanc. 1995 ISBN 0-7923-3559-7
178. M. Marion and R.S. Cohen (eds.): *Québec Studies in the Philosophy of Science.* Part II: Biology, Psychology, Cognitive Science and Economics. Essays in Honor of Hugues Leblanc. 1996 ISBN 0-7923-3560-0
    Set (177–178) ISBN 0-7923-3561-9
179. Fan Dainian and R.S. Cohen (eds.): *Chinese Studies in the History and Philosophy of Science and Technology.* 1996 ISBN 0-7923-3463-9
180. P. Forman and J.M. Sánchez-Ron (eds.): *National Military Establishments and the Advancement of Science and Technology.* Studies in 20th Century History. 1996 ISBN 0-7923-3541-4
181. E.J. Post: *Quantum Reprogramming.* Ensembles and Single Systems: A Two-Tier Approach to Quantum Mechanics. 1995 ISBN 0-7923-3565-1
182. A.I. Tauber (ed.): *The Elusive Synthesis: Aesthetics and Science.* 1996 ISBN 0-7923-3904-5
183. S. Sarkar (ed.): *The Philosophy and History of Molecular Biology: New Perspectives.* 1996 ISBN 0-7923-3947-9
184. J.T. Cushing, A. Fine and S. Goldstein (eds.): *Bohmian Mechanics and Quantum Theory: An Appraisal.* 1996 ISBN 0-7923-4028-0
185. K. Michalski: *Logic and Time.* An Essay on Husserl's Theory of Meaning. 1996 ISBN 0-7923-4082-5
186. G. Munévar (ed.): *Spanish Studies in the Philosophy of Science.* 1996 ISBN 0-7923-4147-3
187. G. Schubring (ed.): *Hermann Günther Graßmann (1809–1877): Visionary Mathematician, Scientist and Neohumanist Scholar.* Papers from a Sesquicentennial Conference. 1996 ISBN 0-7923-4261-5
188. M. Bitbol: *Schrödinger's Philosophy of Quantum Mechanics.* 1996 ISBN 0-7923-4266-6
189. J. Faye, U. Scheffler and M. Urchs (eds.): *Perspectives on Time.* 1997 ISBN 0-7923-4330-1
190. K. Lehrer and J.C. Marek (eds.): *Austrian Philosophy Past and Present.* Essays in Honor of Rudolf Haller. 1996 ISBN 0-7923-4347-6
191. J.L. Lagrange: *Analytical Mechanics.* Translated and edited by Auguste Boissonade and Victor N. Vagliente. Translated from the *Mécanique Analytique, novelle édition* of 1811. 1997 ISBN 0-7923-4349-2
192. D. Ginev and R.S. Cohen (eds.): *Issues and Images in the Philosophy of Science.* Scientific and Philosophical Essays in Honour of Azarya Polikarov. 1997 ISBN 0-7923-4444-8

Boston Studies in the Philosophy of Science

193. R.S. Cohen, M. Horne and J. Stachel (eds.): *Experimental Metaphysics.* Quantum Mechanical Studies for Abner Shimony, Volume One. 1997 ISBN 0-7923-4452-9
194. R.S. Cohen, M. Horne and J. Stachel (eds.): *Potentiality, Entanglement and Passion-at-a-Distance.* Quantum Mechanical Studies for Abner Shimony, Volume Two. 1997 ISBN 0-7923-4453-7; Set 0-7923-4454-5
195. R.S. Cohen and A.I. Tauber (eds.): *Philosophies of Nature: The Human Dimension.* 1997 ISBN 0-7923-4579-7
196. M. Otte and M. Panza (eds.): *Analysis and Synthesis in Mathematics.* History and Philosophy. 1997 ISBN 0-7923-4570-3
197. A. Denkel: *The Natural Background of Meaning.* 1999 ISBN 0-7923-5331-5
198. D. Baird, R.I.G. Hughes and A. Nordmann (eds.): *Heinrich Hertz: Classical Physicist, Modern Philosopher.* 1999 ISBN 0-7923-4653-X
199. A. Franklin: *Can That be Right?* Essays on Experiment, Evidence, and Science. 1999 ISBN 0-7923-5464-8
200. D. Raven, W. Krohn and R.S. Cohen (eds.): *The Social Origins of Modern Science.* 2000 ISBN 0-7923-6457-0
201. Reserved
202. Reserved
203. B. Babich and R.S. Cohen (eds.): *Nietzsche, Theories of Knowledge, and Critical Theory.* Nietzsche and the Sciences I. 1999 ISBN 0-7923-5742-6
204. B. Babich and R.S. Cohen (eds.): *Nietzsche, Epistemology, and Philosophy of Science.* Nietzsche and the Science II. 1999 ISBN 0-7923-5743-4
205. R. Hooykaas: *Fact, Faith and Fiction in the Development of Science.* The Gifford Lectures given in the University of St Andrews 1976. 1999 ISBN 0-7923-5774-4
206. M. Fehér, O. Kiss and L. Ropolyi (eds.): *Hermeneutics and Science.* 1999 ISBN 0-7923-5798-1
207. R.M. MacLeod (ed.): *Science and the Pacific War.* Science and Survival in the Pacific, 1939-1945. 1999 ISBN 0-7923-5851-1
208. I. Hanzel: *The Concept of Scientific Law in the Philosophy of Science and Epistemology.* A Study of Theoretical Reason. 1999 ISBN 0-7923-5852-X
209. G. Helm; R.J. Deltete (ed./transl.): *The Historical Development of Energetics.* 1999 ISBN 0-7923-5874-0
210. A. Orenstein and P. Kotatko (eds.): *Knowledge, Language and Logic.* Questions for Quine. 1999 ISBN 0-7923-5986-0
211. R.S. Cohen and H. Levine (eds.): *Maimonides and the Sciences.* 2000 ISBN 0-7923-6053-2
212. H. Gourko, D.I. Williamson and A.I. Tauber (eds.): *The Evolutionary Biology Papers of Elie Metchnikoff.* 2000 ISBN 0-7923-6067-2
213. S. D'Agostino: *A History of the Ideas of Theoretical Physics.* Essays on the Nineteenth and Twentieth Century Physics. 2000 ISBN 0-7923-6094-X
214. S. Lelas: *Science and Modernity.* Toward An Integral Theory of Science. 2000 ISBN 0-7923-6303-5
215. E. Agazzi and M. Pauri (eds.): *The Reality of the Unobservable.* Observability, Unobservability and Their Impact on the Issue of Scientific Realism. 2000 ISBN 0-7923-6311-6
216. P. Hoyningen-Huene and H. Sankey (eds.): *Incommensurability and Related Matters.* 2001 ISBN 0-7923-6989-0
217. A. Nieto-Galan: *Colouring Textiles.* A History of Natural Dyestuffs in Industrial Europe. 2001 ISBN 0-7923-7022-8

Boston Studies in the Philosophy of Science

218. J. Blackmore, R. Itagaki and S. Tanaka (eds.): *Ernst Mach's Vienna 1895–1930.* Or Phenomenalism as Philosophy of Science. 2001 ISBN 0-7923-7122-4
219. R. Vihalemm (ed.): *Estonian Studies in the History and Philosophy of Science.* 2001 ISBN 0-7923-7189-5
220. W. Lefèvre (ed.): *Between Leibniz, Newton, and Kant.* Philosophy and Science in the Eighteenth Century. 2001 ISBN 0-7923-7198-4
221. T.F. Glick, M.Á. Puig-Samper and R. Ruiz (eds.): *The Reception of Darwinism in the Iberian World.* Spain, Spanish America and Brazil. 2001 ISBN 1-4020-0082-0
222. U. Klein (ed.): *Tools and Modes of Representation in the Laboratory Sciences.* 2001 ISBN 1-4020-0100-2
223. P. Duhem: *Mixture and Chemical Combination.* And Related Essays. Edited and translated, with an introduction, by Paul Needham. 2002 ISBN 1-4020-0232-7
224. J.C. Boudri: *What was Mechanical about Mechanics.* The Concept of Force Betweem Metaphysics and Mechanics from Newton to Lagrange. 2002 ISBN 1-4020-0233-5
225. B.E. Babich (ed.): *Hermeneutic Philosophy of Science, Van Gogh's Eyes, and God.* Essays in Honor of Patrick A. Heelan, S.J. 2002 ISBN 1-4020-0234-3
226. D. Davies Villemaire: *E.A. Burtt, Historian and Philosopher.* A Study of the Author of The Metaphysical Foundations of Modern Physical Science. 2002 ISBN 1-4020-0428-1
227. L.J. Cohen: *Knowledge and Language.* Selected Essays of L. Jonathan Cohen. Edited and with an introduction by James Logue. 2002 ISBN 1-4020-0474-5
228. G.E. Allen and R.M. MacLeod (eds.): *Science, History and Social Activism: A Tribute to Everett Mendelsohn.* 2002 ISBN 1-4020-0495-0
229. O. Gal: *Meanest Foundations and Nobler Superstructures.* Hooke, Newton and the "Compounding of the Celestiall Motions of the Planetts". 2002 ISBN 1-4020-0732-9
230. R. Nola: *Rescuing Reason.* A Critique of Anti-Rationalist Views of Science and Knowledge. 2003 Hb: ISBN 1-4020-1042-7; Pb ISBN 1-4020-1043-5
231. J. Agassi: *Science and Culture.* 2003 ISBN 1-4020-1156-3
232. M.C. Galavotti (ed.): *Observation and Experiment in the Natural and Social Science.* 2003 ISBN 1-4020-1251-9
233. A. Simões, A. Carneiro and M.P. Diogo (eds.): *Travels of Learning.* A Geography of Science in Europe. 2003 ISBN 1-4020-1259-4
234. A. Ashtekar, R. Cohen, D. Howard, J. Renn, S. Sarkar and A. Shimony (eds.): *Revisiting the Foundations of Relativistic Physics.* Festschrift in Honor of John Stachel. 2003 ISBN 1-4020-1284-5
235. R.P. Farell: *Feyerabend and Scientific Values.* Tightrope-Walking Rationality. 2003 ISBN 1-4020-1350-7
236. D. Ginev (ed.): *Bulgarian Studies in the Philosophy of Science.* 2003 ISBN 1-4020-1496-1
237. C. Sasaki: *Descartes Mathematical Thought.* 2003 ISBN 1-4020-1746-4
238. K. Chemla (ed.): *History of Science, History of Text.* 2004 ISBN 1-4020-2320-0
239. C.R. Palmerino and J.M.M.H. Thijssen (eds.): *The Reception of the Galilean Science of Motion in Seventeenth-Century Europe.* 2004 ISBN 1-4020-2454-1

Boston Studies in the Philosophy of Science

240. J. Christianidis (ed.): *Classics in the History of Greek Mathematics*. 2004 ISBN 1-4020-0081-2
241. R.M. Brain and O. Knudsen (eds.): *Hans Christian Ørsted and the Romantic Quest for Unity.* Ideas, Disciplines, Practices. 2005 ISBN 1-4020-2979-9
242. D. Baird, E. Scerri and L. McIntyre (eds.): *Philosophy of Chemistry.* Synthesis of a New Discipline. 2006 ISBN 1-4020-3256-0
243. D.B. Zilberman, H. Gourko and R.S. Cohen (eds.): *Analogy in Indian and Western Philosophical Thought.* 2005 ISBN 1-4020-3339-7
244. G. Irzik and G. Güzeldere (eds.): *Turkish Studies in the History and Philosophy of Science.* 2005 ISBN 1-4020-3332-X
245. H.E. Gruber and K. Bödeker (eds.): *Creativity, Psychology and the History of Science*. 2005 ISBN 1-4020-3491-1

Also of interest:

R.S. Cohen and M.W. Wartofsky (eds.): *A Portrait of Twenty-Five Years Boston Colloquia for the Philosophy of Science, 1960-1985.* 1985 ISBN Pb 90-277-1971-3

Previous volumes are still available.

springer.com